Problems and Solutions in

Mathematical Olympiad

Secondary 3

Related Titles

Problems and Solutions in Mathematical Olympiad
High School 1
by Bin Xiong and Zhi-Gang Feng
translated by Tian-You Zhou
ISBN: 978-981-122-985-5
ISBN: 978-981-123-142-1 (pbk)

Problems and Solutions in Mathematical Olympiad
High School 2
by Shi-Xiong Liu
translated by Jiu Ding
ISBN: 978-981-122-988-6
ISBN: 978-981-123-143-8 (pbk)

Problems and Solutions in Mathematical Olympiad
High School 3
by Hong-Bing Yu
translated by Fang-Fang Lang and Yi-Chao Ye
ISBN: 978-981-122-991-6
ISBN: 978-981-123-144-5 (pbk)

Problems and Solutions in

Mathematical Olympiad

Secondary 3

Editors-in-Chief

Zun Shan *Nanjing Normal University, China*

Bin Xiong *East China Normal University, China*

Original Author

Jun Ge *High School Affiliated To Nanjing Normal University, China*

English Translator

Huan-Xin Xie *No.2 High School of East China Normal University, China*

Copy Editors

Ming Ni *East China Normal University Press, China*

Ling-Zhi Kong *East China Normal University Press, China*

Lei Rui *East China Normal University Press, China*

Published by

East China Normal University Press
3663 North Zhongshan Road
Shanghai 200062
China

and

World Scientific Publishing Co. Pte. Ltd.
5 Toh Tuck Link, Singapore 596224
USA office: 27 Warren Street, Suite 401-402, Hackensack, NJ 07601
UK office: 57 Shelton Street, Covent Garden, London WC2H 9HE

British Library Cataloguing-in-Publication Data
A catalogue record for this book is available from the British Library.

PROBLEMS AND SOLUTIONS IN MATHEMATICAL OLYMPIAD
Secondary 3

ISBN 978-981-122-982-4 (hardcover)
ISBN 978-981-123-141-4 (paperback)
ISBN 978-981-122-983-1 (ebook for institutions)
ISBN 978-981-122-984-8 (ebook for individuals)

For any available supplementary material, please visit
https://www.worldscientific.com/worldscibooks/10.1142/12086#t=suppl

Desk Editor: Tan Rok Ting

Typeset by Stallion Press
Email: enquiries@stallionpress.com

Printed in Singapore

Editorial Board

Preface

It is said that in many countries, especially the United States, children are afraid of mathematics and regard mathematics as an "unpopular subject." But in China, the situation is very different. Many children love mathematics, and their math scores are also very good. Indeed, mathematics is a subject that the Chinese are good at. If you see a few Chinese students in elementary and middle schools in the United States, then the top few in the class of mathematics are none other than them.

At the early stage of counting numbers, Chinese children already show their advantages.

Chinese people can express integers from 1 to 10 with one hand, whereas those in other countries would have to use two.

The Chinese have long had the concept of digits, and they use the most convenient decimal system (many countries still have the remnants of base 12 and base 60 systems).

Chinese characters are all single syllables, which are easy to recite. For example, the multiplication table can be quickly mastered by students, and even the "stupid" people know the concept of "three times seven equals twenty one." But for foreigners, as soon as they study multiplication, their heads get bigger. Believe it or not, you could try and memorise the multiplication table in English and then recite it, it is actually much harder to do so in English.

It takes the Chinese one or two minutes to memorize $\pi = 3.14159\cdots$ to the fifth decimal place. However, in order to recite these digits, the Russians wrote a poem. The first sentence contains three words and the second sentence contains one $\cdots$ To recite π, recite poetry first. In our opinion, this just simply asks for trouble, but they treat it as a magical way of memorization.

Application problems for the four arithmetic operations and their arithmetic solutions are also a major feature of Chinese mathematics. Since ancient times, the Chinese have compiled a lot of application questions, which has contact or close relations with reality and daily life. Their solutions are simple and elegant as well as smart and diverse, which helps increase students' interest in learning and enlighten students'. For example:

"There are one hundred monks and one hundred buns. One big monk eats three buns and three little monks eat one bun. How many big monks and how many little monks are there?"

Most foreigners can only solve equations, but Chinese have a variety of arithmetic solutions. As an example, one can turn each big monk into 9 little monks, and 100 buns indicate that there are 300 little monks, which contain 200 added little monks. As each big monk becomes a little monk 8 more little monks are created, so $200/8 = 25$ is the number of big monks, and naturally there are 75 little monks. Another way to solve the problem is to group a big monk and three little monks together, and so each person eats a bun on average, which is exactly equal to the overall average. Thus the big monks and the little monks are not more and less after being organized this way, that is, the number of the big monks is $100/(3+1) = 25$.

The Chinese are good at calculating, especially good at mental arithmetic. In ancient times, some people used their fingers to calculate (the so-called "counting by pinching fingers"). At the same time, China has long had computing devices such as counting chips and abaci. The latter can be said to be the prototype of computers.

In the introductory stage of mathematics – the study of arithmetic, our country has obvious advantages, so mathematics is often the subject that our smart children love.

Geometric reasoning was not well-developed in ancient China (but there were many books on the calculation of geometric figures in our country), and it was slightly inferior to the Greeks. However, the Chinese are good at learning from others. At present, the geometric level of middle school students in our country is far ahead of the rest of the world. Once a foreign education delegation came to a junior high school class in our country. They thought that the geometric content taught was too in-depth for students to comprehend, but after attending the class, they had to admit that the content was not only understood by Chinese students, but also well mastered.

The achievements of mathematics education in our country are remarkable. In international mathematics competitions, Chinese contestants have won numerous medals, which is the most powerful proof. Ever since our country officially sent a team to participate in the International Mathematical Olympiad in 1986, the Chinese team has won 14 team championships, which can be described as very impressive. Professor Shiing-Shen Chern, a famous contemporary mathematician, once admired this in particular. He said, "One thing to celebrate this year is that China won the first place in the international math competition ⋯ Last year it was also the first place." (Shiing-Shen Chern's speech, *How to Build China into a Mathematical Power*, at Cheng Kung University in Taiwan in October 1990)

Professor Chern also predicted: "China will become a mathematical power in the 21st century."

It is certainly not an easy task to become a mathematical power. It cannot be achieved overnight. It requires unremitting efforts. The purpose of this series of books is: (1) To further popularize the knowledge of mathematics, to make mathematics be loved by more young people, and to help them achieve good results; (2) To enable students who love mathematics to get better development and learn more knowledge and methods through the series of books.

"The important things in the world must be done in detail." We hope and believe that the publication of this series of books will play a role in making our country a mathematical power. This series was first published in 2000. According to the requirements of the curriculum reform, each volume is revised to different degrees.

Well-known mathematician, academician of the Chinese Academy of Sciences, and former chairman of the Chinese Mathematical Olympiad Professor Yuan Wang, served as a consultant to this series of books and wrote inscriptions for young math enthusiasts. We express our heartfelt thanks. We would also like to thank East China Normal University Press, and in particular Mr. Ming Ni and Mr. Lingzhi Kong. Without them, this series of books would not have been possible.

Zun Shan and Bin Xiong
May 2018

Contents

Chapter 1

Quadratic Equations

The standard form of a quadratic equation is

$$ax^2 + bx + c = 0 \ (a \neq 0),$$

and its simple form is $Ax^2 = B$ $(A \neq 0)$.

If the root of the above equation is x_0, then

$$ax_0^2 + bx_0 + c = 0, \qquad (*)$$

and vice versa.

Using the formula (*), we can easily get the relationship between a, b, c and x_0.

The quadratic formula for solving the equation $ax^2+bx+c = 0$ $(a \neq 0)$ is

$$x_{1,2} = \frac{-b \pm \sqrt{b^2 - 4ac}}{2a}.$$

Example 1. Observe the following equations:

$$\begin{aligned} 225 &= 15^2 \\ 625 &= 25^2 \\ 1225 &= 35^2 \\ 2025 &= 45^2 \\ &\cdots. \end{aligned}$$

Given that $21025 = x^2$, try to find the value of x according to the above rules.

Solution Let $x = 10n + 5$, $n > 0$, so

$$\begin{aligned} 21025 &= (10n+5)^2, \\ n^2 + n - 210 &= 0, \\ (n+15)(n-14) &= 0, \\ n = 14 \quad &or \quad n = -15. \end{aligned}$$

As $n > 0$, we discard $n = -15$.

Thus $n = 14$, $x = 10 \times 14 + 5 = 145$.

Example 2. Given that b and c are two roots of the equation $x^2+bx+c=0$, where $c \neq 0$, $b \neq c$, find the values of b and c.

Analysis Two identities are obtained by substituting b and c into the equation respectively. Then, the two identities can be properly added or subtracted, which is the basic method of identity deformation. In order to simplify the problem, we can observe the characteristics of two equations and try to subtract them.

Solution Because b and c are two roots of the equation $x^2 + bx + c = 0$, we can obtain that

$$\begin{cases} b^2 + b^2 + c = 0, & (1) \\ c^2 + bc + c = 0. & (2) \end{cases}$$

From (1)–(2), we get $(b - c)[(b + c) + b] = 0$.

Since $b - c \neq 0$, the above equation can be reduced to

$$c + 2b = 0. \tag{3}$$

Since $c \neq 0$, the equation (2) can be reduced to

$$c + b + 1 = 0. \tag{4}$$

Combine the equations (3) and (4) to get the solution $b = 1$, $c = -2$.

Think about it Do you have any other solutions?

Example 3. As shown in Figure 1.1, it is planned to lay a rectangular carpet in the middle of the floor of a rectangular conference room with a length of 16m and a width of 12m. The width of the unpaved floor around is the same. If the carpet area accounts for half of the floor area of the entire conference room, find the length and width of the carpet.

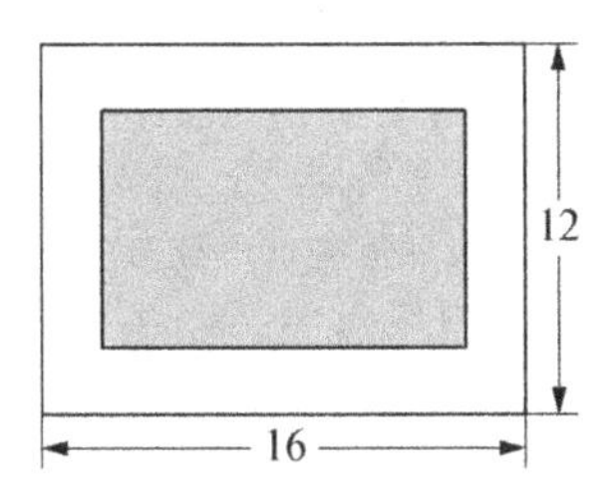

Fig. 1.1

Solution Let the width of the unpaved floor around be xm, then $0 < 2x < 12$, i.e. $0 < x < 6$.

Thus the length of the carpet is $(16 - 2x)$m and the width of the carpet is $(12 - 2x)$m.

From the given condition,

$$(16 - 2x)(12 - 2x) = \tfrac{1}{2} \times 16 \times 12,$$

$$x^2 - 14x + 24 = 0,$$

$$x = 2 \text{ or } x = 12.$$

As $0 < x < 6$, we discard $x = 12$. Thus $x = 2$.

Therefore,

$$16 - 2x = 16 - 2 \times 2 = 12,$$

$$12 - 2x = 12 - 2 \times 2 = 8.$$

So the length of the carpet is 12m, and the width of the carpet is 8m.

Example 4. Suppose that the following three quadratic equations of x

$$ax^2 + bx + c = 0, \quad bx^2 + cx + a = 0, \quad cx^2 + ax + b = 0$$

have exactly one common real root, find the value of $\frac{a^2}{bc} + \frac{b^2}{ca} + \frac{c^2}{ab}$.

Analysis Refer to Example 2, we can consider adding these three equations to simplify the problem.

Solution Let x_0 be the common real root, then

$$ax_0^2 + bx_0 + c = 0, \quad bx_0^2 + cx_0 + a = 0, \quad cx_0^2 + ax_0 + b = 0.$$

Adding up the above three equations, we can obtain that

$$(a + b + c)(x_0^2 + x_0 + 1) = 0.$$

Since $x_0^2 + x_0 + 1 = (x_0 + \frac{1}{2})^2 + \frac{3}{4} > 0$, the above equation can be reduced to

$$a + b + c = 0.$$

Therefore, $\dfrac{a^2}{bc} + \dfrac{b^2}{ca} + \dfrac{c^2}{ab} = \dfrac{a^3 + b^3 + c^3}{abc} = \dfrac{a^3 + b^3 - (a+b)^3}{abc} = \dfrac{-3ab(a+b)}{abc} = 3.$

Remark The values of these two algebraic expressions $x^2 + x + 1$ and $x^2 - x + 1$ are always greater than 0. This basic fact is often used.

Example 5. Given that m and n are rational numbers, and one root of the equation $x^2 + mx + n = 0$ is $\sqrt{5} - 2$, find the value of $m + n$.

Solution From the given condition,

$$(\sqrt{5} - 2)^2 + m(\sqrt{5} - 2) + n = 0,$$

i.e. $(9 - 2m + n) + (m - 4)\sqrt{5} = 0$.

Since m and n are rational numbers, and $\sqrt{5}$ is an irrational number, we have that

$$\begin{cases} 9 - 2m + n = 0, \\ m - 4 = 0. \end{cases}$$

So $\begin{cases} m = 4, \\ n = -1. \end{cases}$

Hence $m + n = 3$.

Think about it (1) Can you prove that $\sqrt{5}$ is an irrational number?

(2) If $\sqrt{5} - 2$ in the question is replaced by $\sqrt{c} - 2$ and $\sqrt{c}$ is an irrational number (c is a constant), what is the answer?

Example 6. Given that one root of the equation $(a-1)x^2 - 4x - 1 + 2a = 0$ is $x = 3$,

(1) find the value of a and the other root of the equation;
(2) if the lengths of all three sides of a triangle are the roots of this equation, find the perimeter of this triangle.

Solution (1) From the given condition,

$$9(a-1) - 4 \times 3 - 1 + 2a = 0.$$

Thus $a = 2$.

So the equation is $x^2 - 4x + 3 = 0$, the other root of the equation is $x = 1$.

(2) Since all three sides of the triangle are the roots of this equation, we know that at least two sides of the triangle are equal. Then there are two situations as follows:

(1) All three sides are equal. The lengths of the three sides are 1, 1, 1 or 3, 3, 3. So the perimeter of this triangle is 3 or 9.
(2) Only two sides are equal. Since $1 + 1 < 3$, the length of the three sides can just be 3, 3, 1. So the perimeter of this triangle is 7.

In summary, the perimeter of this triangle is 3 or 7 or 9.

Think about it If two sides of a triangle are the roots of this equation, try to find the range of the length of the third side.

Example 7. Suppose that x_1 and x_2 are two roots of the equation $x^2 + x - 3 = 0$, find the value of $x_1^3 - 4x_2^2 + 19$.

Solution From the given condition,

$$x_1^2 = 3 - x_1,$$
$$x_2^2 = 3 - x_2.$$

Therefore,

$$\begin{aligned} x_1^3 - 4x_2^2 + 19 &= x_1(3 - x_1) - 4(3 - x_2) + 19 \\ &= 3x_1 - (3 - x_1) + 4x_2 + 7 \\ &= 4(x_1 + x_2) + 4. \end{aligned}$$

From the quadratic formula we know that the two roots of the equation $x^2 + x - 3 = 0$ are $\frac{-1+\sqrt{13}}{2}$ and $\frac{-1-\sqrt{13}}{2}$, so $x_1 + x_2 = -1$.

Hence $x_1^3 - 4x_2^2 + 19 = 0$.

Remark The calculation of $x_1 + x_2$ can also be obtained by the factorization of two equations $x_1^2 = 3 - x_1$, $x_2^2 = 3 - x_2$ after subtraction. It can also be directly obtained by using the relationship between the roots and the coefficients in Chapter 4 (Vieta's Theorem). Using the equations satisfied by the roots of the equation to deal with the calculation of some algebraic expressions gives the role of "reducing the power", so as to simplify the calculation process.

Example 8. Given that $x = \frac{1}{1+\sqrt{2}}$, find the value of $\sqrt{x^3 + 2x^2 - x + 8}$.

Solution From the given condition,

$$x = \frac{1}{1+\sqrt{2}}$$
$$= \sqrt{2} - 1,$$

i.e. $x + 1 = \sqrt{2}$.
So $(x+1)^2 = (\sqrt{2})^2$,
i.e. $x^2 + 2x - 1 = 0$.
Therefore $\sqrt{x^3 + 2x^2 - x + 8} = \sqrt{x(x^2 + 2x - 1) + 8} = 2\sqrt{2}$.

Reading

Learn to Try

When solving a mathematical problem, we should be confident and try hard.

A little poem "I'll try" by C. Rossetti (1830–1894) is worth reading:

The little boy who says "I'll try",
Will climb to the hill-top.
The little boy who says "I can't",
Will at the bottom stop.
"I'll try" does great things every day,
"I can't" gets nothing done.
Be sure then that you say "I'll try",
And let "I can't" alone.

(Excerpted from: Shan Zun. *Problem solving research.* Shanghai: Shanghai Education Press, 2016, 4–5 and 218.)

Exercises

1. Let b and c be integers. When x is taken as 1, 3, 6 and 11 in turn, Xiao Cong calculates the values of the polynomials $x^2 + bx + c$ as 3, 5, 21

and 93 respectively. It is verified that only one result is wrong. Which one is wrong? ( )

(A) When $x = 1$, $x^2 + bx + c = 3$.
(B) When $x = 3$, $x^2 + bx + c = 5$.
(C) When $x = 6$, $x^2 + bx + c = 21$.
(D) When $x = 11$, $x^2 + bx + c = 93$.

2. Solve the equation: $(2x - 1)(x - 1) = (3x + 2)(x - 1)$.
3. Solve the equation: $x^2 - |x| - 1 = 0$.
4. Let k be a real number. Given that a root of the quadratic equation $x^2 - 3kx + k^2 - k = 0$ is -1, find the value of k.
5. Let b be a real number. Given that the equation $x^2 + bx + 1 = 0$ and the equation $x^2 - x - b = 0$ have a common real root, find the value of b.
6. Let a and b be integers. Given that a root of the equation $x^2 + ax + b = 0$ is $\sqrt{4 - 2\sqrt{3}}$, find the value of $a + b$.
7. Given that the real number a is a root of the equation $x^2 - 5x + 1 = 0$, find the last digit number of $a^4 + a^{-4}$.
8. Let c be a real number. Given that the opposite number of a root of the equation $x^2 - 3x + c = 0$ is a root of the equation $x^2 + 3x - c = 0$, find the roots of the equation $x^2 - 3x + c = 0$.
9. Given that $x = \frac{4-\sqrt{7}}{3}$, find the value of $\frac{x^4+x^2+1}{x^2}$.
10. Let m and n be real numbers. It is known that $m^2 = m+1$, $n^2 = n+1$, $m \neq n$, find the value of $m^5 + n^5$.

Chapter 2

Equations that can be Transformed into Quadratic Equations

In middle schools, for the integral equation whose degree is more than 2 (called higher-order equation), the factorization or the substitution is usually used to convert it into a quadratic equation with one unknown or a linear equation with one unknown to solve it.

Example 1. Solve the equation: $(x-1)(x^2+3x+2)=0$.

Solution The original equation is converted to

$$x-1=0 \quad \text{or} \quad x^2+3x+2=0.$$

Solving the equation $x-1=0$ gives $x_1=1$.
Solving the equation $x^2+3x+2=0$ gives $x_2=-1$ or $x_3=-2$.
After checking, we know that the roots of the original equation are

$$x_1=1, \quad x_2=-1, \quad x_3=-2.$$

Example 2. Given that the real number $x=\sqrt[3]{20+14\sqrt{2}}+\sqrt[3]{20-14\sqrt{2}}$, find the last digit of x^{2018}.

Solution From the given condition,

$$x^3=\left(\sqrt[3]{20+14\sqrt{2}}+\sqrt[3]{20-14\sqrt{2}}\right)^3.$$

So $x^3=40+3x\left(\sqrt[3]{20+14\sqrt{2}}\cdot\sqrt[3]{20-14\sqrt{2}}\right)^3,$

i.e. $x^3-6x-40=0.$

Observing the characteristics of the above equation, we can see that the value of x must be greater than 0, and x^3 must be at least greater than 40. It is obvious that the nearest number is $x = 4$, and it satisfies the equation. Thus the equation $x^3 - 6x - 40 = 0$ has a root of 4, that is, the expression $x^3 - 6x - 40$ has a factor of $x - 4$. Then, $x^3 - 6x - 40$ is factorized into

$$(x-4)(x^2+4x+10).$$

Therefore, $(x-4)(x^2+4x+10) = 0$.

Since $x^2+4x+10 = (x^2+4x+4)+6 = (x+2)^2+6 > 0$, we can obtain that $x = 4$.

As $4^{2018} = (16)^{1009}$, the last digit of it is 6. So the last digit of x^{2018} is 6.

Remark If there is a rational root for the equation with the form $ax^3 + bx^2+cx+d = 0$ (a, b, c, d are integers), the root is the divisor of the integer d (including negative integers), or the quotient of the divisor of d and the divisor of a (not equal to 1). (Please try to prove it) For example, if the equation $x^3 - 6x - 40 = 0$ has a rational root, its root is the divisor of 40, then we can try to put $x = -1, 1, 2, -2, -5, 5, 4, -4, -8, 8, 10, -10, 20, -20, 40, -40$ into the equation and get the rational root. Here we need to use the observation method to examine the value characteristics of the equation $x^3 - 6x - 40 = 0$.

Example 3. Solve the equation: $|x| - \frac{4}{x} = \frac{3|x|}{x}$.

Analysis The absolute value is removed, the fractional equation is obtained, and then it can be transformed into a quadratic equation.

Solution From the given condition, $x \neq 0$.

When $x > 0$, the original equation is converted to

$$x - \frac{4}{x} = 3,$$

i.e. $x^2 - 3x - 4 = 0$.

So $x = 4$ or $x = -1$.

As $x > 0$, we discard $x = -1$. Thus $x = 4$.

When $x < 0$, the original equation is converted to

$$-x - \frac{4}{x} = -3,$$

i.e. $x^2 - 3x + 4 = 0$.

Since $x^2 - 3x + \frac{9}{4} + \frac{7}{4} = \left(x - \frac{3}{2}\right)^2 + \frac{7}{4} > 0$, the equation $x^2 - 3x + 4 = 0$ has no real root.

In summary, the root of the original equation is $x = 4$.

Example 4. Solve the equation: $\dfrac{1}{x^2 + x - 2} + \dfrac{1}{x^2 + 7x + 10} = 2.$

Solution The original equation is converted to

$$\frac{1}{(x-1)(x+2)} + \frac{1}{(x+2)(x+5)} = 2,$$

$$\frac{1}{3}\left[\frac{1}{x-1} - \frac{1}{x+2}\right] + \frac{1}{3}\left[\frac{1}{x+2} - \frac{1}{x+5}\right] = 2,$$

i.e. $\frac{1}{x-1} - \frac{1}{x+5} = 6$.

Transform it into a quadratic equation $x^2 + 4x - 6 = 0$.

Solving the equation gives $x = -2 \pm \sqrt{10}$.

After checking, we know that the roots of the original equation are

$$x_1 = -2 + \sqrt{10}, x_2 = -2 - \sqrt{10}.$$

Remark

(1) For irrational equations, the basic idea of its solution is to transform it into rational equations. The methods commonly used are: completing the square, substitution, and factorization.

(2) According to Example 4, can you easily solve the equation $\frac{1}{x^2-5x+4} + \frac{1}{x^2+x-2} + \frac{1}{x^2+7x+10} = 2$? From this can you solve a similar equation? Try to give examples of similar equations.

Example 5. Let m be a real number. If the equation $\sqrt{x^2 - m} + 2\sqrt{x^2 - 1} = x$ has only one real root, find the value range of m.

Solution Mark the given equation as equation (1).

From the given condition,

$$x^2 - m \geq 0, \quad x^2 - 1 \geq 0, \quad x \geq 0.$$

So $x^2 \geq m$ and $x \geq 1$. If $m < 0$, then $\sqrt{x^2 - m} + 2\sqrt{x^2 - 1} > x$. At this time, the equation (1) has no root, which does not conform to the question, so $m \geq 0$. The equation (1) is converted to $2\sqrt{x^2 - 1} = x - \sqrt{x^2 - m}$. Square on both sides, $2x^2 + m - 4 = -2x\sqrt{x^2 - m}$. Square on both sides again, we get $8(2 - m)x^2 = (m - 4)^2$. So $(2 - m)x^2 \geq 0$, i.e. $2 - m \geq 0$.

Since $m \neq 2$, $0 \leq m < 2$. At this time, the equation (1) can only have the root $x = \frac{4-m}{\sqrt{8(2-m)}}$.

By substituting $x = \dfrac{4-m}{\sqrt{8(2-m)}}$ into the equation (1), we can obtain

$$\sqrt{\frac{(m-4)^2}{8(2-m)} - m} + 2\sqrt{\frac{(m-4)^2}{8(2-m)} - 1} = \frac{4-m}{\sqrt{8(2-m)}}.$$

Simplying and equation with the rearrangements, we obtain

$$|3m - 4| = 4 - 3m.$$

Thus, $0 \leq m \leq \frac{4}{3}$, at this time, the equation (1) has only one root $x = \frac{4-m}{\sqrt{8(2-m)}}$.

In summary, the value range of the real number m is $0 \leq m \leq \frac{4}{3}$.

Example 6. Let k be a real number. If the equation $x^2 - 2x = k|x|$ has exactly three different real roots, find the value range of k.

Solution It's obvious that $x = 0$ is a root of the equation $x^2 - 2x = k|x|$.

If the equation $x^2 - 2x = k|x|$ has a root when $x < 0$, then

$$x^2 - 2x = -kx, i.e. x - 2 = -k.$$

Thus $-k + 2 < 0$, $k > 2$.

If the equation $x^2 - 2x = k|x|$ has a root when $x > 0$, then

$$x^2 - 2x = kx, \text{ i.e. } x - 2 = k.$$

Thus $k + 2 > 0$, $k > -2$.

In summary, the value range of the real number k is $k > 2$.

Example 7. Given that a and b are negative real numbers satisfying $\frac{1}{a} + \frac{1}{b} - \frac{1}{a-b} = 0$, find the value of $\frac{b}{a}$.

Solution From $\frac{1}{a} + \frac{1}{b} - \frac{1}{a-b} = 0$, we can obtain $a^2 - b^2 - ab = 0$, i.e.

$$\left(\frac{b}{a}\right)^2 + \frac{b}{a} - 1 = 0.$$

Since a and b are negative real numbers, $\frac{b}{a}$ can be regarded as a positive root of the equation $x^2 + x - 1 = 0$.

Solving this equation gives $x = \frac{-1-\sqrt{5}}{2}$ or $x = \frac{-1+\sqrt{5}}{2}$. Hence, $\frac{b}{a} = \frac{-1+\sqrt{5}}{2}$.

Think about it

(1) Whether there exist any real numbers a and b, which satisfy one of the following equations: (i) $\frac{1}{a}-\frac{1}{b}-\frac{1}{a+b}=0$, (ii) $\frac{1}{a}+\frac{1}{b}+\frac{1}{a+b}=0$, (iii) $\frac{1}{a}-\frac{1}{b}+\frac{1}{a-b}=0$.
(2) Whether there exist any real numbers a and b, which satisfy $\frac{1}{a}+\frac{1}{b}-\frac{1}{a+b}=0$?
(3) Furthermore, whether there exist any real numbers a, b and c, which satisfy $\frac{1}{a}+\frac{1}{b}+\frac{1}{c}=\frac{1}{a+b+c}$?

Example 8. Given that the real numbers a, b and c satisfy the equation $a+\frac{1}{b}=b+\frac{1}{c}=c+\frac{1}{a}=t$, where a, b and c are not equal to each other, find the value of t.

Solution From $a+\frac{1}{b}=t$, we can obtain $b=\frac{1}{t-a}$. Substituting this into the equation $b+\frac{1}{c}=t$, we have

$$\frac{1}{t-a}+\frac{1}{c}=t.$$

Rearrange the equation and get

$$ct^2-(ac+1)t+(a-c)=0.$$

From $c+\frac{1}{a}=t$, we can obtain $ac+1=at$. Substituting this into the above equation, and we get

$$ct^2-at^2+(a-c)=0,$$

i.e.

$$(c-a)(t^2-1)=0.$$

Since $c\neq a$, $t^2-1=0$. Thus, $t=\pm 1$. Check: When $b=\frac{1}{1-a}$ and $c=\frac{a-1}{a}$, $t=1$. When $b=-\frac{1}{1+a}$ and $c=-\frac{a+1}{a}$, $t=-1$. Hence, $t=\pm 1$.

Think about it

(1) Is there any other solution?
(2) Under the condition of Example 8, find the value of abc. Furthermore, you can also refer to Example 8 of Chapter 5.

Example 9. As shown in Figure 2.1, in Rt$\triangle ABC$, the length of the hypotenuse AB is 35. The square $CDEF$ is inscribed in $\triangle ABC$, and the length of its side is 12. Find the perimeter of $\triangle ABC$.

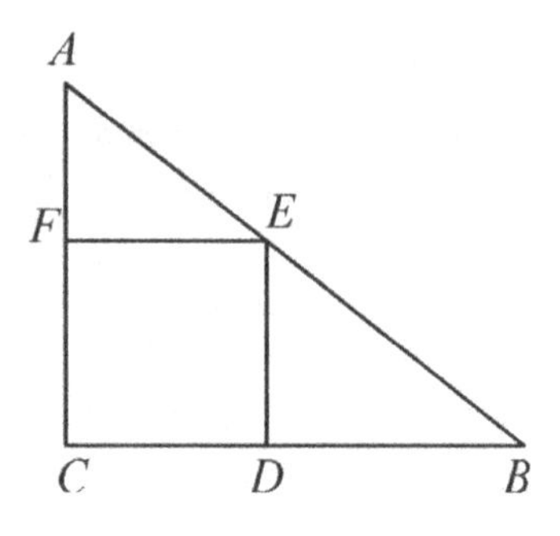

Fig. 2.1

Solution As shown in Figure 2.1, suppose that $BC=a$, $AC=b$, then

$$a^2+b^2=35^2=1225. \tag{1}$$

Since $\text{Rt}\triangle AFE \backsim \text{Rt}\triangle ACB$, we have that $\frac{FE}{CB}=\frac{AF}{AC}$, i.e. $\frac{12}{a}=\frac{b-12}{b}$. Thus,

$$12(a+b)=ab. \tag{2}$$

From the equations (1) and (2),

$$(a+b)^2=a^2+b^2+2ab=1225+24(a+b),$$

i.e. $(a+b)^2-24(a+b)-1225=0$.

Solving the quadratic equation gives $a+b=49$ or $a+b=-25$ (discarded).

Hence,

$$a+b+c=49+35=84.$$

Reading

Do Not Use the Skill Just for Skillful Solution.

Skill is quite important, however, the purpose of using skill is to solve the problem rather than for just using the skills.

Example Solve the fractional equation: $\dfrac{2}{x^2-4}-\dfrac{1}{x(x-1)}+\dfrac{x-4}{x(x+2)}=0$.

Solution 1 Multiplying both sides of the equation by $x(x-2)(x+2)$ gives

$$2x-(x+2)+(x-2)(x-4)=0,$$

i.e. $x^2-5x+6=0$.

Solving the quadratic equation gives $x_1 = 1$, $x_2 = 3$. After checking, we know that only $x = 3$ is the root of the original equation.

Solution 2 Since $\frac{1}{x^2-4} = \frac{1}{2}\left(\frac{1}{x-2} - \frac{1}{x+2}\right)$,

$$\frac{1}{x(x-2)} = \frac{1}{2}\left(\frac{1}{x-2} - \frac{1}{x}\right),$$

$$\frac{x-4}{x(x+2)} = \frac{1}{x+2} - \frac{4}{x(x+2)} = \frac{3}{x+2} - \frac{2}{x},$$

the original equation can be transformed into

$$\frac{1}{2}\left(\frac{1}{x-2} - \frac{1}{x+2}\right) - \frac{1}{2}\left(\frac{1}{x-2} - \frac{1}{x}\right) + \left(\frac{3}{x+2} - \frac{2}{x}\right) = 0,$$

i.e. $\frac{1}{2}\left(\frac{5}{x+2} - \frac{3}{x}\right) = 0$.

Solving the above equation gives $x = 3$.

The second solution takes advantage of the skill by transforming fraction to partial fraction, which looks quite smart, however it does not show a good result.

(Excerpted from: Shan Zun. *Problem solving research.* Shanghai: Shanghai Education Press, 2016, $188 \sim 189$.)

Exercises

1. Solve the equation: $(x-2)(x^2-3x-4) = 0$.
2. Solve the equation: $x^2 - |2x-1| - 2 = 0$.
3. Solve the equation: $\frac{x+1}{x+2} + \frac{x+8}{x+9} = \frac{x+2}{x+3} + \frac{x+7}{x+8}$.
4. Solve the equation: $x^2 + \frac{1}{x^2} - \frac{7}{2}\left(x - \frac{1}{x}\right) + 1 = 0$.
5. Solve the equation: $2\left(x^2 + \frac{1}{x^2}\right) - 3\left(x + \frac{1}{x}\right) = 1$.
6. Solve the equation: $\left(\frac{x^2-1}{x}\right)^2 - \frac{7}{6}\left(\frac{x^2-1}{x}\right) - 4 = 0$.
7. Solve the equation of x: $\frac{a+x}{b+x} + \frac{b+x}{a+x} = \frac{5}{2}$.
8. Solve the equation: $\sqrt{x+8} - \sqrt{5x+20} + 2 = 0$.

9. Solve the equation: $x^2 + x + 2x\sqrt{x+2} = 14$.
10. Solve the equation: $x^2 + 18x + 30 = 2\sqrt{x^2 + 18x + 45}$.
11. Given that the real numbers a and b satisfy the equation $\frac{1}{a} - \frac{1}{b} + \frac{1}{a-b} = 0$, find the value of $\frac{b}{a}$.
12. Solve the equation: $4x^2 + 2x\sqrt{3x^2 + x} + x - 9 = 0$.
13. Solve the equation: $x^3 + 2x^2 - 5x + 2 = 0$.

Chapter 3

Discriminant of Quadratic Equation

As we know, to solve the quadratic equation $ax^2 + bx + c = 0 (a \neq 0)$ by completing the square, it can be transformed into

$$a\left(x + \frac{b}{2a}\right)^2 = \frac{b^2 - 4ac}{4a},$$

i.e.

$$\left(x + \frac{b}{2a}\right)^2 = \frac{b^2 - 4ac}{4a^2}.$$

According to the definition of square root, when $b^2 - 4ac \geq 0$, the equation has real roots; when $b^2 - 4ac < 0$, the equation has no real roots. The quantity $b^2 - 4ac$ is called the discriminant of a quadratic equation. The symbol delta Δ is used to represent the discriminant, so $\Delta = b^2 - 4ac$.

When $b^2 - 4ac = 0$, then $\left(x + \frac{b}{2a}\right)^2 = 0$. So the equation has two identical real roots which is call a repeated root. When $b^2 - 4ac > 0$, the equation has two distinct real roots which are $x_1 = \frac{-b+\sqrt{\Delta}}{2a}$ and $x_2 = \frac{-b-\sqrt{\Delta}}{2a}$ respectively. Thus we know that the sign of Δ is associated with the nature of the solutions of the quadratic equation.

Example 1. Let k be a real number. Try to judge whether the equation $(k-1)x^2 + 3kx + k + 1 = 0$ has real roots or not?

Prove From the given condition, when $k = 1$, the original equation is a linear equation $3x + 2 = 0$. So $x = -\frac{2}{3}$ is the root of the equation.

When $k \neq 1$, the original equation is a quadratic equation. Then its discriminant is

$$\Delta = (3k)^2 - 4(k-1)(k+1) = 5k^2 + 4.$$

Since $5k^2 \geq 0$, $5k^2 + 4 > 0$, i.e. $\Delta > 0$, there are two distinct real roots.

In summary, when $k = 1$, the equation has only one real root; when $k \neq 1$, the equation has two distinct real roots.

Remark The discriminant Δ can only be used for quadratic equations.

Example 2. Given that a, b and c are three sides of a triangle, try to judge whether the equation $b^2x^2 + (b^2 + c^2 - a^2)x + c^2 = 0$ has real roots or not?

Solution The discriminant of the equation $b^2x^2+(b^2+c^2-a^2)x+c^2 = 0$ is

$$\Delta = (b^2 + c^2 - a^2)^2 - 4b^2c^2.$$

Factoring the discriminant, we get

$$\begin{aligned}\Delta &= (b^2 + c^2 - a^2 - 2bc)(b^2 + c^2 - a^2 + 2bc) \\ &= [(b-c)^2 - a^2][(b+c)^2 - a^2] \\ &= (b-c-a)(b-c+a)(b+c-a)(b+c+a).\end{aligned}$$

Since $b + c > a$, $a + b > c$, $c + a > b$, we can obtain

$$b - c - a < 0, \quad b - c + a > 0, \quad b + c - a > 0.$$

Thus $\Delta < 0$, the equation has no real roots.

Example 3. Consider the equation $|x^2 - 5x| = a$. Find the value of a for which the equation has and only has two distinct real roots.

Solution From the given condition, $a \geq 0$.

When $a = 0$, $|x^2 - 5x| = 0$. Solve the equation and get $x_1 = 0$, $x_2 = 5$, which satisfy the given condition.

When $a > 0$, the original equation is transformed into

$$x^2 - 5x - a = 0, \tag{1}$$

$$\text{and } x^2 - 5x + a = 0. \tag{2}$$

Consider the equation (1). Because its discriminant is $25 + 4a > 0$, the equation (1) has two distinct real roots.

Therefore, it is known from the given condition that the equation (2) has no real roots or its roots are the same as that of the equation (1).

Obviously, if the equations (1) and (2) have the same roots, $a = 0$.

Since $a > 0$, the equation (2) must have no real roots. Thus, the discriminant is $25 - 4a < 0$, i.e. $a > \frac{25}{4}$.

In summary, the value range of a is $a = 0$ or $a > \frac{25}{4}$.

Think about it For the equation $|x^2 - 5x| = a$ in Example 3, try to discuss the case of real roots and find the corresponding real roots.

Example 4. For real numbers u and v, an operation "*" is defined as $u * v = uv + v$. If the equation $x * (a * x) = -\frac{1}{4}$ with regard to x has two distinct real roots, find the range of the real number a.

Solution From the definition of "*", we can know that

$$a * x = ax + x, \quad x * (a * x) = x(ax + x) + (ax + x).$$

Since $x * (a * x) = -\frac{1}{4}$,

$$(a+1)x^2 + (a+1)x + \frac{1}{4} = 0.$$

From the given condition,

$$\begin{cases} a + 1 \neq 0, \\ \Delta = (a+1)^2 - (a+1) > 0. \end{cases}$$

So $a(a+1) > 0$, then $a > 0$ or $a < -1$.

Think about it Following the think given in Example 3 you try to discuss the case of real roots of the equation $x * (a * x) = -\frac{1}{4}$ about x.

Example 5. Find the real roots of the equation $x + y = x^2 - xy + y^2 + 1$.

Solution The equation $x+y = x^2 - xy + y^2 + 1$ is regarded as a quadratic equation with regard to x, that is,

$$x^2 - (y+1)x + (y^2 - y + 1) = 0.$$

To make sure that this equation has real roots, there must have

$$\Delta = (y+1)^2 - 4(y^2 - y + 1) = -3(y-1)^2 \geq 0.$$

Since $-3(y-1)^2 \leq 0$, we can obtain $y - 1 = 0$, i.e. $y = 1$. Substituting $y = 1$ into the original equation, we get $x^2 - 2x + 1 = 0$, i.e. $x = 1$.

After checking, we know that $x = 1$, $y = 1$ satisfy the given equation. So the real roots of the equation are $x = 1$, $y = 1$.

Remark When solving the quadratic equations of two variables (or the equations with letter coefficients), one of the basic ideas is to use discriminant to separate variables and discuss one by one according to the nature of the real solutions. Of course, we can also use the method of completing the square to get the result. Transform the equation in Example 5 into

$$2x + 2y = 2x^2 - 2xy + 2y^2 + 2,$$

i.e. $(x-1)^2 + (y-1)^2 + (x-y)^2 = 0$.

Thus, $x = 1$, $y = 1$, satisfying the given condition. It can be seen that completing the square is also one of the basic ideas to deal with this kind of problem.

We know that from the equation $ax^2 + bx + c = 0 (a \neq 0)$, we will think of the discriminant Δ. On the contrary, when the form of $\Delta = b^2 - 4ac$ is encountered, it is easy to solve the problem by associating with the corresponding quadratic equation.

Example 6. Given that $(x-z)^2 - 4(x-y)(y-z) = 0$, find the relationship of $x + z$ and y.

Analysis Seeing "$(x-z)^2 - 4(x-y)(y-z) = 0$", we think of "$b^2 - 4ac = 0$", and strive to solve the problem of a quadratic equation with two identical real roots.

Solution 1 When $x - y \neq 0$ *i.e.* $x \neq y$, construct an equation

$$(x-y)t^2 - (x-z)t + (y-z) = 0.$$

Because $\Delta = (x-z)^2 - 4(x-y)(y-z) = 0$, the above equation has two identical real roots. Since $(x-y) - (x-z) + (y-z) = 0$, the real roots of the equation are $t_1 = t_2 = 1$.

From the quadratic formula,

$$t_1 + t_2 = \frac{(x-z) + \sqrt{\Delta}}{2(x-y)} + \frac{(x-z) - \sqrt{\Delta}}{2(x-y)} = \frac{x-z}{x-y} = 1 + 1 = 2,$$

$$t_1 \cdot t_2 = \frac{(x-z) + \sqrt{\Delta}}{2(x-y)} \cdot \frac{(x-z) - \sqrt{\Delta}}{2(x-y)} = \frac{y-z}{x-y} = 1 \times 1 = 1.$$

Thus $x - z = 2(x-y)y - z = x - y$, i.e. $x + z = 2y$.

When $x - y = 0$ *i.e.* $x = y$, from the given condition,

$$x - z = 0, \text{ i.e. } x = z.$$

So $x + z = 2y$. In summary, $x + z = 2y$.

Solution 2 From the given condition,

$$x^2 + 2xz + z^2 - 4xy - 4yz + 4y^2 = 0,$$

i.e. $(x + z)^2 - 4y(x + z) + 4y^2 = 0$.
So $[(x + z) - 2y]^2 = 0$, i.e. $x + z = 2y$.

Solution 3 Let $x - y = a$, $y - z = b$. Then the given equation is transformed into

$$(x - y + y - z)^2 = 4(x - y)(y - z),$$

i.e. $(a + b)^2 = 4ab$. Then we have $(a - b)^2 = 0$, so $a = b$. Hence, $x - y = y - z$, i.e. $x + z = 2y$.

Remark It can be seen that it is rather simple to solve this problem by using identical transformation, especially overall substitution. The solution of constructing a quadratic equation presented in this example is only to expand the readers' thinking, not to encourage the readers to "go for the abstruse and forget the obvious" when solving problems. We should highly recommend the natural and simple solutions.

Example 7. Given that $m \geq -1$, the equation

$$x^2 + 2(m - 2)x + m^2 - 3m + 3 = 0$$

has two distinct real roots x_1 and x_2, find the maximum value of $\frac{mx_1^2}{1-x_1} + \frac{mx_2^2}{1-x_2}$.

Solution From the given condition, $\Delta > 0$, i.e.

$$\Delta = [2(m - 2)]^2 - 4(m^2 - 3m + 3) = -4m + 4 > 0.$$

So $m < 1$. Since $m \geq -1$, we get $-1 \leq m < 1$.

From the quadratic formula, the roots of the equation are

$$x_1 = 2 - m + \sqrt{1 - m}, \quad x_2 = 2 - m - \sqrt{1 - m}.$$

Thus,

$$\begin{aligned}\frac{x_1^2}{1-x_1} &= \frac{-(1-x_1^2)+1}{1-x_1} \\ &= \frac{1}{1-x_1} - 1 - x_1 \\ &= \frac{1}{m-1-\sqrt{1-m}} + m - 3 - \sqrt{1-m} \\ &= \frac{m-1+\sqrt{1-m}}{(m-1)^2-(\sqrt{1-m})^2} + m - 3 - \sqrt{1-m} \\ &= \frac{m-1+\sqrt{1-m}}{m(m-1)} + m - 3 - \sqrt{1-m}.\end{aligned}$$

Similarly,

$$\frac{x_2^2}{1-x_2} = \frac{m-1-\sqrt{1-m}}{m(m-1)} + m - 3 + \sqrt{1-m}.$$

Let $S = \frac{mx_1^2}{1-x_1} + \frac{mx_2^2}{1-x_2}$. Then

$$\begin{aligned}S &= \frac{m \cdot 2(m-1)}{m(m-1)} + m \cdot 2(m-3) \\ &= 2(m^2 - 3m + 1) \\ &= 2\left(m - \frac{3}{2}\right)^2 - \frac{5}{2}.\end{aligned}$$

Since $-1 \leq m < 1$, we know that when $m = -1$, the maximum value of S is 10.

Remark We can also use the knowledge of Chapter 4 to deal with this problem. Try to compare the advantages and disadvantages of both methods.

Reading

This is mathematics: she makes you realize that the soul is invisible; she gives life to mathematical discovery; she wakes up the mind and enlightens the wisdom; she lights up the ideas in our hearts; she eliminates our innate fatuity and ignorance······

——Proclus (ancient Greece)

Why should we do the exercises? First, it is to deepen our understanding of the basic concepts, definitions and theorems in books. This is the main thing. Second, it is also to train our operational skills and logical thinking. Although this is secondary, it is essential.

Doing exercises is beneficial to deepen understanding and improve operation skills and logical thinking. However, it must be pointed out that it is not good to learn mathematics if we neglect the basic concepts, definitions and theorems (including proofs) in the textbook of deep learning only by calculating exercises. Therefore, when solving a problem, we should first see what basic knowledge this problem contains and which formulas or theorems will be used, and then solve the problem step by step from a formula or theorem.

(Excerpted from: Su Buqing. *Magic symbols.* Changsha: Hunan children's publishing house, 2010, 186.)

Exercises

1. If the equation $9x^2 - (k+6)x + k + 1 = 0$ has two identical real roots, find the value of k.
2. If the equation $ax^2 + x + 2 = 0$ of x has no real roots, find the value range of a.
3. Consider the equation $kx^2 - (2k+1)x + k = 0$. Find the value of k for which the equation has:

   (1) two distinct real roots;
   (2) a repeated root;
   (3) no real roots.
4. Given that the equation $x^2 + 2x = n - 1$ has no real roots. where n is a real number, prove that the equation $x^2 + nx = 1 - 2n$ must have two distinct real roots.
5. Given that the equation $(a^2 - 1)x^2 - 2(a+1)x + 1 = 0$ has exactly one real root, where a is a real number, find the value of a.
6. Given that the equation $2x(kx - 4) - x^2 + 6 = 0$ has no real roots, where k is a real number find the value of k.
7. Given that the equation $x^2 - 2\sqrt{-a}x + \frac{(a-1)^2}{4} = 0$ has a real root x_0, where a is a real number, find the value of $a^5 - x_0^5$.
8. Consider the equation $2(m+1)x^2 + 2\sqrt{6}mx + 3m - 2 = 0$, where m is a real number. Find the value of m for which the equation has two distinct real roots.

9. Suppose that a and b are real numbers. Try to decide the number of real roots of the equation $(x-a)(x-a-b)=1$.
10. Consider the equation $|x^2+3x|=a$ where a is a real number. Find the value of a for which the equation has exactly three distinct real roots.
11. Suppose that a, b and c are real numbers which are unequal to each other. Prove that the quadratic equations $ax^2+2bx+c=0$, $bx^2+2cx+a=0$ and $cx^2+2ax+b=0$ can't have two identical real roots at the same time.
12. Find the real roots of the following equation of x and y:
$$5x^2-3xy+\frac{1}{2}y^2-2x+\frac{1}{2}y+\frac{1}{4}=0.$$
13. Suppose that a is a real number. Consider the equation $\frac{x+1}{x-1}+\frac{x-1}{x+1}+\frac{2x+a+2}{x^2-1}=0$. Find the sum of all the values of a for which the equation has only one real root.
14. Suppose that a is a real number. If the equation $x^2+2ax+7a-10=0$ has no real roots, then which of the following equations must has real roots? ( ).

    (A) $x^2+2ax+3a-2=0$
    (B) $x^2+2ax+5a-6=0$
    (C) $x^2+2ax+10a-21=0$
    (D) $x^2+2ax+2a+3=0$

Chapter 4

Relationship Between Roots and Coefficients and Application

If the quadratic equation $ax^2+bx+c=0(a \neq 0)$ has two real roots x_1 and x_2, then

$$ax^2+bx+c=a(x-x_1)(x-x_2).$$

Comparing the coefficients of the corresponding terms on both sides of the equation, we conclude that

$$\begin{cases} x_1+x_2=-\dfrac{b}{a}, & (1) \\ x_1 \cdot x_2=\dfrac{c}{a}. & (2) \end{cases}$$

Formulas (1) and (2) can also be obtained by using the quadratic formula.

Formulas (1) and (2) are called Vieta's theorem, that is, the relationship between roots and coefficients.

Therefore, given the quadratic equation $ax^2+bx+c=0$, there must be formulas (1) and (2). Conversely, if there are two numbers x_1 and x_2 satisfying formulas (1) and (2), then these two numbers x_1 and x_2 must be the roots of a quadratic equation $ax^2+bx+c=0$. Using this basic knowledge can often solve the problems easily.

Example 1. Given that the equation $2x^2 - 3x - 5 = 0$ has roots x_1 and x_2, find the values of:

$$(1)\ x_1^2 + x_2^2, \quad (2)\ x_1^3 + x_2^3, \quad (3)\ x_1^5 + x_2^5.$$

Solution From Vieta's Formula, we recognize that

$$x_1 + x_2 = \frac{3}{2},$$
$$x_1 \cdot x_2 = -\frac{5}{2}.$$

(1)
$$\begin{aligned} x_1^2 + x_2^2 &= (x_1 + x_2)^2 - 2x_1x_2 \\ &= \frac{9}{4} - 2 \times \left(-\frac{5}{2}\right) \\ &= \frac{29}{4}. \end{aligned}$$

(2) From the solution of (1), we get

$$\begin{aligned} x_1^3 + x_2^3 &= (x_1 + x_2)(x_1^2 + x_2^2) - x_1x_2^2 - x_2x_1^2 \\ &= (x_1 + x_2)(x_1^2 + x_2^2) - x_1x_2(x_1 + x_2) \\ &= \frac{3}{2} \times \frac{29}{4} - \left(-\frac{5}{2}\right) \times \frac{3}{2} \\ &= \frac{117}{8}. \end{aligned}$$

(3) From the solutions of (1) and (2), we get

$$\begin{aligned} x_1^4 + x_2^4 &= (x_1 + x_2)(x_1^3 + x_2^3) - x_1x_2(x_1^2 + x_2^2) \\ &= \frac{3}{2} \times \frac{117}{8} - \left(-\frac{5}{2}\right) \times \frac{29}{4} \\ &= \frac{641}{16}. \end{aligned}$$

So,

$$\begin{aligned} x_1^5 + x_2^5 &= (x_1 + x_2)(x_1^4 + x_2^4) - x_1x_2(x_1^3 + x_2^3) \\ &= \frac{3}{2} \times \frac{641}{16} - \left(-\frac{5}{2}\right) \times \frac{117}{8} \\ &= \frac{3093}{32}. \end{aligned}$$

Remark

(1) For Example 1, there are other ways for readers to try.
(2) By observing the solution process of Example 1, we find that we can calculate the value of the general expression $x_1^n + x_2^n$ (n is a positive integer). If we have already known $A_{n-2} = x_1^{n-2} + x_2^{n-2}$ and $A_{n-1} = x_1^{n-1} + x_2^{n-1}$, we can use

$$A_n = x_1^n + x_2^n = (x_1 + x_2)A_{n-1} - x_1x_2A_{n-2}$$

to calculate the value of $A_n = x_1^n + x_2^n$.
(3) In fact, since you can find the value of $x_1^n + x_2^n$ when n is a positive integer, you can also find it when n is an integer. Please think about the reasons.

Example 2. Suppose that k is a real number. Given that x_1 and x_2 are two distinct real roots of the equation $x^2 - 2(k+1)x + k^2 + 2 = 0$ with $(x_1+1)(x_2+1) = 8$, find the value of k.

Solution From Vieta's Formula, we recognize that

$$x_1 + x_2 = 2(k+1),$$
$$x_1 \cdot x_2 = k^2 + 2.$$

So $(x_1 + 1)(x_2 + 1) = x_1x_2 + (x_1 + x_2) + 1 = k^2 + 2k + 5$. Since $(x_1+1)(x_2+1) = 8$, we have

$$k^2 + 2k - 3 = 0.$$

Solving this equation, we get $k = -3$ or $k = 1$. Because the equation has two distinct real roots,

$$\Delta = 4(k+1)^2 - 4(k^2+2) = 8k - 4 > 0,$$

i.e. $k > \frac{1}{2}$. Hence, we get $k = 1$.

Example 3. Find the largest integer which is not more than $(\sqrt{5}+\sqrt{3})^6$.

Solution $(\sqrt{5}+\sqrt{3})^6 = (8+2\sqrt{15})^3$.

Let $8 + 2\sqrt{15} = a$, $8 - 2\sqrt{15} = b$. Then we have $ab = 4$, $a + b = 16$. So a and b are the roots of the equation $x^2 - 16x + 4 = 0$. Substituting a and b in the equation, we get

$$a^2 = 16a - 4, \quad b^2 = 16b - 4.$$

Then we have $a^3 = 16a^2 - 4a$, $b^3 = 16b^2 - 4b$.

Hence,

$$\begin{aligned} a^3 + b^3 &= 16(a^2 + b^2) - 4(a + b) \\ &= 16[(a + b)^2 - 2ab] - 4(a + b) \\ &= 16 \times (16^2 - 2 \times 4) - 4 \times 16 \\ &= 3904. \end{aligned}$$

Since $0 < b < 1$, we know $0 < b^3 < 1$. Thus the largest integer which is not more than a^3 is 3903. So the largest integer which is not more than $(\sqrt{5} + \sqrt{3})^6$ is 3903.

Remark We always associate the expression in the form of $c + \sqrt{d}$ (or $c - \sqrt{d}$) with the expression $c - \sqrt{d}$ (or $c + \sqrt{d}$). Using the equations $(c + \sqrt{d}) + (c - \sqrt{d}) = 2c$ and $(c + \sqrt{d}) \cdot (c - \sqrt{d}) = c^2 - d$ to construct a quadratic equation with one unknown can solve the related problems simply.

Example 4. Suppose a and b are real numbers satisfying $b > 0$ and $a^2 \geq 4b$. Try to explain that there must be non-zero real numbers x and y such that $a + 2\sqrt{b} = (\sqrt{x} + \sqrt{y})^2$.

Solution From $a + 2\sqrt{b} = (\sqrt{x} + \sqrt{y})^2$, we have $a + 2\sqrt{b} = x + y + 2\sqrt{x}\sqrt{y}$.

Suppose that $x + y = a$, $xy = b$, (1). Then we try to explain whether there are real numbers x and y satisfying (1).

From Vieta's Formula, we know that x and y can be the roots of the equation

$$t^2 - at + b = 0.$$

Since $a^2 \geq 4b$, the above equation has two real roots x and y.

From the quadratic formula, we get

$$x = \frac{a + \sqrt{a^2 - 4b}}{2}, \quad y = \frac{a - \sqrt{a^2 - 4b}}{2}.$$

So there must be real numbers x and y that satisfy $a + 2\sqrt{b} = (\sqrt{x} + \sqrt{y})^2$.

Remark Using the conclusion of Example 4, we can simplify $\sqrt{11 + 2\sqrt{18}}$ to $3 + \sqrt{2}$.

Example 5. Suppose that a, b, c and d are real numbers. Given that the quadratic equation $x^2 + cx + d = 0$ has the roots a and b, and the quadratic equation $x^2 + ax + b = 0$ has the roots c and d, find all arrays (a, b, c, d) that satisfy the conditions.

Solution From Vieta's Formula, we get

$$a + b = -c, \quad ab = d, \quad c + d = -a, \quad cd = b.$$

So $b = -a - c = d$.

When $b = d = 0$, we have $c = -a$. Thus $(a, b, c, d) = (t, 0, -t, 0)$ where t is any real number. When $b = d \neq 0$, we have $a = \frac{d}{b} = 1$, $c = \frac{d}{b} = 1$. Thus $b = -2$, $d = -2$. So $(a, b, c, d) = (1, -2, 1, -2)$.

After checking, we get that the arrays $(1, -2, 1, -2)$ and $(t, 0, -t, 0)$ (where t is any real number) satisfy the conditions.

Example 6. Given that real numbers x and y satisfy the expression $xy - x - y = 1$, then which of the following numbers is the minimum value of $x^2 + y^2$?

(A) $3 - 2\sqrt{2}$ (B) $6 - 4\sqrt{2}$ (C) 1 (D) $6 + 4\sqrt{2}$

Solution Suppose $x + y = t$. Then from the given condition, we know that $xy = x + y + 1 = t + 1$. So x and y are the real roots of the quadratic equation $m^2 - tm + t + 1 = 0$.

Therefore, the discriminant of this quadratic equation is $\Delta = t^2 - 4(t + 1) \geq 0$.

Solving the inequality gives $t \geq 2 + 2\sqrt{2}$ or $t \leq 2 - 2\sqrt{2}$. Since $x^2 + y^2 = (x + y)^2 - 2xy = t^2 - 2(t + 1) = (t - 1)^2 - 3$, we know that $x^2 + y^2$ gets the minimum value when $t = 2 - 2\sqrt{2}$ (i.e. $x = y = 1 - \sqrt{2}$). The minimum value of $x^2 + y^2$ is $(2 - 2\sqrt{2} - 1)^2 - 3 = 6 - 4\sqrt{2}$. So the answer is B.

Example 7. As shown in Figure 4.1, the straight line DE is parallel to BC, and the area of ΔBDE is $S_{\Delta BDE} = k^2$ (constant value). What's the relation between k^2 and $S_{\Delta ABC}$ when such lines DE exist? And how many lines can we get?

Solution Suppose $\frac{AD}{AB} = x$, $S_{\Delta ABC} = S$.

Since $DE//BC$, we know that $\frac{AE}{AC} = \frac{AD}{AB} = x$. Thus,

$$S_{\Delta ADE} = x^2 S, \quad S_{\Delta ABE} = xS.$$

Because $S_{\Delta ABE} = S_{\Delta ADE} + S_{\Delta BDE}$, we have

$$xS = x^2 S + k^2,$$

i.e. $x^2 S - xS + k^2 = 0$.

To make this equation have real roots, then

$$\Delta = S^2 - 4k^2 S \geq 0,$$

i.e. $S \geq 4k^2$.

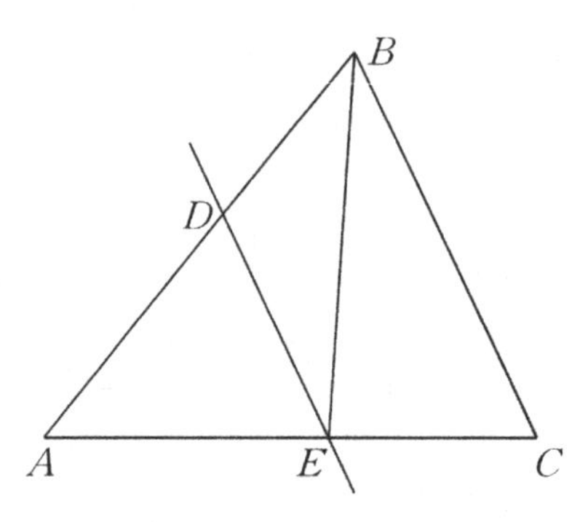

Fig. 4.1

Let the real roots of the equation be x_1 and x_2. From Vieta's Formula, we get

$$x_1 + x_2 = 1, \quad x_1 x_2 = \frac{k^2}{S}.$$

Because both 1 and $\frac{k^2}{S}$ are positive numbers, x_1 and x_2 are non-negative numbers. And because $S \geq 4k^2$, we get

$$0 < x_1, x_2 < 1.$$

Hence, when $S_{\Delta ABC} = S = 4k^2$, $x_1 = x_2$, there is only one straight line DE to meet the requirements. When $S_{\Delta ABC} = S > 4k^2$, $x_1 \neq x_2$, there are two such lines to meet the requirements.

Example 8. For all the natural numbers n which are not less than 2, the roots of the quadratic equation $x^2 - (n+2)x - 2n^2 = 0$ are recorded as a_n and $b_n (n \geq 2)$. Find the value of

$$\frac{1}{(a_2 - 2)(b_2 - 2)} + \frac{1}{(a_3 - 2)(b_3 - 2)} + \cdots + \frac{1}{(a_{101} - 2)(b_{101} - 2)}.$$

Solution From Vieta's Formula, we get

$$a_n + b_n = n + 2, \quad a_n b_n = -2n^2.$$

Thus,

$$\begin{aligned}(a_n - 2)(b_n - 2) &= a_n b_n - 2(a_n + b_n) + 4 \\ &= -2n^2 - 2(n+2) + 4 \\ &= -2n(n+1).\end{aligned}$$

Then we have $\dfrac{1}{(a_n-2)(b_n-2)} = -\dfrac{1}{2n(n+1)} = -\dfrac{1}{2}\left(\dfrac{1}{n} - \dfrac{1}{n+1}\right).$

Hence,

$$\begin{aligned}
&\frac{1}{(a_2-2)(b_2-2)} + \frac{1}{(a_3-2)(b_3-2)} + \cdots + \frac{1}{(a_{101}-2)(b_{101}-2)} \\
&= -\frac{1}{2}\left[\left(\frac{1}{2}-\frac{1}{3}\right) + \left(\frac{1}{3}-\frac{1}{4}\right) + \cdots + \left(\frac{1}{101}-\frac{1}{102}\right)\right] \\
&= -\frac{1}{2}\left(\frac{1}{2}-\frac{1}{102}\right) \\
&= -\frac{25}{102}.
\end{aligned}$$

Reading

Solve the quadratic equation of one unknown by using the linear equations of two unknowns

Zhao Shuang, a mathematician in Chinese Three Kingdoms period, first proposed the formula for finding the roots of a quadratic equation in the world. But we don't know how he deduced it. Here is a unique derivation method: Suppose that the quadratic equation $x^2 + px + q = 0$ has two real roots x_1 and x_2. From the relationship between roots and coefficients, we have

$$\begin{cases} x_1 + x_2 = -p, & (1) \\ x_1 x_2 = q. & (2) \end{cases}$$

Using the above formulas and the identities $(x_1 - x_2)^2 = (x_1 + x_2)^2 - 4x_1x_2$, we can get $(x_1 - x_2)^2 = p^2 - 4q$, thus

$$x_1 - x_2 = \pm\sqrt{p^2 - 4q}. \qquad (3)$$

Combining (1) and (3), we can find the values of x_1 and x_2. This method is to convert the quadratic equation of one unknown into the linear equations of two unknowns.

(Excerpted from: Zhang Jingzhong and Ren hongshuo. *Talk about mathematics*. Beijing: China children's publishing house, 2003,55)

Exercises

1. Suppose that a and b are the roots of the quadratic equation $x^2 - x - 1 = 0$. Find the value of $3a^3 + 4b + \frac{2}{a^2}$.
2. Suppose that a and b are two different real numbers that satisfy
$$a^2 + 1 = 3a, \quad b^2 + 1 = 3b.$$
Find the value of $\dfrac{1}{a^2} + \dfrac{1}{b^2}$.
3. Suppose that real numbers a and b satisfy
$$19a^2 + 99a + 1 = 0, \quad b^2 + 99b + 19 = 0,$$
and $ab \neq 1$. Find the value of $\dfrac{ab + 4a + 1}{b}$.
4. Suppose that the roots of the equation $x^2 - 5x + 2 = 0$ are x_1 and x_2. Find a quadratic equation whose roots are $-x_1^2$ and $-x_2^2$.
5. Suppose that the roots of the equation $(2x + 7)(x - 3) = 1$ are x_1 and x_2. Find a quadratic equation whose roots are $x_1 + x_2$ and x_1x_2.
6. Let a, b and c be real numbers, and $a \neq 0$. Suppose that the roots of the equation $ax^2 + bx + c = 0$ are x_1 and x_2. Find a quadratic equation whose roots are $\frac{1}{x_1} + 1$ and $\frac{1}{x_2} + 1$.
7. Let m be a real number. Given that one root of the equation $2x^2 - 3x + m = 0$ is twice the other root, find the value of m.
8. Let a be a real number. Given that the difference between two roots of the equation $2x^2 - (a - 1)x + a + 1 = 0$ is 1, find the value of a.
9. Let a be a real number. Given that the roots of the equation $x^2 + 2ax = 3$ are x_1 and x_2, and $x_1^2 + x_2^2 = 10$, find the value of $|a|$.
10. Solve the equation:
$$\frac{x^2 + 3x}{2x^2 + 2x - 8} + \frac{x^2 + x - 4}{3x^2 + 9x} = \frac{11}{12}.$$
11. Let a, b and c be real numbers, and $a \neq 0$. Given that the quadratic equation $ax^2 + bx + c = 0$ has no real roots. Student A misread the coefficient of the quadratic term, and found that two roots are 2 and 4. Student B misread the sign of one of the coefficients, and found that two roots are -1 and 4. Find the value of $\dfrac{2b + 3c}{a}$.
12. Let a be a real number. Given that the quadratic equation $x^2 - 4ax + 5a^2 - 6a = 0$ has two real roots, and the absolute value of the difference between these two roots is 6, find the value of a.

13. Let m be a real number. Given that the roots of the equation $x^2 - 2mx + (m^2 + 2m + 3) = 0$ are x_1 and x_2, find the minimum value of $x_1^2 + x_2^2$.
14. Given that one root of the equation $x^2 - (2 + \sqrt{3})x + 2\sqrt{3} = 0$ is the length of hypotenuse c of a right triangle, and the other root is the length of leg a, find the length of the third side b of the right triangle.
15. Let k be a real number. Given that the roots of the quadratic equation $x^2 + kx + k + 1 = 0$ are x_1 and x_2, and $x_1 + 2x_2^2 = k$, find the value of k.
16. Let a and b be real numbers. Given that the roots of the quadratic equation $x^2 + ax + b = 0$ are x_1 and x_2, and $x_1^3 + x_2^3 = x_1^2 + x_2^2 = x_1 + x_2$, how many ordered binary arrays (a, b) are there?

Chapter 5

Simultaneous Quadratic Equations with Two Unknowns

The basic ways to solve the simultaneous quadratic equations with two unknowns are "elimination" and "order reduction".

Example 1. Solve the simultaneous equations

$$\begin{cases} x + 2y = 5, \\ x^2 + y^2 - 2xy - 1 = 0. \end{cases}$$

Solution From $x + 2y = 5$, we see

$$x = 5 - 2y.$$

Substituting $x = 5 - 2y$ in $x^2 + y^2 - 2xy - 1 = 0$, we have

$$3y^2 - 10y + 8 = 0.$$

The solutions of the quadratic equation are $y_1 = 2$ and $y_2 = \frac{4}{3}$. Thus we find that $x_1 = 1$ and $x_2 = \frac{7}{3}$.

So the solutions of the original simultaneous equations are

$$\begin{cases} x_1 = 1, \\ y_1 = 2, \end{cases} \quad \text{and} \quad \begin{cases} x_2 = \dfrac{7}{3}, \\ y_2 = \dfrac{4}{3}. \end{cases}$$

Think about it Do you have any other solutions?

When solving simultaneous quadratic equations with two unknowns or special simultaneous equations we often transform them into the basic equations

$$\begin{cases} x+y=a, \\ xy=b. \end{cases}$$

Especially for the simultaneous equations involving a symmetrical equation in which the equation remains unchanged after the exchange of unknowns x and y, one of the basic solutions is to change the simultaneous equations into the above basic simultaneous equations.

Example 2. Given that $x+y=3$ and $x^2+y^2-xy=4$, find the value of $x^4+y^4+x^3y+xy^3$.

Solution From $x^2+y^2-xy=4$, we see that

$$(x^2+2xy+y^2)-3xy=4,$$

i.e. $(x+y)^2-3xy=4$.

Since $x+y=3$, we get

$$3xy=(x+y)^2-4=5,$$

i.e. $xy=\dfrac{5}{3}$.

So the original equations becomes the basic simultaneous equations

$$\begin{cases} x+y=3, \\ xy=\dfrac{5}{3}. \end{cases}$$

Since

$$\begin{aligned} &x^4+y^4+x^3y+xy^3 \\ &=(x+y)(x^3+y^3) \\ &=(x+y)^2[(x+y)^2-3xy], \end{aligned}$$

we find that $x^4+y^4+x^3y+xy^3=3^2\times\left(3^2-3\times\frac{5}{3}\right)=36$.

Example 3. Solve the simultaneous equations

$$\begin{cases} xy+x+y=-13, & (1) \\ x^2+y^2=29. & (2) \end{cases}$$

Analysis It is found that the simultaneous equations are symmetrical simultaneous equations, so they can be transformed into the basic simultaneous equations.

Solution From (1) $\times 2+$ (2), we have

$$(x+y)^2+2(x+y)-3=0$$

i.e. $[(x+y)+3][(x+y)-1]=0$.

So $(x+y)+3=0$ or $(x+y)-1=0$.

Substituting them in (1), we have

$$\begin{cases} x+y=-3, \\ xy=-10, \end{cases} \quad \text{or} \quad \begin{cases} x+y=1, \\ xy=-14. \end{cases}$$

Solving the simultaneous equations

$$\begin{cases} x+y=-3, \\ xy=-10, \end{cases}$$

we get that $\begin{cases} x_1=-5, \\ y_1=2 \end{cases}$ and $\begin{cases} x_2=2, \\ y_2=-5. \end{cases}$

Solving the simultaneous equations

$$\begin{cases} x+y=1, \\ xy=-14. \end{cases}$$

we get that $\begin{cases} x_3=\frac{1+\sqrt{57}}{2}, \\ y_3=\frac{1-\sqrt{57}}{2} \end{cases}$ and $\begin{cases} x_4=\frac{1-\sqrt{57}}{2}, \\ y_4=\frac{1+\sqrt{57}}{2}. \end{cases}$

So the solutions of the original simultaneous equations are

$$\begin{cases} x_1=-5, \\ y_1=2, \end{cases} \quad \begin{cases} x_2=2, \\ y_2=-5, \end{cases} \quad \begin{cases} x_3=\dfrac{1+\sqrt{57}}{2}, \\ y_3=\dfrac{1-\sqrt{57}}{2}, \end{cases} \quad \begin{cases} x_4=\dfrac{1-\sqrt{57}}{2}, \\ y_4=\dfrac{1+\sqrt{57}}{2}. \end{cases}$$

Remark There are two basic ways to solve the basic simultaneous equations

$$\begin{cases} x+y=a, \\ xy=b. \end{cases}$$

One method is elimination by substitution, and the other is to transform the question into finding the roots of a quadratic equations.

Example 4. Solve the simultaneous equations

$$\begin{cases} \dfrac{1}{x} + \dfrac{1}{y} = 5, \\ xy = \dfrac{1}{6}. \end{cases}$$

Solution Let $\frac{1}{x} = u$ and $\frac{1}{y} = v$. Then we have

$$\begin{cases} u + v = 5, \\ uv = 6. \end{cases}$$

Solving the simultaneous equations, we get

$$\begin{cases} u_1 = 2, \\ v_1 = 3, \end{cases} \quad \begin{cases} u_2 = 3, \\ v_2 = 2. \end{cases}$$

Thus we know that

$$\begin{cases} x_1 = \dfrac{1}{2}, \\ y_1 = \dfrac{1}{3}, \end{cases} \quad \begin{cases} x_2 = \dfrac{1}{3}, \\ y_2 = \dfrac{1}{2}. \end{cases}$$

After checking, we get that

$$\begin{cases} x_1 = \dfrac{1}{2}, \\ y_1 = \dfrac{1}{3}, \end{cases} \quad \begin{cases} x_2 = \dfrac{1}{3}, \\ y_2 = \dfrac{1}{2}, \end{cases}$$

are the solutions of the original simultaneous equations.

Remark When dealing with a quadratic equation with two unknowns, we often regard it as a quadratic equation of one variable with letters. By using the discriminant of the quadratic equation of one variable and other basic knowledge to "divide and conquer", we achieve the purpose of solving the simultaneous quadratic equations with two unknowns.

Example 5. Given that the upper base, the height and the lower base of a trapezoid $ABCD$ are three consecutive positive integers from small to large, and these three positive integers make the values of $x^3 - 30x^2 + ax$ (a is a constant) be three consecutive positive integers in the same order, find the area of trapezoid $ABCD$.

Solution Let the height of the trapezoid be t. Then the length of upper base is $t-1$, the length of the lower base is $t+1$, where $t>1$.

Thus the area of trapezoid $ABCD$ is

$$\frac{1}{2}[(t-1)+(t+1)]\cdot t=t^2.$$

From the given condition,

$$\begin{aligned}&(t-1)^3-30(t-1)^2+a(t-1)+1\\&=t^3-30t^2+at\\&=(t+1)^3-30(t+1)^2+a(t+1)-1.\end{aligned}$$

Then we have

$$\begin{cases}3t^2-3t-60t+30+a=0,\\3t^2+3t-60t-30+a=0.\end{cases}$$

Solving the simultaneous equations, we get that $t=10,\quad a=300$.
So the area of trapezoid $ABCD$ is $t^2=100$.

Remark For three consecutive integers, if the sum of the three numbers is involved, it is usually assumed that the middle number is a and the other two numbers are $a-1$ and $a+1$, which can simplify the calculation.

Example 6. Solve the simultaneous equations

$$\begin{cases}ab+c+d=3,\\bc+a+d=5,\\cd+a+b=2,\\da+b+c=6.\end{cases}$$

Solution Adding $ab+c+d=3$ and $cd+a+b=2$, we get

$$ab+cd+a+b+c+d=5.$$

Adding $bc+a+d=5$ and $da+b+c=6$, we get

$$ad+bc+a+b+c+d=11.$$

Then we have $(ad+bc)-(ab+cd)=6$, i.e.

$$(a-c)(d-b)=6.$$

So $a\neq c$, $b\neq d$.

Since $(da+b+c)-(bc+a+d)=1=(ab+c+d)-(cd+a+b)$, we get

$$(a+c-2)(b-d)=0.$$

Thus $a+c=2$.

Since $(cd+a+b)+(da+b+c)=8$, we get

$$(a+c)d+(a+c)+2b=8.$$

Thus $b+d=3$.

Similarly, From

$$(da+b+c)-(cd+a+b)=2\times(bc+a+d)-2\times(ab+d+d),$$

we get $(a-c)(d+2b-3)=0$. So $d+2b=3$.

Hence, from $b+d=3$ and $d+2b=3$, we obtain that $b=0$, $d=3$. Substituting them into $(a-c)(d-b)=6$ gives $a-c=2$.

Thus from $a-c=2$ and $a+c=2$, we obtain that $a=2$, $c=0$.

So the solution of the original simultaneous equations is $a=2$, $b=0$, $c=0$, $d=3$.

Example 7. Given that three different real numbers a, b and c satisfy $a-b+c=3$, the equations $x^2+ax+1=0$ and $x^2+bx+c=0$ have an identical real root, and the equations $x^2+x+a=0$ and $x^2+cx+b=0$ have an identical real root too. Find the values of a, b and c.

Solution Let x_1 be the identical real root of the equations $x^2+ax+1=0$ and $x^2+bx+c=0$. Then we have

$$\begin{cases} x_1^2+ax_1+1=0, \\ x_1^2+bx_1+c=0. \end{cases}$$

By subtracting two equations, we can get $x_1=\frac{c-1}{a-b}$. Let x_2 be the identical real root of the equations $x^2+x+a=0$ and $x^2+cx+b=0$. Then we have

$$\begin{cases} x_2^2+x_2+a=0, \\ x_2^2+cx_2+b=0. \end{cases}$$

By subtracting two equations, we can get $x_2=\frac{a-b}{c-1}$. Thus, $x_1x_2=1$.

Since the product of two roots of the equation $x^2+ax+1=0$ is equal to 1, x_2 is also the root of equation. Then we have $x_2^2+ax_2+1=0$. By subtracting two equations $x_2^2+x_2+a=0$ and $x_2^2+ax_2+1=0$, we get $(a-1)x_2=a-1$. If $a=1$, the equation $x^2+ax+1=0$ has no real roots. So $a\neq 1$, then $x_2=1$.

Thus $a = -2$, $b + c = -1$.

And since $a - b + c = 3$, we get $b = -3$, $c = 2$.

Hence, $a = -2$, $b = -3$, $c = 2$.

Example 8. Given that real numbers a, b, c and d are not equal to each other, and

$$a + \frac{1}{b} = b + \frac{1}{c} = c + \frac{1}{d} = d + \frac{1}{a} = x,$$

try to find the value of x.

Solution From $a + \frac{1}{b} = x$, we have $b = \frac{1}{x-a}$. From $b + \frac{1}{c} = x$ and $b = \frac{1}{x-a}$, we get

$$c = \frac{x-a}{x^2 - ax - 1}.$$

From $c + \frac{1}{d} = x$ and $c = \frac{x-a}{x^2-ax-1}$, we get

$$\frac{x-a}{x^2 - ax - 1} + \frac{1}{d} = x,$$

i.e. $dx^3 - (ad+1)x^2 - (2d-a)x + ad + 1 = 0$.

Since $d + \frac{1}{a} = x$, i.e. $ad + 1 = ax$, we have

$$(d-a)(x^3 - 2x) = 0.$$

From the given condition $d - a \neq 0$, we can obtain that $x = 0$ or $x = \pm\sqrt{2}$. If $x = 0$, from $c = \frac{x-a}{x^2-ax-1}$ we get $a = c$, which conflicts with the given condition. Hence, the value of x is $x = \pm\sqrt{2}$.

Reading

The solution should be simple and natural.

To solve the problem, we should try to be simple and natural. We should grasp the essence of the problem, directly dissect the core, and do not drag the mud and water, circle around, and use many "useless moves".

Example If the equations $2x^2 - (3m+2)x + 12 = 0$ and $4x^2 - (9m-2)x + 36 = 0$ have an identical root, find the value of m.

If we solve these two quadratic equations respectively, and then compare the roots, we will spend a lot of efforts. The essence of this problem is to think x of the two equations as a same number and solve the following simultaneous equations:

$$\begin{cases} 2x^2-(3m+2)x+12=0, & (1) \\ 4x^2-(9m-2)x+36=0. & (2) \end{cases}$$

By eliminating x, we can get the value of m.

Eliminate the items of x^2, and then get

$$(m-2)x=4. \quad (3)$$

Substituting (3) into (1) $\times(m-2)^2$ gives

$$2\times 4^2-4(3m+2)(m-2)+12(m-2)^2=0.$$

Solving the equation, we obtain

$$m=3.$$

It's not hard to verify that when $m=3$, the equations (1) and (2) have the identical root $x=4$.

(Excerpted from: Shan Zun. *Problem solving research.* Shanghai: Shanghai Education Press, 2016,48.)

Exercises

1. Solve the simultaneous equations:

$$\begin{cases} x+y=5, \\ xy=4. \end{cases}$$

2. Given that the real numbers a and b satisfy the equations $a^2+b^2-11=0$ and $a^2-5b-5=0$, find the value of b.
3. If the simultaneous equations

$$\begin{cases} x+y=4-a, \\ xy+a(x+y)=5 \end{cases}$$

have real roots, then which is the value range of a? ( )

(A) $a\geq\frac{2}{3}$ (B) $a\leq 2$ (C) $\frac{2}{3}<a<2$ (D) $\frac{2}{3}\leq a<2$

4. Try to judge whether there are any real roots of the simultaneous equations

$$\begin{cases} x^2-4x-2y+1=0, \\ y-2=kx. \end{cases}$$

5. Solve the simultaneous equations:

$$\begin{cases} x^2 + 2xy - 10x = 0, \\ y^2 + 2xy - 10y = 0. \end{cases}$$

6. Solve the simultaneous equations:

$$\begin{cases} x^2 + y^2 = 5, \\ 2x^2 - 3xy - 2y^2 = 0. \end{cases}$$

7. Solve the simultaneous equations:

$$\begin{cases} x + y = 2, \\ xy - z^2 = 1. \end{cases}$$

8. If the real numbers x, y and Z satisfy the equations $x = y + \sqrt{2}$ and $2xy + 2\sqrt{2}z^2 + 1 = 0$, find the value of $x + y + Z$.
9. There is a team moving at a constant speed, with a total length of 40 meters. The messenger at the end of the line rushed to the front of the line to deliver the order and immediately returned to the end at the same speed. If the team march forward 30 meters, how much more distance does the messenger travel than the team?
10. There are three kinds of goods: A, B and C. If you buy 3 pieces of A, 7 pieces of B and 1 piece of C, it will cost 3.15 yuan in total. If you buy 4 pieces of A, 10 pieces of B and 1 piece of C, it will cost 4.20 yuan in total. Then how much does it cost to buy 1 piece of A, B and C respectively?
11. Given that the real numbers x and y satisfy the equation $x^2 - 2x - 4y = 5$, find the range of $x - 2y$.
12. Xiaoming rowed 15 kilometers. If he rowed at his usual speed, it would take five hours less to go downstream than to go upstream. If he rowed at twice his usual speed, the time required for downstream was only 1 hour less than that for upstream. Find the speed of water flow.
13. If the real numbers x, y and z satisfy the simultaneous equations

$$\begin{cases} x = 6 - 3y, \\ x + 3y - 2xy + 2z^2 = 0, \end{cases}$$

find the value of x^{2y+z}.

Chapter 6

Integer Roots of Quadratic Equation

For the quadratic equation $ax^2 + bx + c = 0$, using the quadratic formula or completing the square, we can get $\sqrt{b^2 - 4ac} = \pm(2ax + b)$. If a, b and x are all rational numbers, then $2ax + b$ is also a rational number, so $\sqrt{b^2 - 4ac}$ is a rational number too. Let m be the value of $\sqrt{b^2 - 4ac}$. Then $\Delta = b^2 - 4ac = m^2$. Especially if a, b and c are integers and the equation has integer roots, the discriminant of the equation $\Delta = b^2 - 4ac$ must be a perfect square number.

Example 1. Let n be a rational number. Given that the equation $x^2 - (\sqrt{3} + 1)x + \sqrt{3}n - 6 = 0$ has an integer root, find the value of n.

Analysis $\sqrt{3}$ can be associated with the relationship between rational numbers and irrational numbers.

Solution From the given condition, we let an integer root of the equation be α, then we have

$$\alpha^2 - (\sqrt{3} + 1)\alpha + \sqrt{3}n - 6 = 0,$$

i.e. $\alpha^2 - \alpha - 6 = \sqrt{3}(\alpha - n)$.

Because $\sqrt{3}$ is an irrational number and $\alpha - n$, $\alpha^2 - \alpha - 6$ are integers, we can get $\alpha = n$. Thus,

$$n^2 - n - 6 = 0.$$

Solving the equation gives $n = -2$ or $n = 3$.

So the value of n is -2 or 3.

Example 2. Suppose that the integer a makes the two roots of the quadratic equation $5x^2 - 5ax + 26a - 143 = 0$ be integers, find the value of a.

Solution From the given condition, the discriminant of the equation should be a perfect square number, i.e.

$\Delta = 25a^2 - 4 \times 5 \times (26a - 143) = k^2$, where k is a non-negative integer.

Then we get $k^2 - (5a - 52)^2 = 156$, i.e.

$$\begin{aligned}&(k + 5a - 52)(k - 5a + 52)\\&= 156 \times 1 = 78 \times 2 = 39 \times 4 = 26 \times 6\\&= 13 \times 12 = 3 \times 52.\end{aligned}$$

Because $k + 5a - 52$ and $k - 5a + 52$ are both even or both odd, and

$$(k + 5a - 52) + (k - 5a + 52) = 2k \geq 0,$$

we can obtain that

$$\begin{cases} k + 5a - 52 = 78, \\ k - 5a + 52 = 2; \end{cases} \quad \begin{cases} k + 5a - 52 = 26, \\ k - 5a + 52 = 6; \end{cases} \quad \begin{cases} k + 5a - 52 = 2, \\ k - 5a + 52 = 78; \end{cases}$$

$$\text{or} \quad \begin{cases} k + 5a - 52 = 6, \\ k - 5a + 52 = 26. \end{cases}$$

Then we get $k = 40$, $a = 18$, which accord with the given condition. So the value of a is 18.

Example 3. Let a be an integer. Given that there exist integers b and c, for any real number x, the equation $(x + a)(x - 15) - 25 = (x + b)(x + c)$ holds, find the possible values of a.

Solution From $(x + a)(x - 15) - 25 = (x + b)(x + c)$, we have

$$(b + c - a + 15)x + bc + 25 + 15a = 0.$$

By the arbitrariness of x, we get

$$\begin{cases} b + c = a - 15, \\ bc = -15a - 25. \end{cases}$$

Then $bc + 15(b + c) + 15^2 = -25$, i.e. $(b + 15)(c + 15) = -1 \times 25 = -5 \times 5 = 1 \times (-25)$.

Thus, we have

$$\begin{cases} b+15=-1, \\ c+15=25; \end{cases} \quad \begin{cases} b+15=25, \\ c+15=-1; \end{cases} \quad \begin{cases} b+15=-5, \\ c+15=5; \end{cases}$$

$$\begin{cases} b+15=5, \\ c+15=-5; \end{cases} \quad \begin{cases} b+15=1, \\ c+15=-25; \end{cases} \quad \text{or} \quad \begin{cases} b+15=-25, \\ c+15=1. \end{cases}$$

That is, $b+c=-6$ or $b+c=-30$ or $b+c=-54$.
When $b+c=-6$, $a=9$.
When $b+c=-30$, $a=-15$.
When $b+c=-54$, $a=-39$.
Therefore, the value of a that meets the requirements is 9 or -15 or -39.

Remark Example 3 can also be solved by the discriminant.
In the above process, we have found that

$$\begin{cases} b+c=a-15, \\ bc=-15a-25. \end{cases}$$

Then b and c are the integer roots of the quadratic equation $t^2-(a-15)t+(-15a-25)=0$, so

$$\Delta=(15-a)^2-4\times(-15a-25)=m^2,$$

where m is an integer. Thus,

$$m^2-(a+15)^2=100.$$

Since $m-(a+15)$ and $m+(a+15)$ are both even or both odd, we have

$$[m-(a+15)][m+(a+15)]=2\times 50=10\times 10.$$

Therefore,

$$\begin{cases} m-(a+15)=2, \\ m+(a+15)=50; \end{cases} \quad \begin{cases} m-(a+15)=50, \\ m+(a+15)=2; \end{cases}$$

$$\begin{cases} m-(a+15)=-2, \\ m+(a+15)=-50; \end{cases} \quad \begin{cases} m-(a+15)=-50, \\ m+(a+15)=-2; \end{cases}$$

$$\begin{cases} m-(a+15)=10, \\ m+(a+15)=10; \end{cases} \quad \text{or} \quad \begin{cases} m-(a+15)=-10, \\ m+(a+15)=-10. \end{cases}$$

Then we get $a+15=24$, $a+15=-24$, or $a+15=0$, i.e. $a=9$, -15 or -39.

Example 4. Let a be a positive integer. If the equation of x

$$x^2 + 4x + \sqrt{10-a} + 2 = 0$$

has two rational roots, find all the values of a.

Solution Since two roots of the equation are rational numbers, the discriminant of the equation $\Delta \geq 0$ and it is a perfect square number.

From $\Delta = 16 - 4(\sqrt{10-a} + 2) = 4(2 - \sqrt{10-a}) \geq 0$, we get

$$2 \geq 2 - \sqrt{10-a} \geq 0.$$

Thus, $0 \leq \sqrt{10-a} \leq 2$.

Since Δ is a perfect square number, $2 - \sqrt{10-a}$ is a non-negative integer, i.e. $\sqrt{10-a}$ is a non-negative integer, then we can get $\sqrt{10-a} = 0$, 1 or 2.

When $\sqrt{10-a} = 0$, $a = 10$.

When $\sqrt{10-a} = 1$, $a = 9$.

When $\sqrt{10-a} = 2$, $a = 6$.

After checking, we know that $a = 6$ and $a = 9$ satisfy the given condition.

So $a = 6$ or 9.

Example 5. Are there prime numbers p and q, which make the quadratic equation $px^2 - qx + p = 0$ have rational roots?

Solution Because the equation has rational roots, the discriminant should be a square number.

Suppose $\Delta = q^2 - 4p^2 = n^2$, where n is a non-negative integer, then $(q-n)(q+n) = 4p^2$.

Since $1 \leq q-n \leq q+n$, and $q-n$ and $q+n$ are both even or both odd, both $q-n$ and $q+n$ are even.

Thus, there are several possible situations as follows:

$$\begin{cases} q-n=2, \\ q+n=2p^2; \end{cases} \quad \begin{cases} q-n=4, \\ q+n=p^2; \end{cases} \quad \begin{cases} q-n=p, \\ q+n=4p; \end{cases}$$

$$\begin{cases} q-n=2p, \\ q+n=2p; \end{cases} \quad \text{or} \quad \begin{cases} q-n=p^2, \\ q+n=4. \end{cases}$$

Eliminating n gives $q = p^2 + 1$, $q = 2 + \frac{p^2}{2}$, $q = \frac{5p}{2}$, or $q = 2p$. Because p and q are prime numbers, $q = 2p$ is imposible. To make $q = p^2 + 1$, $q = 2 + \frac{p^2}{2}$, or $q = \frac{5p}{2}$, there must be $p = 2$. At this time, $q = 2^2 + 1 = 5$, or $q = 4$. Since $q = 4$ is not prime, we can get $p = 2$, $q = 5$. When $p = 2$ and $q = 5$, the equation is $2x^2 - 5x + 2 = 0$, and its roots are $x_1 = \frac{1}{2}$ and $x_2 = 2$, both of which are rational numbers.

In summary, there exist prime numbers $p = 2$ and $q = 5$, which satisfy the given condition.

Example 6. Given that the sum of three integers a, b and c is 13, and $\frac{b}{a} = \frac{c}{b}$, find the maximum and minimum values of a, and find the corresponding values of b and c too.

Solution Let $\frac{b}{a} = \frac{c}{b} = x$. Then $b = ax$, $c = ax^2$.

Therefore, $a + b + c = 13$ is transformed into $a(x^2 + x + 1) = 13$.

Since $a \neq 0$, we get

$$x^2 + x + 1 - \frac{13}{a} = 0. \tag{1}$$

We know that a, b and c are integers, then the equation (1) has roots and the roots must be rational numbers, so

$$\Delta = 1 - 4\left(1 - \frac{13}{a}\right) = \frac{52}{a} - 3 > 0.$$

Then we can obtain that $1 \leq a < \frac{52}{3}$, thus $1 \leq a \leq 17$.

When $a = 17$, $\sqrt{\Delta} = \sqrt{\frac{52}{17} - 3} = \frac{\sqrt{17}}{17}$ is not a rational number, then

$$1 \leq a \leq 16.$$

And when $a = 1$, equation (1) is transformed into $x^2 + x - 12 = 0$, the roots are $x_1 = -4$, $x_2 = 3$. So $a_{\min} = 1$, $b = -4$, $c = 16$; or $a_{\min} = 1$, $b = 3$, $c = 9$.

And when $a = 16$, the equation (1) is transformed into $x^2 + x + \frac{3}{16} = 0$, the roots are $x_1 = -\frac{3}{4}, x_2 = -\frac{1}{4}$. So $a_{\max} = 16, b = -12, c = 9$; or $a_{\max} = 16$, $b = -4$, $c = 1$. Hence, the minimum value of a is 1, at this time, $b = -4$, $c = 16$ or $b = 3$, $c = 9$; the maximum value of a is 16, at this time, $b = -12$, $c = 9$ or $b = -4$, $c = 1$.

Example 7. If $[x]$ is used to represent the largest integer not greater than x, then the number of roots of equation $x^2 - 2[x] - 3 = 0$ is ________.

Solution From the given condition, $[x] \leq x < [x] + 1, 2[x] + 3 = x^2 \geq 0$, then we get

$$[x] \geq -\frac{3}{2}.$$

It's obvious that $[x] \neq 0$.

When $[x] > 0, ([x])^2 \leq x^2 < ([x]+1)^2$, we obtain that

$$([x])^2 \leq 2[x]+3 < ([x]+1)^2.$$

Thus, from $[x] \geq -\frac{3}{2}$ and the above inequality, we know that $[x] = 2$ or 3. If $[x] = 2$, then $x^2 = 2[x]+3 = 2\times 2+3 = 7$, $x = \sqrt{7}$. If $[x] = 3$, then $x^2 = 2\times 3+3 = 9$, $x = 3$.

When $[x] < 0, ([x]+1)^2 < x^2 \leq ([x])^2$, we obtain that

$$([x]+1)^2 < 2[x]+3 \leq ([x])^2.$$

And since $[x] \geq -\frac{3}{2}$, we get $[x] = -1$. Thus,

$$x^2 = 2\times(-1)+3 = 1, x = -1.$$

So the roots of the equation are $x = -1$, $x = \sqrt{7}$ or $x = 3$, that is, there are three roots of the given equation.

Remark To solve the question about integral roots of a quadratic equation, we usually consider the relationship between rational numbers and real numbers directly, or factorize, or use the discriminant and the quadratic formula, or use the relationship between roots and coefficients, or convert it to discuss the value range of the roots of the quadratic function, or complete the square, or use some basic properties of integer. However, in specific operation, according to the characteristics of specific problems, we should choose the proper solution.

Reading

A famous odd question

There is such a story, which once circulates in some international mathematician gathering. They call the question raised in the story "seemingly unanswerable question".

Now, I will write down the story and make some analysis. There is a quadratic equation, the roots of the equation are two positive integers greater than 1, and the sum of the two roots is not more than 40. The equation is written as follows:

$$x^2 - px + q = 0$$

(p and q on the paper are numbers.)

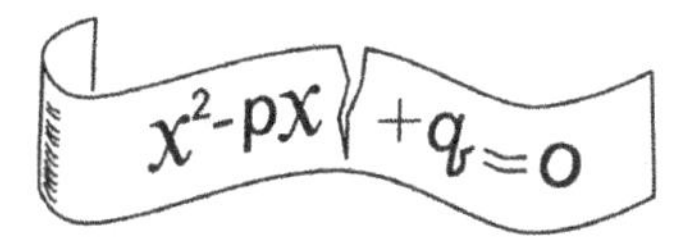

Someone tore the paper with this equation from the middle, gave half paper with the number p to mathematician A, and the other half paper with the number q to mathematician B from other places. Therefore, mathematician A knows the sum p of the two roots, mathematician B knows the product q of the two roots.

After a while, A called B and said, "I'm sure you don't know the number p in my hand." A moment later, B called back and said, "but I have already known what is the number p you have." And after a while, A called back and said, "I know the number q you have, too."

What are the two roots of this equation? (Answer: 4 and 13.)

(Excerpted from: Zhang Jingzhong. *Help you with your math.* Beijing: China children's publishing house, 2002, 86∼87.)

Exercises

1. Suppose that $\sqrt{27-10\sqrt{2}} = a+b$, where a is a positive integer and b is between 0 and 1, find the value of $\dfrac{a+b}{a-b}$.
2. Three consecutive positive integers a, b and c are filled into three boxes of $\Box x^2 + \Box x + \Box = 0$ in any order (different order is regarded as a different group) as the coefficient of quadratic term, coefficient of linear term and constant term of the quadratic equation. Find the value group of a, b and c so that the equation has at least one integer root.
3. Let r be an integer. If the roots of the equation $rx^2 - (2r+7)x + (r+7) = 0$ are positive integers, find the value of r.
4. Let a and b be integers. If the roots of the quadratic equation $ax^2 + bx + c = 0$ are a and b, find the value of b.
5. How many pairs of integers (a, b) which satisfy $|ab| + |a+b| = 1$? ( ).
   (A) 4 (B) 7 (C) 6 (D) 8
6. Suppose that x and y are positive integers and satisfy the equation $\dfrac{1}{x} - \dfrac{1}{y} = \dfrac{1}{100}$, find the maximum value of y.
7. Given that the equation $(x-a)(x-8) = 1$ has two integer roots, find the value of a.

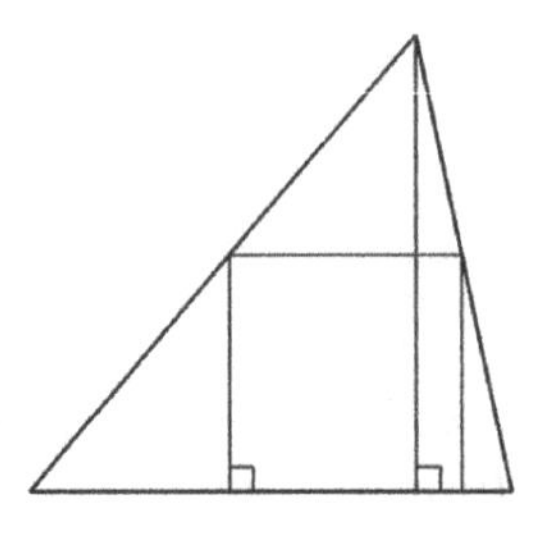

Fig. 6.1

8. As shown in Figure 6.1, square $EFGH$ inscribed in $\triangle ABC$. Let $BC = \overline{ab}$ where $\overline{ab}$ is a double digit composed of tens digit a and units digit b, $EF = c$, the height of the triangle $AD = d$. Given that a, b, c and d are four consecutive positive integers from small to large, try to find the area of $\triangle ABC$.
9. Given that the lengths of three sides of a right triangle are integers. If the lengths of two legs of the triangle are the roots of equation $x^2 - (k+2)x + 4k = 0$, find the value of k and determine the lengths of three sides of the right triangle.
10. If m and n are integers, try to determine whether the equation $x^2 + 10mx - 5n + 3 = 0$ has integer roots or not, and explain the reason.
11. For the equation $ax^2 + bx + c = 0$, if the coefficients a, b and c are all odd, try to judge that this equation has no integer roots and explain the reason.

Chapter 7

Perfect Square Numbers

It is easy to see that

$$(2n)^2 = 4n^2,$$
$$(2n+1)^2 = 4n^2 + 4n + 1 = 4n(n+1) + 1.(n \in \mathbf{Z})$$

This shows that the square numbers have the following characteristics: if the square of an even number is divided by 4, the remainder is 0; if the square of an odd number is divided by 4, the remainder is 1. If we use the notation of congruence to represent this, i.e

$$(2n)^2 \equiv 0(\text{mod } 4),$$
$$(2n+1)^2 \equiv 1(\text{mod } 4)(n \in \mathbf{Z}).$$

In other words, when an integer is divided by 4, the remainder is neither 0 nor 1, then it is not a square number.

We know that all the numbers such as $100, 1000, 10000, \ldots$ are multiples of 4. So the remainder of a positive integer divided by 4 is the remainder of its last two digits divided by 4.

Example 1. Consider a sequence of numbers: $11, 111, 1111, 11111, \ldots$ (the n-th term is a $(n+1)$-digit number consisting of $n+1$ digits of ones). Try to prove that each number in the sequence is not a perfect square number.

Solution It is easy to know that 11 is not a perfect square number.

$$\underbrace{11\ldots11}_{(n+1)\text{ digits}} = \underbrace{11\ldots1}_{(n-1)\text{ digit}} \times 100 + 11 (n \geq 2),$$

The remainder of this number divided by 4 is 3, not 0 or 1. Therefore, each number in this sequence is not a perfect square number.

Remark By using the method of Example 1, it is easy to get that all the numbers such as $22, 222, 2222, \ldots$; $55, 555, 5555, \ldots$; $66, 666, 6666, \ldots$; $99, 999, 9999, \ldots$ are not perfect square numbers.

Although we know that when a square number is divided by 4, the remainder is 0 or 1, we can't say that a positive integer leaving a remainder of 0 or 1 when divided by 4 must be a perfect square number. For example, 33 is not a square number. To this end, we often use the numerical characteristics of the last digit of the square number. It is easy to calculate that the last digit of the square numbers must be 0, 1, 4, 5, 6 or 9. Therefore, It is easy to know that all the numbers such as $33, 333, 3333, \ldots$; $77, 777, 7777, \ldots$; $88, 888, 8888, \ldots$ are not perfect square numbers.

As for the question whether $44, 444, 4444, \ldots$ are square numbers, we can also use the characteristic of prime factorization of square numbers, that is, the number of each different prime factor must be even or it must be the product of some perfect square numbers. Since $44 = 4 \times 11, 444 = 4\times 111, 4444 = 4\times 1111, \ldots$, where $11, 111, 1111, \ldots$ are not square numbers, $44, 444, 4444, \ldots$ are not perfect square numbers.

Example 2. Prove that whatever integers a and b are, $15a - 35b + 3$ cannot be a perfect square number.

Solution Since $15a - 35b + 3 = 5 \times (3a - 7b) + 3$, considering that the number $5 \times (3a - 7b)$ is a multiple of 5, we know that the last digit of this number must be 0 or 5. So the last digit of $15a - 35b + 3$ must be 3 or 8. This is not consistent with the fact that the units digit of a perfect square number can only be 0, 1, 4, 5, 6, 9. So $15a - 35b + 3$ is not a perfect square number.

Think about it Can you solve this problem by using the characteristic of dividing the square numbers by 4?

Example 3. Prove that any number in the form of $(4n + 2)$ cannot be expressed as the difference of two square numbers.

Solution It is obvious that $a^2 - b^2 = (a - b)(a + b)$, and $(a - b)$ and $(a + b)$ are both even or both odd.

If both are even, $a^2 - b^2$ must be a multiple of 4. If both are odd, $a^2 - b^2$ is odd.

So $4n + 2$ cannot be expressed as the difference of two square numbers.

Example 4. Prove that if the tens digit of a perfect square number is odd, its units digit number must be equal to 6.

Solution Let the perfect square number be $A = (10m + n)^2$, where m is a natural number and n is one of 0, 1, 2, ..., 9. Then

$$A = (10m + n)^2 = 100m^2 + 20mn + n^2.$$

Since the tens digit number of $100m^2$ is 0, and the tens digit number of $20mn$ (i.e. $2mn \times 10$) is the units digit number of $2mn$ which is even, the tens digit number of A is equal to the sum of the tens digit number of n^2 and an even number. To make the tens digit number of A be odd, the tens digit number of n^2 must be odd. Therefore, n cannot be an odd number, and $n \neq 0$, 2 or 8. So n can only be 4 or 6, and we can know that the last digit of n^2 is 6. Hence, the units digit of a perfect square number whose tens digit is odd must be 6.

Remark This example tells us that if the tens digit of a number is an odd number and the units digit is not equal to 6, then the number must not be a square number. For example, $34, 334, 3334, \ldots$ are not square numbers.

Example 5. Let a be an integer. If $100a + 64$ and $201a + 64$ are four-digit numbers, and both are perfect square numbers, find the value of a.

Solution From the given condition, $1000 < 100a+64 < 10000$ and $1000 < 201a + 64 < 10000$, then we get $10 \leq a < 50$.

Let $100a + 64 = m^2$, $201a + 64 = n^2$. Then $32 \leq m < n < 100$. Thus we can obtain that $64 < n + m < 200$, $0 < n - m < 100 < 101$, and

$$101a = n^2 - m^2 = (n - m)(n + m).$$

Because 101 is prime, $(n - m)(n + m)$ is divisible by 101. And since $0 < n - m < 101$, we can know that $(n + m)$ is divisible by 101, that is, $n + m$ is a multiple of 101. Moreover, from $n + m < 2 \times 101$, we can get that $n + m = 101$. Therefore, $a = n - m$, and we obtain $a = 2n - 101$. Substituting this into the equation $201a + 64 = n^2$ gives

$$n^2 - 402n + 20237 = 0.$$

Solving the equation gives $n = 59$ or $n = 343$. As $n < 100$, we discard $n = 343$.

Hence, $a = 2n - 101 = 2 \times 59 - 101 = 17$.

Example 6. Let n be a positive integer. If $2^n + 256$ is a perfect square number, find the value of n.

Solution When $n \leq 8$, we know that $2^n + 256 = 2^n(1 + 2^{8-n})$. If it is a perfect square number, n must be even.

If $n = 2$, $2^n + 256 = 2^2 \times 65$.

If $n = 4$, $2^n + 256 = 2^4 \times 17$.

If $n = 6$, $2^n + 256 = 2^6 \times 5$.

If $n = 8$, $2^n + 256 = 2^8 \times 2$.

Therefore, when $n \leq 8$, $2^n + 256$ is not a perfect square number.

When $n > 8$, we know that $2^n + 256 = 2^8(2^{n-8} + 1)$. If it is a perfect square number, $2^{n-8} + 1$ is the square of an odd number. Suppose that $2^{n-8} + 1 = (2k + 1)^2$ (where k is a positive integer), then $2^{n-10} = k(k + 1)$.

Since k and $k + 1$ are one odd and one even, we can get $k = 1$. Then we have $2^{n-10} = 2$. Solving the equation gives $n = 11$. Hence, the value of n is 11.

Example 7. Given that the positive integer n makes $t = 2 + 2\sqrt{1 + 12n^2}$ be a positive integer, prove that t is a perfect square number.

Proof From the given condition, t is a positive integer. Then $2\sqrt{1 + 12n^2}$ is also a positive integer. Let $2\sqrt{1 + 12n^2} = a$, where a is a positive integer. Then $a^2 = 4(1 + 12n^2)$ is an even number, a is an even number. Suppose that $a = 2b$ (b a positive integer). Then $b^2 = 1 + 12n^2$, b is an odd number.

Let $b = 2c + 1$. Substituting it into $b^2 = 1 + 12n^2$ gives $c(c + 1) = 3n^2$.

Since c and $c + 1$ are coprime (i.e. the greatest common divisor of these two numbers is 1), c is a perfect square number while $c + 1$ is three times of a perfect square number; or c is three times of a perfect square number while $c + 1$ is a perfect square number. Hence, there are positive integers x and y, which make $3x^2 - y^2 = 1$ or $x^2 - 3y^2 = 1$ be tenable.

Since the remainder of a perfect square number divided by 3 must be 0 or 1, the remainder of $3x^2 - y^2$ divided by 3 is 0 or 2, which can't be 1. This shows that there are no positive integers x and y which make $3x^2 - y^2 = 1$ be tenable. Hence, we can infer that $x^2 - 3y^2 = 1$, then we get $c + 1 = x^2$ and $c = 3y^2$.

Therefore, $t = 2 + 2\sqrt{1 + 4c(c + 1)} = 2 + 2(2c + 1) = 4(c + 1) = 4x^2$ is a perfect square number.

Example 8. Prove that there are infinitely many such sequences consisting of infinitely many numbers, all terms of the sequence are positive integers that are different from each other, and for each positive integer n, the sum of the first n terms of the sequence is divisible by n.

Solution $1+3=2^2, 1+3+5=3^2, \ldots$ In general,

$$1+3+5+\cdots+(2n-1)=n^2$$

is divisible by n. This shows that the sequence $1, 3, 5, \ldots, 2n-1, \ldots$ meets the requirements in the question.

For the above infinite sequence, each term is multiplied by k, where k is any positive integer, then we get a sequence

$$k\cdot 1, k\cdot 3, k\cdot 5, \ldots$$

which also meets the requirements.

Because k can be infinitely many values, in this way, we can obtain infinitely many sequences satisfying the requirements.

Example 9. Let the positive integers m and n satisfy that the equation $(x+m)(x+n)=x+m+n$ has at least one positive integer root. Prove that $2(m^2+n^2)<5mn$.

Proof Rearranging the equation gives

$$x^2+(m+n-1)x+mn-m-n=0. \quad (1)$$

The discriminant of the equation (1) is

$$\begin{aligned}\Delta &= (m+n-1)^2-4(mn-m-n)\\ &= (m+n)^2-4mn+2(m+n)+1\\ &= (m-n)^2+2(m+n)+1.\end{aligned}$$

Suppose that $m \geq n$. From the given condition, the equation (1) with integral coefficients has at least one positive integer root, so Δ should be a perfect square number.

We have noticed that

$$\Delta=(m-n)^2+2(m+n)+1=(m-n+1)^2+4n>(m-n+1)^2,$$

$$\Delta=(m-n)^2+2(m+n)+1=(m-n+3)^2-(4m-8n+8).$$

If $4m-8n+8>0$, i.e. $m>2n-2$, then $\Delta<(m-n+3)^2$. Thus we have

$$(m-n+1)^2<\Delta<(m-n+3)^2,$$

Δ can only be $(m-n+2)^2$, i.e

$$(m-n)^2+2(m+n)+1=(m-n+2)^2.$$

Rearranging this equation gives $m = 3n - \frac{3}{2}$, which contradicts that m and n are positive integers. Therefore, $m \leq 2n - 2$, then we have $m < 2n$, i.e. $\frac{m}{n} < 2$. Since $\frac{m}{n} > 1 > \frac{1}{2}$, we can get $\left(\frac{m}{n} - \frac{1}{2}\right)\left(\frac{m}{n} - 2\right) < 0$, i.e. $2(m^2 + n^2) < 5mn$. So the conclusion is proved.

Reading

Difficult problem

Bogdanov Beresky's famous painting "difficult problem" (see Figure 7.1) is known by many people, but the people who look at this painting may not have a deep understanding of the content of the subject "difficult problem". This problem is to find out the result of the calculation of the following problem quickly by using mental arithmetic:

$$\frac{10^2 + 11^2 + 12^2 + 13^2 + 14^2}{365}.$$

Fig. 7.1

This problem is not easy in fact, but the students of a teacher can handle it very well.This teacher is the one who was praised by the painting which shows a portrait of him. He is O.A. Radzinski, a professor of natural science, who gave up the chair of university to become an ordinary teacher in a rural school. The wise educator has taught mental arithmetic in his school and skillfully uses the properties of numbers. The numbers 10, 11, 12, 13 and 14 have an interesting property as follow:

$$10^2 + 11^2 + 12^2 = 13^2 + 14^2.$$

Since $100 + 121 + 144 = 365$, it's easy to work out in your mind that the formula in the picture is equal to 2.

The above interesting characteristics lead us to go a little further by using the knowledge of equations, that is, to think about the following problems:

Question 1 Find all these five consecutive integers, where the quadratic sum of the first three integers is equal to the quadratic sum of the last two integers.

Question 2 Find all these four consecutive integers, where the cubes of the largest integer is equal to the sum of the cubes of the other three integers. I believe you can easily solve the above two problems.

(Answer Question 1: 10, 11, 12, 13, 14; or −2, −1, 0, 1, 2. Question 2: 3, 4, 5, 6.)

(Excerpted from: Tserel'man. *Entertaining Algebra.* Kaiming Book Company, 1950, 127 ∼ 129.)

Exercises

1. Prove that

(1) $\underbrace{11\ldots1}_{n \text{ digits}}\underbrace{55\ldots56}_{(n-1) \text{ digits}}$ is a perfect square number,

(2) $\underbrace{899\ldots99}_{(n-1) \text{ digits}}\underbrace{400\ldots001}_{(n-1) \text{ digits}}$ is a perfect square number.

2. Given that the integer P is equal to the quadratic sum of two adjacent natural numbers.

   (1) What is the minimum three-digit number P?
   (2) Try to explore the characteristics of the last digit number of P.

3. Suppose that $a > 0$ and $\sqrt{41 - 2a}$ is an integer, find the value of a.
4. If the last digit number of the quadratic sum of three consecutive positive integers a, b and c is 2, find the units digit number of b.
5. Consider two conclusions: (1) the sum of the last two digits of an odd perfect square number is odd; (2) the sum of the last two digits of an even perfect square number is even. Which of the following sentence is correct?( )

   (A) Only (1) is correct
   (B) Only (2) is correct
   (C) Both (1) and (2) are correct
   (D) Both (1) and (2) are incorrect

6. If the sum of 75 consecutive positive integers starting from a certain number is a perfect square number, find the minimum value of this number.
7. Prove that the product of the units digit and the tens digit of any perfect square number B must be even.
8. Is there such a two-digit number $\overline{ab}$ that the sum of $\overline{ab}$ and $\overline{ba}$ obtained by exchanging the two-digit numbers of $\overline{ab}$ is a perfect square number? If so, how many numbers are there? If not, explain the reason.
9. If a positive integer is added or subtracted by the same odd number, can the sum and the difference be perfect square numbers? Try to prove your conclusion.
10. Let m be a positive integer and satisfy that $5 \times 2^m + 1$ is a perfect square number. Find all the values of m.
11. A double-digit number is the same as the last two digits of its square. How many of such double-digit numbers are there?
12. The sequence $1, 2, 3, \ldots, n$ is divided into two groups, so that the sum of any two different numbers in each group is not a square number. What is the maximum value of n that can be grouped in this way?
13. The sequence $2, 3, \ldots, n$ is divided into two groups, and the sum of any two different numbers in each group is not a square number. What is the maximum value of n?

14. Given an algebraic expression $\pm 1^2 \pm 2^2 \pm 3^2 \pm \cdots \pm n^2$ (n is a natural number), where the signs "+" and "−" can be selected arbitrarily. What is the minimum absolute value of the resulting algebraic sum?
15. Find the integer solutions of the equation $x^2 + y^2 = 2019$.
16. Prove that all the square numbers are divided into two groups, then there must be two numbers in a group, the sum of which is a square number.

Chapter 8

Quadratic Functions

A quadratic function has the form $y = ax^2 + bx + c$ where $a \neq 0$. The domain of the function is I where I is a set of numbers.

The following three basic methods can be used flexibly to find the expression of a quadratic function in the given domain.

(1) Using three points

If we know that the coordinates of three points on the graph of the quadratic function (or three groups of corresponding values of the function) are (x_1, y_1), (x_2, y_2) and (x_3, y_3), then the simultaneous equations

$$\begin{cases} y_1 = ax_1^2 + bx_1 + c, \\ y_2 = ax_2^2 + bx_2 + c, \\ y_3 = ax_3^2 + bx_3 + c \end{cases}$$

can uniquely determine the constants a, b and c. Thus we can get that the expression of the function is $y = ax^2 + bx + c$.

(2) Using the vertex

As we know, $y = ax^2+bx+c = a\left(x + \frac{b}{2a}\right)^2 + \frac{4ac-b^2}{4a}$. If we know the vertex of the graph is $\left(-\frac{b}{2a}, \frac{4ac-b^2}{4a}\right)$, we can suppose that the quadratic function has the form $y = a\left(x + \frac{b}{2a}\right)^2 + \frac{4ac-b^2}{4a}$. By using the other conditions, the expression of this function can be easily obtained. Besides, the line $x = -\frac{b}{2a}$ is called the axis of symmetry for the quadratic function.

(3) Using the roots of the quadratic equation
As we know,

$$y = ax^2 + bx + x = a\left(x + \frac{b}{2a}\right)^2 + \frac{4ac - b^2}{4a} = a(x - x_1)(x - x_2),$$

where x_1 and x_2 are two roots of the quadratic equation $ax^2 + bx + c = 0$. If the graph of a quadratic function has intersections with x-axis (or the equation $ax^2 + bx + c = 0$ has real roots), we can suppose that the quadratic function has the form $y = a(x - x_1)(x - x_2)$, then the expression is obtained.

The above basic methods are often used to solve the problem of finding the expression of a quadratic function.

Example 1. As shown in Figure 8.1, given that the vertex of the graph of the quadratic function $y = ax^2 + bx + c$ is on the y-axis, $c - b = 2$, and the graph passes through the point $P(2, 8)$, find the expression of the quadratic function.

Analysis Use the vertex to determine the expression of the quadratic function.

Solution Since $y = a\left(x + \frac{b}{2a}\right)^2 + \frac{4ac-b^2}{4a}$, and the vertex of the graph of the quadratic function is on the y-axis, we get that $-\frac{b}{2a} = 0$, i.e. $b = 0$. Substituting $b = 0$ into $c - b = 2$ gives $c = 2$. Therefore, the quadratic function is $y = ax^2 + 2$.

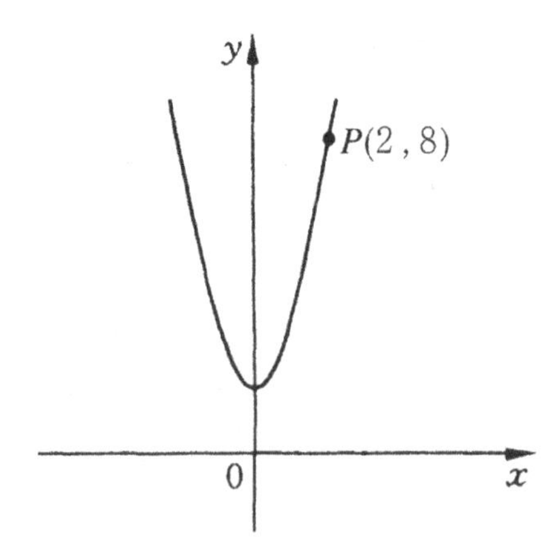

Fig. 8.1

Since the graph of the quadratic function passes through the point $P(2, 8)$, we have

$$8 = a \times 2^2 + 2,$$

then $a = \frac{3}{2}$.

So the expression of the quadratic function is

$$y = \frac{3}{2}x^2 + 2.$$

Remark According to the graph of quadratic functions, it is easy to get the following basic properties of quadratic functions:

(1) Symmetry
The graph of the quadratic function $y = ax^2 + bx + c$ is symmetrical about the line $x = -\frac{b}{2a}$.

(2) Monotonicity
When $a > 0$, the quadratic function $y = ax^2 + bx + c$ is monotonically decreasing on the interval $\left(-\infty, -\frac{b}{2a}\right]$, and monotonically increasing on the interval $\left(-\frac{b}{2a}, +\infty\right)$.
When $a < 0$, the quadratic function $y = ax^2 + bx + c$ is monotonically increasing on the interval $\left(-\infty, -\frac{b}{2a}\right]$, and monotonically decreasing on the interval $\left(-\frac{b}{2a}, +\infty\right)$.

(3) Maximum and minimum values
When $a > 0$, the quadratic function $y = ax^2 + bx + c$ has the minimum value $\frac{4ac-b^2}{4a}$ in the domain of $(-\infty, +\infty)$.
When $a < 0$, the quadratic function $y = ax^2 + bx + c$ has the maximum value $\frac{4ac-b^2}{4a}$ in the domain of $(-\infty, +\infty)$.
Notice that the maximum and minimum values of a quadratic function are related to the given domain.

(4) The x-intercepts
For a quadratic function $y = f(x) = ax^2 + bx + c$, we consider the discriminant $\Delta = b^2 - 4ac$. If $\Delta \geq 0$, the graph of this function intersects the x-axis. When $\Delta > 0$, the graph cuts the x-axis twice; when $\Delta = 0$, the graph touches the x-axis.

Let x_1 and x_2 be the x-intercepts of the quadratic function, and $x_1 < x_2$.

As shown in Figure 8.2(1), we know that $a > 0$ and $\Delta > 0$. When $x < x_1$, $f(x) > 0$; when $x_1 < x < x_2$, $f(x) < 0$; when $x > x_2$, $f(x) > 0$. As shown in Figure 8.2(2), we know that $a < 0$ and $\Delta > 0$. When $x < x_1$ or $x > x_2$, $f(x) < 0$; when $x_1 < x < x_2$, $f(x) > 0$.

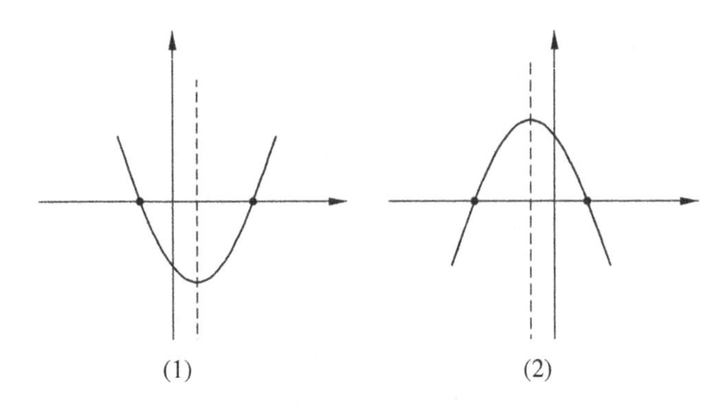

Fig. 8.2

Example 2. Let k be a real number. Given that the vertex of the graph of a quadratic function is $(1, 2)$, and the curve and the line $y = 2x + k$ intersect at point $(2, -1)$,

(1) find the expression of the quadratic function;
(2) find the value of k;
(3) find the coordinates of the other point at which the graph of the quadratic function and the line $y = 2x + k$ intersect.

Analysis It is easy to use the vertex to determine the expression of the quadratic function.

Solution

(1) Since the vertex of the graph is $(1, 2)$, suppose that the expression of the quadratic function is $y = a(x-1)^2 + 2$. And because point $(2, -1)$ is on the graph of the function, we have

$$-1 = a(2-1)^2 + 2.$$

Then we get $a = -3$.
So the expression of the quadratic function is $y = -3(x-1)^2 + 2$.

(2) Since the point $(2, -1)$ is also on the line $y = 2x + k$, we have

$$-1 = 2 \times 2 + k.$$

So $k = -5$.

(3) From the given conditions,

$$2x - 5 = -3(x-1)^2 + 2,$$

i.e. $3x^2 - 4x - 4 = 0$.

Solving the equation gives $x = 2$ or $x = -\frac{2}{3}$. Substituting x into the equation of the line, we get that when $x = 2$, $y = -1$ and when $x = -\frac{2}{3}$, $y = -\frac{19}{3}$. Hence, the coordinates of the other point which the graph of the quadratic function and the line $y = 2x + k$ intersect at are $\left(-\frac{2}{3}, -\frac{19}{3}\right)$.

Example 3. Let k be a real number. If the graph of the function $y = (k^2 - 1)x^2 - (k+1)x + 1$ and the x-axis have no intersection points, find the value range of k.

Solution From the given condition, when the function is a quadratic function, we have $k^2 - 1 \neq 0$ and the equation

$$(k^2 - 1)x^2 - (k+1)x + 1 = 0$$

has no real roots. Thus we get $\Delta = (k+1)^2 - 4(k^2 - 1) < 0$, i.e. $3k^2 - 2k - 5 > 0$. Solving the inequality gives $(3k-5)(k+1) > 0$, then $k > \frac{5}{3}$ or $k < -1$.

When the function is a linear function, we have $k^2 - 1 = 0$, $k + 1 \neq 0$. Thus we get $k = 1$. At this time, the line $y = -2x + 1$ and the x-axis intersect at point $(\frac{1}{2}, 0)$, which does not meet the given requirements.

When the function is a constant function, we have $k^2 - 1 = 0$ and $k + 1 = 0$. Thus we get $k = -1$. At this time, the line $y = 1$ and the x-axis have no intersection points.

Hence, the value range of k satisfying the requirements is $k > \frac{5}{3}$ or $k \leq -1$.

Example 4. Let b and c be real numbers. Given that in the plane rectangular coordinate system xOy, the vertex of the graph of the quadratic function $y = x^2 + bx + c$ is point D. The curve intersects the positive direction of x-axis at points A and B from left to right, and intersects the positive direction of y-axis at point C. If ΔABD and ΔOBC are isosceles right triangles, try to judge whether the point $(1, 0)$ is on the graph of this function.

Solution From the given condition, $b^2 - 4c > 0$, $c > 0$ and the coordinates of the points are $C(0, c)$, $A\left(\frac{-b-\sqrt{b^2-4c}}{2}, 0\right)$, $B\left(\frac{-b+\sqrt{b^2-4c}}{2}, 0\right)$, $D\left(-\frac{b}{2}, -\frac{b^2-4c}{4}\right)$, then we have

$$OC = c, \quad OB = \left|\frac{-b+\sqrt{b^2-4c}}{2}\right|, \quad AB = \sqrt{b^2-4c}.$$

Since ΔABD is an isosceles right triangles, we know that $AD = BD$. Draw a line DE through D which is perpendicular to AB and intersects AB at E, then we have $2DE = AB$, i.e. $2 \times \frac{b^2-4c}{4} = \sqrt{b^2-4c}$, thus we get $b^2 - 4c = 2\sqrt{b^2-4c}$. Since $b^2 - 4c > 0$, i.e. $\sqrt{b^2-4c} > 0$, we find that $\sqrt{b^2-4c} = 2$.

Since ΔOCB is an isosceles right triangles and $\angle COB = 90°$, we know that $OC = OB$, i.e.

$$c = \left|\frac{-b+\sqrt{b^2-4c}}{2}\right| = \left|\frac{-b+2}{2}\right|.$$

Therefore, we get $\begin{cases} c > 0, \\ 2c = |-b+2|, \\ \sqrt{b^2-4c} = 2. \end{cases}$

Solving the simultaneous equations gives $b = -4$ and $c = 3$. Then we have $b+c+1 = -4+3+1 = 0$. Hence, from $1 \cdot 1^2 + b \cdot 1 + c = 1 + b + c = 0$, we can know that the graph of the function $y = x^2 + bx + c$ passes through the point $(1, 0)$.

Think about it

(1) In Example 4, if "the curve intersects the positive direction of y-axis at point C" is changed to "the curve intersects the y-axis at point C", is the conclusion also true?

(2) Is there any other way to solve this problem?

Example 5. Let a be a real number. Given that the coordinates of points A and B are $(1, 0)$ and $(2, 0)$ respectively. If the graph of the quadratic function $y = x^2 + (a-3)x + 3$ and line AB have exactly one intersection point, find the value range of a.

Solution There are two situations as follow:

(1) Since the graph of the quadratic function $y = x^2 + (a-3)x + 3$ and line AB have exactly one intersection point, and the coordinates of points A and B are $(1, 0)$ and $(2, 0)$ respectively, we know that

$$[1^2 + (a-3) \times 1 + 3] \times [2^2 + (a-3) \times 2 + 3] < 0.$$

Solving the inequality gives $-1 < a < -\frac{1}{2}$.
When $1^2 + (a-3) \times 1 + 3 = 0$, i.e. $a = -1$, we get that $x_1 = 1$ and $x_2 = 3$, which satisfy the given requirements.
When $2^2 + (a-3) \times 2 + 3 = 0$, i.e. $a = -\frac{1}{2}$, we get that $x_1 = 2$ and $x_2 = \frac{3}{2}$, which do not satisfy the given requirements.

(2) Let $x^2 + (a-3)x + 3$ be 0. Since the discriminant $\Delta = 0$, we get that $a = 3 \pm 2\sqrt{3}$. When $a = 3 + 2\sqrt{3}$, we get $x_1 = x_2 = -\sqrt{3}$, which do not satisfy the given requirements. When $a = 3 - 2\sqrt{3}$, we get $x_1 = x_2 = \sqrt{3}$, which satisfy the given requirements. In summary, the value range of a is $-1 \le a < -\frac{1}{2}$ or $a = 3 - 2\sqrt{3}$.

Example 6. Given that m, n and p are positive integers with $m < n$. Let O be the coordinate origin and the coordinates of points A, B and C be $(-m, 0)$, $(n, 0)$ and $(0, p)$ respectively. If $\angle ACB = 90°$ and $OA^2 + OB^2 + OC^2 = 3(OA + OB + OC)$, find the expression of a quadratic function of which the graph passes through the three points A, B and C.

Solution Since $\angle ACB = 90°$ and $OC \perp AB$, we find that $\triangle AOC \backsim \triangle COB$. Then we have $OA \cdot OB = OC^2$, i.e. $mn = p^2$.

From $OA^2 + OB^2 + OC^2 = 3(OA + OB + OC)$, we get that

$$m^2 + n^2 + p^2 = 3(m + n + p).$$

Besides, we know that

$$\begin{aligned} m^2 + n^2 + p^2 &= (m+n+p)^2 - 2(mn + np + mp) \\ &= (m+n+p)^2 - 2(p^2 + np + mp) \\ &= (m+n+p)^2 - 2p(m+n+p) \\ &= (m+n+p)(m+n-p). \end{aligned}$$

Since $m + n + p \neq 0$, we obtain that $m + n - p = 3$, i.e. $m + n = p + 3$.

Because $m + n = p + 3$ and $mn = p^2$, m and n are two distinct positive integer roots of the quadratic equation

$$x^2 - (p+3)x + p^2 = 0. \tag{1}$$

Then we have $\Delta = [-(p+3)]^2 - 4p^2 > 0$. Solving the inequality gives $-1 < p < 3$.

Since p is a positive integer, $p = 1$ or $p = 2$.

When $p = 1$, the equation (1) is $x^2 - 4x + 1 = 0$, which has no integer roots.

When $p = 2$, the equation (1) is $x^2 - 5x + 4 = 0$. Since $m < n$, the roots are $m = 1$, $n = 4$.

In summary, $m = 1$, $n = 4$ and $p = 2$.

Let the expression of the quadratic function of which the graph passes through the points A, B and C be $y = k(x+1)(x-4)$. Substituting the coordinates of the point C $(0, 2)$ into the expression gives $2 = k \cdot 1 \cdot (-4)$. Then we get $k = -\frac{1}{2}$. Hence, the expression of the quadratic function of which the graph passes through points A, B and C is

$$y = -\frac{1}{2}(x+1)(x-4) = -\frac{1}{2}x^2 + \frac{3}{2}x + 2.$$

Example 7. Let b and c be real numbers and $c < 0$. Given that the graph of the quadratic function $y = x^2 + bx + c$ intersects the x-axis at the points A and B, and intersects the y-axis at the point C, let the center of the circumscribed circle of $\triangle ABC$ be the point P.

(1) Prove that the other intersection of $\odot P$ and the y-axis is a fixed point.
(2) IfAB is exactly the diameter of $\odot P$ and $S_{\triangle ABC} = 2$, find the values of b and c.

Solution

(1) From the given condition, the coordinates of the point C are $(0,\ c)$. Suppose that the coordinates of the points A and B are $(x_1,\ 0)$ and $(x_2,\ 0)$. Then we have

$$x_1 + x_2 = -b, \quad x_1x_2 = c.$$

Let the other intersection of $\odot P$ and the y-axis be the point D. Since AB and CD are two intersecting chords of $\odot P$ and the point of their intersection is O, according to the power of a point theorem (see Chapter 19), we get $OA \times OB = OC \times OD$. Then

$$OD = \frac{OA \times OB}{OC} = \frac{|x_1x_2|}{|c|} = \frac{|c|}{|c|} = 1.$$

The graph of the quadratic function opens upwards, and from $c < 0$ we know that the point C is below the x-axis, so the point D is on the positive half of the y-axis. Thus the point D is a fixed point, its coordinates are $(0, 1)$.

(2) Since $AB \perp CD$ and AB is exactly the diameter of $\odot P$, the points C and D are symmetric about the point O. Thus the coordinates of point C are $(0,\ -1)$, i.e. $c = -1$.

From the quadratic formula, we have

$$AB = |x_1 - x_2| = \frac{2\sqrt{b^2 - 4c}}{2 \times 1} = \sqrt{b^2 - 4c} = \sqrt{b^2 + 4}.$$

Therefore,

$$S_{\triangle ABC} = \frac{1}{2}AB \cdot OC = \frac{1}{2}\sqrt{b^2+4} \cdot 1 = 2.$$

Solving the equation gives $b = \pm 2\sqrt{3}$. Hence, $b = \pm 2\sqrt{3}$, $c = -1$.

Example 8. Given that the coordinates of the points M and N are $(0, 1)$ and $(0, -1)$ respectively. Point P is a moving point on the parabola $y = \frac{1}{4}x^2$.

(1) Determine the positional relation between the circle with the point P as the center and PM as the radius and the straight line $y = -1$.
(2) Let the other intersection of the straight line PM and the parabola $y = \frac{1}{4}x^2$ be point Q, and connect NP and NQ. Prove that $\angle PNM = \angle QNM$.

Solution

(1) Suppose that the coordinates of point P are $\left(x_0, \frac{1}{4}x_0^2\right)$, then

$$PM = \sqrt{x_0^2 + \left(\frac{1}{4}x_0^2 - 1\right)^2}$$
$$= \sqrt{\left(\frac{1}{4}x_0^2 + 1\right)^2}$$
$$= \frac{1}{4}x_0^2 + 1.$$

Because the distance between point P and line $y = -1$ is $\frac{1}{4}x_0^2 - (-1) = \frac{1}{4}x_0^2 + 1$, the circle with point P as the center and PM as the radius is tangent to line $y = -1$.

(2) As shown in Figure 8.3, draw two perpendicular lines of line $y = -1$ which pass through points P and Q respectively, and the foot points are H and R.

From (1) we know that $PH = PM$. Similarly, we get $QM = QR$.

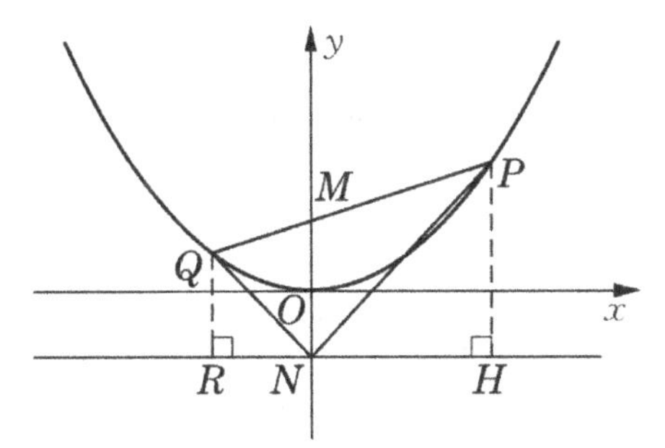

Fig. 8.3

Because PH, MN and QR are all perpendicular to the straight line $y = -1$, we get that $PH // MN // QR$. Then

$$\frac{QM}{RN} = \frac{MP}{NH}.$$

Therefore, we get $\frac{QR}{RN} = \frac{PH}{HN}$, then Rt $\Delta PHN \backsim$ Rt ΔQRN. Hence, $\angle HNP = \angle RNQ, \angle PNM = \angle QNM$.

Reading

Problem is the heart of mathematics

In the article *The Heart of Mathematics* (American Mathematical Monthly, No. 7, 1980), mathematician P.R. Halmos first raised a question:

"What does mathematics consist of?"
"Axiom? Theorem? Proof? Concept? Definition? Theory? Formula? Method?" Halmos thinks that all these are important but not the heart of mathematics. He emphasizes that:
"Problem is the heart of mathematics."

This well-known saying represents the common view of many mathematicians and reflects the real situation of mathematics development.

From ancient times to the present, there are many famous mathematical problems, such as the three major problems of compass-and-straightedge construction (include trisection of an angle, duplication of a cube and squaring the circle, which have been proved impossible to complete); the formula for finding the roots of higher-degree equations (cubic and quartic equations can be solved by using Tartaglia-Cardano formula, the general equation whose degree is higher than four has no roots, which is proved by Abel *et al.*); Brachistochrone Problem (Johann Bernoulli points out that the fastest curve for a particle to fall from one point to another is the cycloid); Fermat's Last Theorem (when $n \geq 3$, the equation $x^n + y^n = z^n$ has no integer roots, this big guess was not solved by Andrew Wiles, 1953 — until 1995) and so on. The research on these problems has greatly promoted the development of mathematics, resulting in many new concepts, new methods and even new branches of mathematics. Stimulated by the problem of squaring the circle, Lindemann proved that π is transcendental number, which became the origin of transcendental number theory. In the process of constructing the concrete and special higher degree equations which can't be solved by radical expression, Galois established significant Group Theory. It is a well-known fact that Brachistochrone Problem led to the birth of calculus of variations. Fermat's Last Theorem is like a chicken laying golden eggs. New researches on Ideal, Cyclotomic Fields, Modular curve and so on are produced by it.

(Excerpted from: Shan Zun. *Problem solving research.* Shanghai: Shanghai Education Press, 2016, 69.)

Exercises

1. Given that the graph of the quadratic function $y = -x^2 + 2mx + 1 - m^2$ passes through the origin, find the coordinates of the vertex.
2. The graph of the quadratic function $y = ax^2 + bx + c$ is shown in Figure 8.4, then in the following six algebraic expressions: ab, ac, $a + b + c$, $a - b + c$, $2a + b$ and $2a - b$, the number of positive expressions is ( ).

   (A) 2 (B) 3 (C) 4 (D) 5

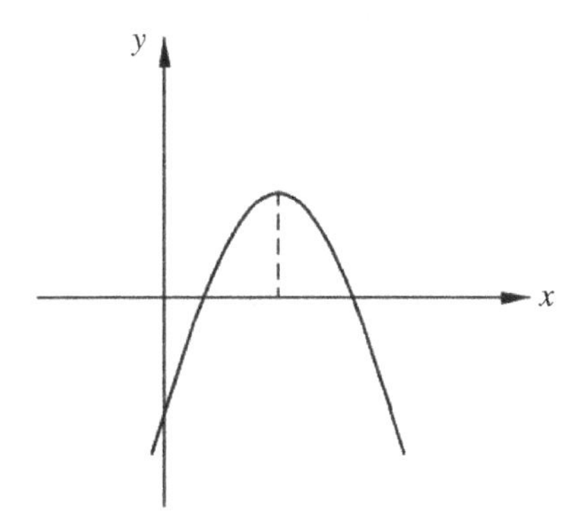

Fig. 8.4

3. The graph of the quadratic function $y = ax^2 + bx + c$ is shown in Figure 8.5, then which quadrant is the point $(a + b, ac)$ in? ( )

   (A) the first quadrant (B) the second quadrant
   (C) the third quadrant (D) the fourth quadrant

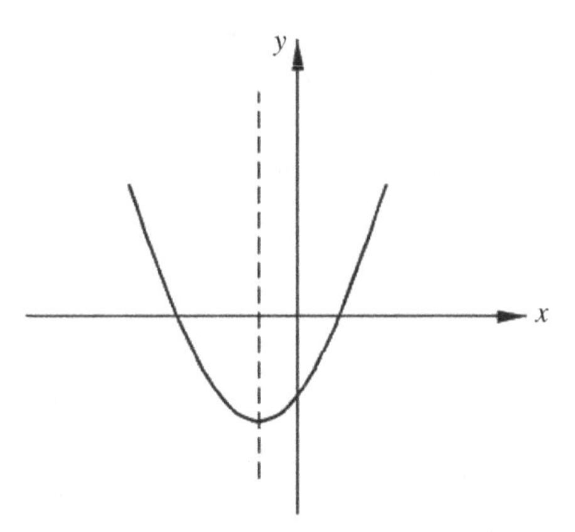

Fig. 8.5

4. Suppose that the points $A(-0.8, 4.132)$, $B(1.2, -1.948)$ and $C(2.8, -3.932)$ are on the graph of the quadratic function $y = ax^2 + bx + c$. When the x-coordinate of point D on the graph is $x = 1.8$, the value of its y-coordinate y is ________.
5. In the plane rectangular coordinate system xOy, a point whose x-coordinate and y-coordinate are integers is called an integral point. The closed figure enclosed by the graph of the quadratic function $y = -x^2 + 6x - \frac{27}{4}$ and x-axis are dyed red, then the number of integral points in the red area and its boundary is ( ).

   (A) 5 (B) 6 (C) 7 (D) 8

6. Let a, b and c be real numbers, and $a \neq 0$. Consider the functions $y = ax + b$ and $y = ax^2 + bx + c$, which of the following figures can be their graphs? ( )

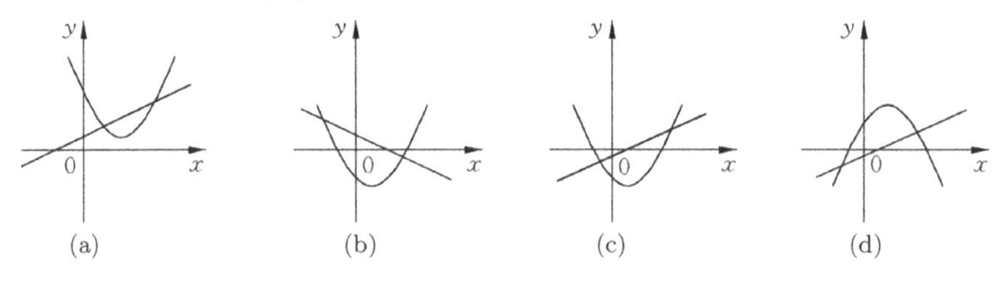

7. The height of a small ball thrown upward from the ground is a quadratic function of its motion time. Xiao Hong throws two little balls up one second apart. Suppose that the two little balls are at the same height

from the ground before they are thrown, and they reach the same maximum height from the ground 1.1 seconds after they are thrown respectively. If t seconds after the first ball is thrown, the first ball has the same height from the ground as the second ball in the air, then $t =$ ________.

8. Suppose that the graph of quadratic function $y_1 = a(x - x_1)(x - x_2)$ (constant $a \neq 0$ and $x_1 \neq x_2$) and the graph of linear function $y_2 = dx + e$ (constant $d \neq 0$, e to be determined) intersect at point $(x_1, 0)$. If the graph of function $y = y_1 + y_2$ and the x-axis have only one intersection point, then the value of $x_2 - x_1$ is ( ).

   (A) $-\frac{d}{a}$ (B) $-\frac{e}{a}$ (C) $\pm\frac{d}{a}$ (D) $\frac{d}{a}$

9. Let a and b be real numbers. Given that the x-coordinate of the intersection of the graph of quadratic function $y = x^2 + ax + b$ and the x-axis is m and n respectively, and $|m| + |n| \leq 1$. Suppose that the maximum and minimum values of b satisfying the above requirements are p and q respectively, then $|p| + |q| =$ ________.

10. The graph of the quadratic function $y = ax^2 + bx + c$ is shown in Figure 8.6.

   (1) Find the signs of a, b, c and $b^2 - 4ac$.
   (2) Find the coordinates of the points A, B, D and M.
   (3) Find the length of $|AB|$.
   (4) Find the value of $|OA| \cdot |OB|$.
   (5) If $|OA| = |OD|$, find the relation of a, b and c.

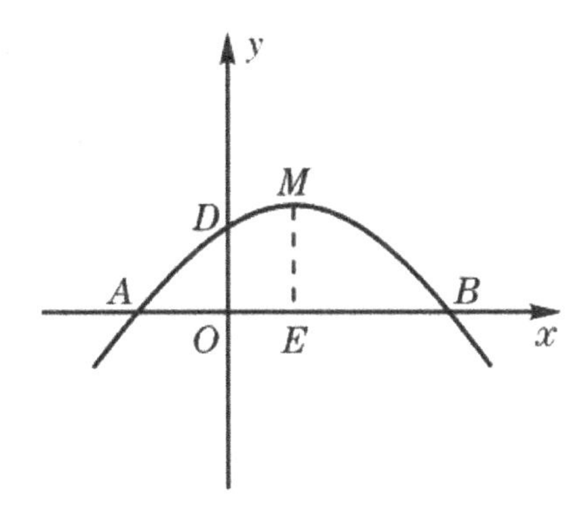

Fig. 8.6

11. The parabola $y = ax^2 + bx + c$ intersects the x-axis at points A and B, and intersects the y-axis at point C. If $\triangle ABC$ is a right triangle, find the value of ac.
12. The graph of linear function y_1 and the graph of quadratic function y_2 intersect at points $A(1, 4)$ and $B(-2, -2)$, and the graph of function y_2 intersects the y-axis at point $C(0, 3)$.
    (1) Find the expressions of these two functions.
    (2) If $y_1 > y_2$, find the value range of x.
    (3) If both y_1 and y_2 increase as x increases, find the value range of x.
13. In the plane rectangular coordinate system xOy, find the coordinates of the points satisfying the following two conditions at the same time:
    (1) Line $y = -2x + 3$ passes through such a point.
    (2) Whatever the value of real number m is, the graph of quadratic function $y = mx^2 + \left(m - \frac{2}{3}\right)x - \left(2m - \frac{3}{8}\right)$ does not pass through such a point.
14. Let a, b and c be real numbers. Given that the graph G of quadratic function $y = ax^2 + bx + c$ and the x-axis have only one intersection point A, the graph G intersects the y-axis at point $B(0, 4)$, and $ac = b$.
    (1) Find the values of a, b and c.
    (2) Translate the graph of the linear function $y = -3x$ to get the graph L which passes through point A, and the other intersection of the graphs L and G is point C. Find the area of $\triangle ABC$.
15. Let a and m be positive real numbers. Given that the vertex of the graph S_1 of quadratic function $y = ax^2 - 2amx + am^2 + 2m + 1$ $(a > 0, m > 0)$ is point A, and the vertex of the graph S_2 of another quadratic function is point B which is on the y-axis. The graphs S_1 and S_2 are centrosymmetric about the point $P(1, 3)$.
    (1) When $a = 1$, find the functional expression of graph S_2 and the value of m.
    (2) Let the intersection of S_2 and the x-axis be point C. When $\triangle ABC$ is an isosceles triangle, find the value of a.
16. Let a, b and c be real numbers, and $a > 0$, $c > 1$. Consider the quadratic function $y = ax^2 + bx + c$, when $x = c$, $y = 0$; when $0 < x < c$, $y > 0$.
    (1) Try to compare ac and 1, and explain the reason.
    (2) When $x > 0$, prove that $\dfrac{a}{x+2} + \dfrac{b}{x+1} + \dfrac{c}{x} > 0$.

Chapter 9

Quadratic Inequalities

Let a, b and c be real numbers. A quadratic inequality has the form $ax^2 + bx + c > 0(<0, \geq 0$ or $\leq 0)$ where $a \neq 0$. The set of values of x satisfying the inequality is called the solution of the quadratic inequality.

The following basic conclusions are often used to solve the quadratic inequalities:

$$AB > 0 \Leftrightarrow A > 0 \text{ and } B > 0 \text{ or } A < 0 \text{ and } B < 0;$$

$$AB \geq 0 \Leftrightarrow A \geq 0 \text{ and } B \geq 0 \text{ or } A \leq 0 \text{ and } B \leq 0;$$

$$AB < 0 \Leftrightarrow A > 0 \text{ and } B < 0 \text{ or } A < 0 \text{ and } B > 0;$$

$$AB \leq 0 \Leftrightarrow A \geq 0 \text{ and } B \leq 0 \text{ or } A \leq 0 \text{ and } B \geq 0.$$

Therefore, we often use factorization to transform $ax^2 + bx + c$ into the form of AB.

Example 1. Solve the following quadratic inequalities:

(1) $x^2 + 4x - 21 \geq 0$;
(2) $x^2 - 3x - 1 < 0$.

Analysis

(1) Factorize directly.
(2) If it is difficult to factorize directly, the inequality in the question can be transformed by the quadratic formula or the method of completing the square.

Solution

(1) The original inequality can be transformed into

$$(x+7)(x-3) \geq 0,$$

i.e $\begin{cases} x+7 \geq 0, \\ x-3 \geq 0; \end{cases}$ or $\begin{cases} x+7 \leq 0, \\ x-3 \leq 0. \end{cases}$

Solving $\begin{cases} x+7 \geq 0, \\ x-3 \geq 0 \end{cases}$ gives $x \geq 3$. Solving $\begin{cases} x+7 \leq 0, \\ x-3 \leq 0 \end{cases}$ gives $x \leq -7$.

So the solution of the original inequality is $x \geq 3$ or $x \leq -7$.

(2) Since $x^2 - 3x - 1 = \left(x - \frac{3+\sqrt{13}}{2}\right)\left(x - \frac{3-\sqrt{13}}{2}\right)$, the inequality $x^2 - 3x - 1 < 0$ can be transformed into

$$\begin{cases} x - \dfrac{3+\sqrt{13}}{2} < 0, \\ x - \dfrac{3-\sqrt{13}}{2} > 0; \end{cases} \quad \text{or} \quad \begin{cases} x - \dfrac{3+\sqrt{13}}{2} > 0, \\ x - \dfrac{3-\sqrt{13}}{2} < 0. \end{cases}$$

Solving

$$\begin{cases} x - \dfrac{3+\sqrt{13}}{2} < 0, \\ x - \dfrac{3-\sqrt{13}}{2} > 0 \end{cases}$$

gives $\frac{3-\sqrt{13}}{2} < x < \frac{3+\sqrt{13}}{2}$.

Solving

$$\begin{cases} x - \dfrac{3+\sqrt{13}}{2} > 0, \\ x - \dfrac{3-\sqrt{13}}{2} < 0 \end{cases}$$

gives no solution.

So the solution of the original inequality is $\frac{3-\sqrt{13}}{2} < x < \frac{3+\sqrt{13}}{2}$.

Remark By summing up the above solving process and combining with the graph of the quadratic function, we can get the following conclusions.

We know that

$$ax^2 + bx + c = a\left(x + \frac{b}{2a}\right)^2 + \frac{4ac - b^2}{4a}.$$

For $a > 0$, when $\Delta = b^2 - 4ac > 0$, the equation $ax^2 + bx + c = 0$ has two distinct real roots α and β, then $ax^2 + bx + c = a(x-\alpha)(x-\beta)$. Suppose that $\alpha < \beta$, then the solution of $ax^2 + bx + c > 0$ is $x < \alpha$ or $x > \beta$; the solution of $ax^2 + bx + c < 0$ is $\alpha < x < \beta$.

When $\Delta = b^2 - 4ac = 0$, the equation $ax^2 + bx + c = 0$ has two identical roots written as $\alpha = \beta$, and $ax^2 + bx + c = a\left(x + \frac{b}{2a}\right)^2 = a(x - \alpha)^2$, then the solution of $ax^2 + bx + c > 0$ is $x \neq \alpha$, i.e. $x \neq -\frac{b}{2a}$; the inequality $ax^2 + bx + c < 0$ has no solution.

When $\Delta = b^2 - 4ac < 0$, From $a\left(x + \frac{b}{2a}\right)^2 \geq 0$ and $\frac{4ac-b^2}{4a} > 0$, it's easy to know that the solutions of inequality $ax^2 + bx + c > 0$ are all real numbers, that is, x can be any real number; the inequality $ax^2 + bx + c < 0$ has no solution.

For $a < 0$, just multiply the two sides of the inequality $ax^2 + bx + c > 0$ (or < 0) by -1, then it becomes the above situation.

Example 2. Solve the inequality:

$$x^2 - 5|x| + 6 > 0.$$

Analysis Notice that $x^2 = (|x|)^2$, factorize it as a whole.

Solution From the given condition,

$$(|x| - 2)(|x| - 3) > 0,$$

i.e. $|x| > 3$ or $|x| < 2$.

Solving $|x| > 3$ gives $x > 3$ or $x < -3$. Solving $|x| < 2$ gives $-2 < x < 2$. So the solution of the original inequality is

$$x < -3, \ -2 < x < 2 \text{ or } x > 3.$$

Remark

(1) In the above solution, $|x|$ is regarded as a whole.
(2) It can also be solved in two cases respectively: $x \geq 0$ and $x < 0$.
(3) The conclusion is easy to know by using the graph of the function.

As shown in Figure 9.1, when $x \geq 0$, consider the graph of the function

$$y = x^2 - 5x + 6 = \left(x - \frac{5}{2}\right)^2 - \frac{1}{4},$$

the vertex is $\left(\frac{5}{2}, -\frac{1}{4}\right)$, and x-intercepts are 3 and 2. Then $x^2 - 5x + 6 > 0$ is a part of the graph on the right side of the y-axis and above the x-axis.

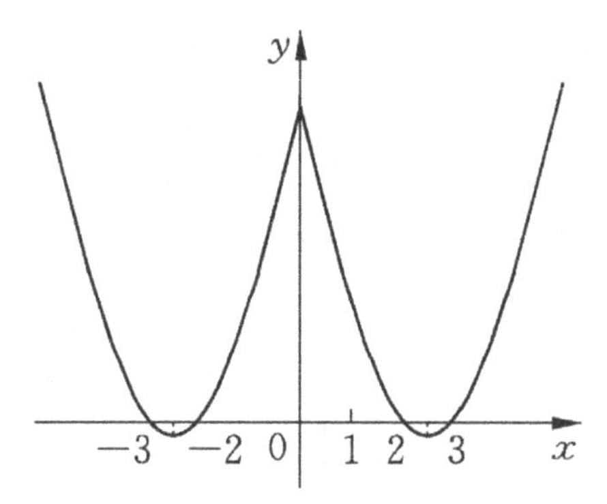

Fig. 9.1

When $x < 0$, consider the graph of the function

$$y = x^2 + 5x + 6 = \left(x + \frac{5}{2}\right)^2 - \frac{1}{4},$$

the vertex is $\left(-\frac{5}{2}, -\frac{1}{4}\right)$, and x-intercepts are -2 and -3. Then $x^2 + 5x + 6 > 0$ is a part of the graph on the left side of the y-axis and above the x-axis. So the solution of the original inequality is

$$x < -3,\ -2 < x < 2 \text{ or } x > 3.$$

Example 3. Let a be a real number. Solve the inequality of x:

$$ax^2 - (a+1)x + 1 < 0.$$

Analysis Notice the discussion of a.

Solution

(1) When $a = 0$, the original inequality is $-x + 1 < 0$, i.e. $x > 1$.
(2) When $a \neq 0$, the original inequality is transformed into

$$(ax - 1)(x - 1) < 0,$$

i.e. $a\left(x - \frac{1}{a}\right)(x - 1) < 0$.

If $a > 0$, we get $\left(x - \frac{1}{a}\right)(x - 1) < 0$. Therefore, when $a > 1$, $\frac{1}{a} < x < 1$; when $a < 1$, $1 < x < \frac{1}{a}$; when $a = 1$, the inequality has no solution. If $a < 0$, we get $\left(x - \frac{1}{a}\right)(x - 1) > 0$, so $x < \frac{1}{a}$ or $x > 1$. Hence, when $a = 0$, the solution of the inequality is $x > 1$; when $a > 1$, the solution of the inequality is $\frac{1}{a} < x < 1$; when $0 < a < 1$, the solution of the inequality is

$1 < x < \frac{1}{a}$; when $a = 1$, the inequality has no solution; when $a < 0$, the solution of the inequality is $x < \frac{1}{a}$ or $x > 1$.

Example 4. Given that three real numbers a, b and c satisfy $c \leq b \leq a$, $a+b+c = 10$, and $abc - 23a = 40$, find the minimum value of $|a| + |b| + |c|$.

Solution From $c \leq b \leq a$ and $a + b + c = 10$, we can get $a > 0$.

Since $b + c = 10 - a$, $bc = 23 + \frac{40}{a}$, b and c are two real roots of the quadratic equation

$$x^2 - (10 - a)x + \left(23 + \frac{40}{a}\right) = 0.$$

Then the discriminant of the quadratic equation is

$$\Delta = (10 - a)^2 - 4\left(23 + \frac{40}{a}\right) \geq 0,$$

so we get $a^3 - 20a^2 + 8a - 160 \geq 0$, i.e. $(a^2 + 8)(a - 20) \geq 0$.

Solving the inequality gives $a \geq 20$.

Therefore, we have $b + c = 10 - a \leq -10$, $|b + c| \geq 10$, then $|b| + |c| \geq |b + c| \geq 10$.

Hence, $|a| + |b| + |c| \geq 30$, when $a = 20$, $b = -5$ and $c = -5$, the equality holds.

Example 5. Given that real numbers x and y satisfy $x^2 + xy + y^2 = 3$, find the maximum value of $x - y$.

Solution Suppose that $x - y = t$, then $x = y + t$. Substituting this into the given equation, we can get

$$(y + t)^2 + (y + t)y + y^2 - 3 = 0,$$

i.e.

$$3y^2 + 3ty + t^2 - 3 = 0.$$

If the above equation is regarded as a quadratic equation of y with real roots, we can know its discriminant is

$$\Delta = (3t)^2 - 4 \times 3 \times (t^2 - 3) \geq 0.$$

Then we get $12 - t^2 \geq 0$. Solving this inequality gives $-2\sqrt{3} \leq t \leq 2\sqrt{3}$. When $t = 2\sqrt{3}$, $\Delta = 0$, then we have $3y^2 + 6\sqrt{3}y + 9 = 0$. Solving the equation gives $y = -\sqrt{3}$, and then we get $x = \sqrt{3}$. Hence, the maximum value of $x - y$ is $2\sqrt{3}$, at this time, $x = \sqrt{3}$ and $y = -\sqrt{3}$.

Example 6. In the rectangular coordinate system xOy, a point whose x-coordinate and y-coordinate are both integers is called an integral point. Try to find all the integral points (x, y) satisfying $y \leq |x|$ on the graph of the quadratic function $y = \frac{x^2}{10} - \frac{x}{10} + \frac{9}{5}$, and explain the reason.

Solution From the given condition, $y = \frac{x^2}{10} - \frac{x}{10} + \frac{9}{5} \leq |x|$, i.e. $x^2 - x + 18 \leq 10|x|$. (1)

When $x \geq 0$, the inequality (1) is transformed into

$$x^2 - 11x + 18 \leq 0.$$

Solving it gives $2 \leq x \leq 9$.

Since $10y = x^2 - x - 2 + 20 = (x-2)(x+1) + 20$, the pairs of numbers satisfying x and y are both integers are $x = 2$, $y = 2$; $x = 4$, $y = 3$; $x = 7$, $y = 6$; $x = 9$, $y = 9$.

When $x < 0$, the inequality (1) is transformed into

$$x^2 + 9x + 18 \leq 0.$$

Solving it gives $-6 \leq x \leq -3$.

Then the pairs of numbers satisfying x and y are both integers are $x = -3$, $y = 3$; $x = -6$, $y = 6$. Hence, the integral points meeting the given requirements are

$$(-3, 3), (-6, 6), (2, 2), (4, 3), (7, 6), (9, 9).$$

Example 7. Let m and n be positive integers, where $m \neq 2$. The distance between two intersections of the graph of quadratic function $y = x^2 + (3 - mt)x - 3mt$ and the x-axis is d_1. The distance between two intersections of the graph of quadratic function $y = -x^2 + (2t - n)x + 2nt$ and the x-axis is d_2. If $d_1 \geq d_2$ is always true for all real numbers t, find the values of m and n.

Solution Since the roots of the quadratic equation $x^2 + (3 - mt)x - 3mt = 0$ are mt and -3, we have $d_1 = |mt + 3|$.

The roots of the quadratic equation $-x^2 + (2t - n)x + 2nt = 0$ are $2t$ and $-n$, so

$$d_2 = |2t + n|.$$

Therefore, $d_1 \geq d_2 \Leftrightarrow |mt + 3| \geq |2t + n| \Leftrightarrow (mt + 3)^2 \geq (2t + n)^2 \Leftrightarrow (m^2 - 4)t^2 + (6m - 4n)t + 9 - n^2 \geq 0$. (1)

From the given condition, $m^2 - 4 \neq 0$, and the inequality (1) is true for all real numbers t, so we have

$$\begin{cases} m^2 - 4 > 0, \\ \Delta = (6m - 4n)^2 - 4(m^2 - 4)(9 - n^2) \leq 0. \end{cases}$$

Rearranging the inequalities gives

$$\begin{cases} m > 2 \text{ or } m < -2, \\ 4(mn - 6)^2 \leq 0, \end{cases} \quad \text{i.e.} \quad \begin{cases} m > 2 \text{ or } m < -2, \\ mn = 6. \end{cases}$$

Since m and n are positive integers, we know that $\begin{cases} m > 2, \\ mn = 6. \end{cases}$

Hence, we get $\begin{cases} m = 3, \\ n = 2; \end{cases}$ or $\begin{cases} m = 6, \\ n = 1. \end{cases}$

Reading

Reflection on the application of quadratic functions

Learn to rethink profoundly by taking an example as a demonstration.

Example Given that $k > a > b > c > 0$, try to prove the following inequality:

$$k^2 - (a + b + c)k + ab + bc + ca > 0. \tag{1}$$

This is an algebra proving problem which junior high school students can solve.

Many students want to apply the knowledge of quadratic functions as soon as they see the inequality (1).

The parabola $y = x^2 - (a + b + c)x + ab + bc + ca$ opens upwards, if the discriminant of the quadratic polynomial

$$x^2 - (a + b + c)x + ab + bc + ca \tag{2}$$

is

$$\Delta = (a + b + c)^2 - 4(ab + bc + ca) \tag{3}$$

which satisfies

$$\Delta < 0, \tag{4}$$

then there is no intersection between the parabola and the x-axis. So the parabola is always above the x-axis, and we always have

$$x^2 - (a + b + c)x + ab + bc + ca > 0. \tag{5}$$

Thus the inequality (1) holds. So the problem turns to prove that the inequality (4) is true.

This plan is also very clear, but it can't be proved that the inequality (4) must be true. When many students come here, they will "doubt whether there is a way out from the endless mountains and rivers". Many of them don't believe in the defects of this plan, and insist on proving the inequality (4) to be true, resulting in futility.

In fact, the problem does not require that the inequality (5) is always true, and it just require that the inequality (5) holds when $x = k$(or $x > a$). Those who insist on proving the inequality (4) do not see it at all, and they even do not use the given condition $k > a > b > c > 0$. Of course, the inequality (4) holds under certain conditions. For example, when

$$a < b + c, \tag{6}$$

we know that

$$\begin{aligned}\Delta &= a^2 + b^2 + c^2 - 2ab - 2bc - 2ca = a(a - b - c) + (b^2 - ab)\\ &\quad + (c^2 - ac) - 2bc < 0.\end{aligned}$$

Therefore, when $a < b + c$, the inequality (1) holds. Then we only need to prove that (1) also holds when $a \geq b + c$.

Students who are familiar with quadratic functions are now in the position of "bright in the dark". It is not difficult to see that the x-coordinate of the vertex is $\frac{a+b+c}{2}$, and the point whose x-coordinate is $a\left(\geq \frac{a+b+c}{2}\right)$ is to the right of the vertex. Corresponding to $x = a$, the value of function y is

$$a^2 - (a + b + c)a + ab + bc + ca = bc > 0.$$

When $x > a$, the function y is monotonically increasing. So $y > 0$, in particular, the inequality (1) holds. Hence no matter $a < b+c$ or $a \geq b+c$, the inequality (1) holds.

Although the solution mentioned above is complicated, it is the method middle school students are good at. The difficulty lies only in being divided into two situations of $a < b + c$ and $a \geq b + c$. However, the above solution is not the best solution, and also not the only solution.

A better solution is as follows.

From the given condition,

$$k - a > 0, \quad k - b > 0, \quad k - c > 0.$$

Then we get that

$$\begin{aligned}0 &< (k-a)(k-b)(k-c)\\ &= k^3-(a+b+c)k^2+(ab+bc+ca)k-abc\\ &< k^3-(a+b+c)k^2+(ab+bc+ca)k.\end{aligned}$$

Since $k>0$, divided by k on both sides of the above inequality gives the inequality (1).

From this solution, we can see that the essence of the problem is a cubic inequality, where $(k-a)(k-b)(k-c) = k^3-(a+b+c)k^2+(ab+bc+ca)$ $k-abc$ is the Vieta's theorem of cubic equation. Applying quadratic function is the reflection of mindset. "The true face of Lushan is lost to my sight, for it is right in this mountain that I reside". Only by gaining more experience can we grasp the essence of the problem.

(Excerpted from: Shan Zun. *Problem solving research.* Shanghai: Shanghai Education Press, 2016, 87~89.)

Exercises

1. Given that $-1<x<0$, then the size relationship between $-x^2$, x and $\frac{1}{x}$ is ( ).
   (A) $-x^2<x<\frac{1}{x}$
   (B) $\frac{1}{x}<-x^2<x$
   (C) $x<-x^2<\frac{1}{x}$
   (D) $\frac{1}{x}<x<-x^2$
2. Solve the inequality: $x-\dfrac{5}{2}>x(x-2)-\dfrac{1}{4}$.
3. Solve the inequality of x: $56x^2+ax>a^2$.
4. If only one value of x satisfies the inequality $0<x^2+ax+5\leq 4$, find the value of a.
5. Solve the inequality: $x^2-2x-3>3|x-1|$.
6. Let k be a real number. Given that the fractional equation

$$\frac{1}{x+2}-\frac{k}{x-2}=1-\frac{4x}{x^2-4}$$

   has two real roots, then k should satisfy ( ).

   (A) $k^2-18k+33>0$
   (B) $k^2-18k+33>0$, and $k\neq 2$

(C) $k \neq 2$

(D) None of the above answers are correct

7. Solve the inequality: $-x < x^2 - 2 < x$.
8. Let α, β and γ be real numbers with $\alpha < \beta < \gamma$. Find the solution of the inequality
$$(x-\alpha)(x-\beta)(x-\gamma) < 0.$$
9. Let a be a real number. Given the quadratic equation $f(x) = x^2 - 2ax + 6$, when $-2 \leq x \leq 2$, $f(x) \geq a$ is always true. Find the value range of a.
10. Let a and b be real numbers. If the solution of inequality $ax^2 + abx + b > 0$ is $1 < x < 2$, try to find the values of a and b.
11. Let a be a real number. In order to make all real numbers x, y and z which satisfy $\frac{x+y}{2} = \frac{y+z}{3} = \frac{z+x}{7}$ to satisfy the inequality $x^2 + y^2 + z^2 + a(x+y+z) > -1$, find the value range of a.
12. Let a, b and c be the three sides of a triangle respectively, and satisfy
$$a^2 - a - 2b - 2c = 0,$$
$$a + 2b - 2c + 3 = 0.$$
Which side of the triangle is longest?

Chapter 10

Distribution of Roots of Quadratic Equation

By using the relation between the roots and coefficients, we can know the positive and negative of the roots of the equation $ax^2 + bx + c = 0$ without finding its roots. Under the condition of $\Delta = b^2 - 4ac \geq 0$, we have the following conclusions.

When $\frac{c}{a} < 0$, the roots of the equation must be one positive and one negative. If $-\frac{b}{a} \geq 0$, the positive root of the equation is not less than the absolute value of the negative root. If $-\frac{b}{a} < 0$, the positive root of the equation is less than the absolute value of the negative root.

When $\frac{c}{a} > 0$, the roots of the equation must be both positive or both negative. If $-\frac{b}{a} > 0$, the roots of the equation are both positive. If $-\frac{b}{a} < 0$, the roots of the equation are both negative.

We can also make more use of the relation between the graph of the function and the roots of the equation, and use a basic conclusion to investigate the distribution of the roots of the quadratic equation. The basic conclusion is as follow: for quadratic function $y = ax^2 + bx + c$, if $f(m) \cdot f(n) < 0 (m < n)$, there must be a number x_0 between m and n, so that $f(x_0) = 0$, that is, equation $f(x) = 0$ must have a real root x_0 between m and n.

Example 1. Let k be an integer. Given that the equation $x^2 + (k+3)x + (2k+3) = 0$ has two roots with opposite signs, and the positive root is less than the absolute value of negative root, find the value of k.

Solution Let x_1 and x_2 be two roots of the equation. From the given condition, we know that the product of the two roots x_1x_2 is less than 0, and the sum of the two roots $x_1 + x_2$ is less than 0.

From the relation between the roots and coefficients and the discriminant of the equation, we get

$$\begin{cases}(k+3)^2 - 4(2k+3) > 0,\\ -(k+3) < 0,\\ 2k+3 < 0,\end{cases}$$

i.e. $\begin{cases}(k-3)(k+1) > 0,\\ -3 < k < -\frac{3}{2}.\end{cases}$

Solving the inequalities gives $-3 < k < -\frac{3}{2}$. Since k is an integer, $k = -2$.

Example 2. Let a and b be real numbers. Given that the quadratic equation $x^2 + ax + b = 0$ has two real roots α and β, prove that if $|\alpha| < 2$ and $|\beta| < 2$, then we have $2|a| < 4 + b$ and $|b| < 4$.

Solution From Vieta's Formula (the relation between the roots and coefficients), we find that $|b| = |\alpha\beta| < 2 \times 2 = 4$.

On the other hand, from the graph of the function $y = x^2 + ax + b$ (see Figure 10.1), it's easy to know that the values of the function are all positive when $x = \pm 2$, that is,

$$(2)^2 + 2a + b > 0,$$

and

$$(-2)^2 + a \cdot (-2) + b > 0.$$

Therefore, we get $4 + b > 2|a|$.

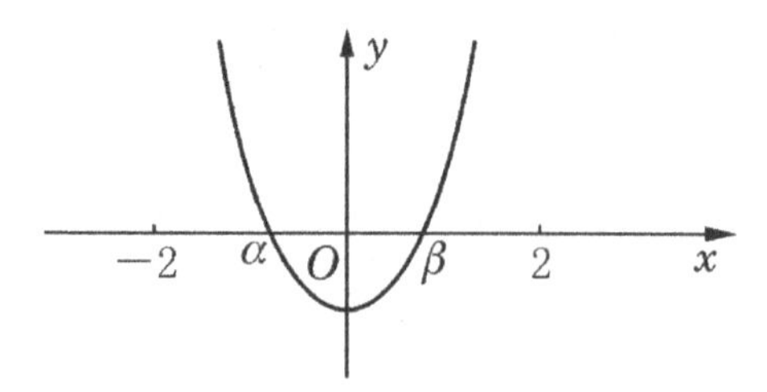

Fig. 10.1

Example 3. Let k be a real number. Given that the equation $7x^2 - (k + 13)x + k^2 - k - 2 = 0$ has two real roots α and β, where $0 < \alpha < 1$, $1 < \beta < 2$, find the value range of k.

Solution 1 Use the relation between the roots and coefficients.

Suppose $x = y + 1$, then the original equation is transformed into

$$7(y+1)^2 - (k+13)(y+1) + k^2 - k - 2 = 0.$$

Rearrange the equation and get $7y^2 - (k-1)y + k^2 - 2k - 8 = 0$.

Since the roots of the original equation are α and β, the roots of the above equation are

$$\alpha_1 = \alpha - 1 \quad \text{and} \quad \beta_1 = \beta - 1.$$

And because $0 < \alpha < 1$, $1 < \beta < 2$, we know that

$$-1 < \alpha - 1 < 0, \quad 0 < \beta - 1 < 1,$$

i.e. $-1 < \alpha_1 < 0$, $0 < \beta_1 < 1$. (1)

Hence, $\alpha_1\beta_1 = \frac{1}{7}(k^2 - 2k - 8) < 0$. (2)

Solving the inequality gives $-2 < k < 4$. (3)

Since $(\alpha_1 + 1)(\beta_1 + 1) = (\alpha_1 + \beta_1) + \alpha_1\beta_1 + 1 > 0$, (4)

we get that $\dfrac{k-1}{7} + \dfrac{k^2 - 2k - 8}{7} + 1 > 0$, i.e. $k^2 - k - 2 > 0$.

Solving the inequality gives $k > 2$ or $k < -1$. (5)

Since $(\alpha_1 - 1)(\beta_1 - 1) = \alpha_1\beta_1 - (\alpha_1 + \beta_1) + 1 > 0$ (6)

we get $\dfrac{k^2 - 2k - 8}{7} - \dfrac{k-1}{7} + 1 > 0$, i.e. $k^2 - 3k > 0$.

Solving the inequality gives $k > 3$ or $k > 0$. (7)

It's easy to prove that α_1 and β_1 satisfying (2), (4) and (6) must satisfy (1). So the range of k is the public part of (3), (5) and (7), that is,

$$-2 < k < -1 \quad \text{or} \quad 3 < k < 4.$$

Solution 2 Use the quadratic formula.

From the quadratic formula, we have

$$x = \frac{(k+13) \pm \sqrt{(k+13)^2 - 4 \times 7 \times (k^2 - k - 2)}}{14},$$

i.e. $x = \dfrac{(k+13) \pm \sqrt{-27k^2 + 54k + 225}}{14}$.

From the given condition, we get that

$$0 < \frac{k+13-\sqrt{-27^2+54k+225}}{14} < 1, \tag{1}$$

$$1 < \frac{k+13+\sqrt{-27^2+54k+225}}{14} < 2. \tag{2}$$

Solving the inequality (1) gives $-2 < k < -1$ or $2 < k < 4$.
Solving the inequality (2) gives $-2 < k < 0$ or $3 < k < 4$.
Hence, $-2 < k < -1$ or $3 < k < 4$.

Solution 3 Use the graph of the function.

Suppose $f(x) = 7x^2 - (k+13)x + k^2 - k - 2$. Then from the given condition, we know that one of the x-intercepts is between 0 and 1, and the other is between 1 and 2. Since $7 > 0$, the graph of the function opens upwards. Hence we get

$$\begin{cases} f(0) = k^2 - k - 2 > 0, & (1) \\ f(1) = 7 - (k+13) + k^2 - k - 2 < 0, & (2) \\ f(2) = 28 - 2(k+13) + k^2 - k - 2 > 0. & (3) \end{cases}$$

Solving the inequality (1) gives $k > 2$ or $k < -1$.
Solving the inequality (2) gives $-2 < k < 4$.
Solving the inequality (3) gives $k > 3$ or $k < 0$.
So the solution set of the simultaneous inequalities is

$$-2 < k < -1 \quad \text{or} \quad 3 < k < 4.$$

Remark This example is about the distribution of roots of a quadratic equation. It is too complex to solve the irrational inequalities by the quadratic formula, and it is also complex to use relation between the roots and coefficients. It is simpler to use the graph of the quadratic function to solve the problem. Generally speaking, to solve the problem about the distribution of roots of quadratic equation, we often transform the problems of equation into the problems of function by using the graph of the quadratic function.

Example 4. Let a be a non-zero real number. Given that one root of the quadratic equation $2ax^2 - 2x - 3a - 2 = 0$ is greater than 1, and the other root is less than 1, find the value range of a.

Solution 1 Use substitution method.

Suppose $y = x - 1$, then $x = y + 1$, the original equation is transformed into

$$2a(y+1)^2 - 2(y+1) - 3a - 2 = 0.$$

Rearrange the equation and get

$$2ay^2 + 2(2a-1)y - a - 4 = 0. \quad (1)$$

Since one root of the original equation is greater than 1 and the other root is less than 1, the roots of the equation (1) must be one positive and one negative. Thus we get

$$-\frac{a+4}{2a} < 0,$$

i.e. $2a(a+4) > 0$.

Solving the inequality gives $a > 0$ or $a < -4$.

Solution 2 Suppose $f(x) = 2ax^2 - 2x - 3a - 2$. Since one root of the original equation is greater than 1 and the other root is less than 1, the intersection points of the quadratic function $f(x)$ and the x-axis is on two sides of the point $(1, 0)$. Observing the graph, we can get

$$\begin{cases} a > 0, \\ f(1) = 2a - 2 - 3a - 2 = -a - 4 < 0, \end{cases} \quad \text{or}$$

$$\begin{cases} a < 0, \\ f(1) = 2a - 2 - 3a - 2 = -a - 4 > 0. \end{cases}$$

Solving these two simultaneous inequalities gives $a > 0$ or $a < -4$.

Think about it Is there any other solution?

Example 5. Let a, b and c which are the coefficients of the quadratic equation $ax^2 + bx + c = 0$ be odd numbers. The two real roots x_1 and x_2 of the equation satisfy $-1 < x_1 < 0$ and $x_2 > 1$. If $b^2 - 4ac = 5$, find the values of x_1 and x_2.

Solution Consider the case of $a > 0$ first.

From the quadratic formula of the equation $ax^2 + bx + c = 0$ and $b^2 - 4ac = 5$, we get

$$x_1 = \frac{-b - \sqrt{5}}{2a}, \quad x_2 = \frac{-b + \sqrt{5}}{2a}.$$

From the given condition, $-1 < x_1 < 0$ and $x_2 > 1$, we know that

$$\begin{cases} -2a + \sqrt{5} < -b < \sqrt{5}, \\ -b > 2a - \sqrt{5}. \end{cases}$$

Then we have $2a - \sqrt{5} < -b < \sqrt{5}$, $a < \sqrt{5}$. Since $a > 0$ and a is an odd number, $a = 1$.

Thus we get $0 < -2 + \sqrt{5} < -b < \sqrt{5}$. Since b is an odd number, $-b = 1$, i.e. $b = -1$.

Hence, $x_1 = \frac{1-\sqrt{5}}{2}$, $x_2 = \frac{1+\sqrt{5}}{2}$.

For the case of $a < 0$, the coefficients of the quadratic equation $-ax^2 - bx - c = 0$ is still odd, the discriminant is $\Delta = (-b)^2 - 4(-a)(-c) = b^2 - 4ac = 5$, and the two real roots are the same as the roots of the equation $ax^2 + bx + c = 0$. Because its coefficient of quadratic term $-a > 0$, according to the results of the previous section, the two real roots of the equation $-ax^2 - bx - c = 0$ are $x_1 = \frac{1-\sqrt{5}}{2}$ and $x_2 = \frac{1+\sqrt{5}}{2}$, which are also the roots of the equation $ax^2 + bx + c = 0$.

Example 6. Let a, b and c be real numbers with $a > 0$. Given a quadratic function $f(x) = ax^2 + bx + c$, if the equation $f(x) = x$ has two roots x_1 and x_2 with $x_2 - x_1 > \frac{1}{a}$, try to compare the size relation between $f(t)$ and x_1 when $0 < t < x_1$.

Solution 1 Rearrange the given equation $f(x) = x$ and get

$$ax^2 + (b-1)x + c = 0.$$

From the quadratic formula and the given condition, we know that

$$x_1 = \frac{-(b-1) - \sqrt{(b-1)^2 - 4ac}}{2a}, \quad x_2 = \frac{-(b-1) + \sqrt{(b-1)^2 - 4ac}}{2a}.$$

Thus, $x_2 - x_1 = \dfrac{\sqrt{(b-1)^2 - 4ac}}{a} > \dfrac{1}{a}$.

Since $a > 0$, we obtain that

$$\sqrt{(b-1)^2 - 4ac} > 1. \qquad (*)$$

Therefore, $x_1 < \dfrac{-(b-1) - 1}{2a} = -\dfrac{b}{2a}$. Because the graph of $f(x)$ opens upwards, the function is monotonically decreasing in $\left(-\infty, -\frac{b}{2a}\right]$, and $0 < t < x_1 < -\frac{b}{2a}$, we find that $f(t) > f(x_1) = x_1$.

Solution 2 The roots of the equation $f(x) = x$ are x_1 and x_2, then we have

$$f(x) - x = a(x - x_1)(x - x_2),$$

i.e. $f(x) = a(x - x_1)(x - x_2) + x$.

Since $f(t) - x_1 = a(t - x_1)(t - x_2) + t - x_1 = (t - x_1)[a(t - x_2) + 1]$, and $0 < t < x_1$, we get that

$$t - x_1 < 0. \tag{1}$$

Since $x_2 - x_1 > \frac{1}{a}$, we have $x_2 - t > \frac{1}{a}$. Thus $a(t - x_2) < -1$, that is,

$$a(t - x_2) + 1 < 0. \tag{2}$$

Combining (1) and (2), we get $f(t) - x_1 > 0$, i.e. $f(t) > x_1$.

Remark It is required to compare the size of $f(t)$ and x_1. Solution 1 is to compare $f(t)$ and $f(x_1)$, $0 < t < x_1 < -\frac{b}{2a}$ is obtained by using the inequalities so as to solve this problem. Solution 2 is to write the expression of function $f(x)$ in the form of $a(x - x_1)(x - x_2) + x$ by using the roots of the given quadratic equation, and to solve the problem by comparing the difference of $f(t)$ and x_1.

Example 7. Let a, b, c and m be real numbers, and satisfy

$$\frac{a}{m+2} + \frac{b}{m+1} + \frac{c}{m} = 0,$$

where $a \geq 0$, $m > 0$. Prove that the equation $ax^2 + bx + c = 0$ has a root x_0 which satisfies $0 < x_0 < 1$.

Solution When $a = 0$, if $b \neq 0$, the given condition becomes $-\dfrac{c}{b} = \dfrac{m}{m+1}$, the root of the equation is $x_0 = -\dfrac{c}{b}$, so $x_0 = \dfrac{m}{m+1}$, i.e. $0 < x_0 < 1$. If $b = 0$, the given condition becomes $c = 0$, the roots of the original equation can be any real number, so there must be a root x_0 satisfying $0 < x_0 < 1$.

When $a > 0$, let $f(x) = ax^2 + bx + c$, then we have

$$f\left(\frac{m}{m+1}\right) = -\frac{ma}{(m+1)^2(m+2)} < 0.$$

If $c > 0$, $f(0) = c > 0$, so there must be a root x_0 satisfying $0 < x_0 < \dfrac{m}{m+1} < 1$.

If $c \leq 0$,

$$f(1) = a + b + c = a + \frac{-(m+1)a}{m+2} + \frac{-(m+1)c}{m} + c = \frac{a}{m+2} - \frac{c}{m} > 0.$$

So there must be a root x_0 satisfying $0 < \dfrac{m}{m+1} < x_0 < 1$.

Example 8. Given that a, b and c are real numbers, $ac < 0$, and $\sqrt{2}a + \sqrt{3}b + \sqrt{5}c = 0$, prove that the quadratic equation $ax^2 + bx + c = 0$ has a root between $\sqrt{\frac{3}{5}}$ and 1.

Solution According to the given conditions, we may assume $a > 0$ (otherwise, we can multiply the polynomial by -1 and replace a, b and c with $-a$, $-b$ and $-c$), then $c < 0$ will be obtained from $ac < 0$.

Considering the characteristics of the graph of quadratic function, we can transform the question as follows: for the function $f(x) = ax^2 + bx + c(a > 0, c < 0)$, prove that $f\left(\sqrt{\frac{3}{5}}\right) < 0, f(1) > 0$.

From $\sqrt{2}a + \sqrt{3}b + \sqrt{5}c = 0(a > 0, c < 0)$, we can get

$$\sqrt{3}a + \sqrt{3}b > \sqrt{2}a + \sqrt{3}b = -\sqrt{5}c > -\sqrt{3}c,$$

i.e. $\sqrt{3}a + \sqrt{3}b + \sqrt{3}c > 0$.

So $f(1) = a + b + c > 0$.

From $-c = \sqrt{\dfrac{2}{5}}a + \sqrt{\dfrac{3}{5}}b > \dfrac{3}{5}a + \sqrt{\dfrac{3}{5}}b$, we can get

$$a\left(\sqrt{\frac{3}{5}}\right)^2 + b\left(\sqrt{\frac{3}{5}}\right) + c < 0,$$

i.e. $f\left(\sqrt{\dfrac{3}{5}}\right) < 0$.

Hence, the graph of the function $f(x)$ and the x-axis must have an intersection point whose x-coordinate is between $\sqrt{\frac{3}{5}}$ and 1, that is, the equation $f(x) = 0$ must have a root between $\sqrt{\frac{3}{5}}$ and 1.

Exercises

1. Try to find the value of real number m, so that the roots of the equation $4x^2 + (m-2)x + m - 5 = 0$ are all negative numbers.
2. Find the value of real number k, so that at least one of the intersections of parabola $y = kx^2 + (k-3)x + 1(k \neq 0)$ and the x-axis is on the right side of the origin.

3. Let m be a real number. Consider the quadratic function $y = (m - 1)x^2 + 8x + m - 7$.

   (1) If there are two intersections of its graph and the x-axis, find the value range of m.

   (2) If there are two intersections of its graph and the x-axis, and these two intersections are on the same side of the origin (excluding the origin), find the value range of m.

4. Let m be a real number. If two roots of the equation $x^2 - 2mx - (m - 12) = 0$ are both greater than 2, find the value of m.
5. Let a be a real number. If the equation $3x^2 - 5x + a = 0$ has two real roots α and β, where $-2 < \alpha < 0$, $1 < \beta < 3$, find the value range of a.
6. Let m be a real number. If one root of the equation $(m-1)x^2 + (3m + 2)x + 2m - 1 = 0$ is greater than 1 and the other root is less than 1, find the value of m.
7. Let m be a real number. Given that the equation $(m + 3)x^2 - 4mx + 2m - 1 = 0$ has two roots with opposite signs, and the positive root is less than the absolute value of negative root, find the value range of m.
8. Let a, b and c be real numbers with $a \neq 0$. If for the quadratic function $f(x) = ax^2 + bx + c$, there is a real number p which makes $af(p) < 0$, prove that the equation $f(x) = 0$ has two distinct real roots, one of which is greater than p, and the other is less than p.
9. Let m and n be positive integers. If the two roots of the equation $4x^2 - 2mx + n = 0$ are both greater than 1 and less than 2, find the values of m and n.
10. Let m be a real number. If the equation $x^2 - 4|x| + 5 = m$ has four distinct real roots, find the value of m.
11. Let m be a real number. Given that the equation $x^2 - mx + 4 = 0$ has roots in the range of $-1 \leq x \leq 1$, find the value range of m.
12. Let b be a real number. If the two intersections of the line $y = \frac{1}{2}x + b$ and the parabola $y = x^2 + 2x - 3$ are in the first and third quadrants respectively, find the value of b.
13. Let b be a real number. Given the function $y = x - b$ and $y = |x^2 - 4x + 3|$, try to solve the following questions by using the graphs of the functions.

    (1) If these two curves have one intersection, find the value of b. What if these two curves have three intersections?

    (2) If these two curves have two intersections, find the value of b.

    (3) If these two curves have four intersections, find the value of b.

14. Let $f_1(x)$ and $f_2(x)$ be two quadratic functions with equal coefficients of quadratic terms. If α_1 and β_1 are two roots of equation $f_1(x) = 0$, α_2 and β_2 are two roots of equation $f_2(x) = 0$, prove that

$$f_1(\alpha_2)f_1(\beta_2) = f_2(\alpha_1)f_2(\beta_1).$$

Chapter 11

Maximum and Minimum Values of Quadratic Functions

The topic about the maximum and minimum values of quadratic functions is an important content in mathematics competition.

Consider the quadratic function

$$y = f(x) = ax^2 + bx + c = a\left(x + \frac{b}{2a}\right)^2 + \frac{4ac - b^2}{4a} \quad (a > 0).$$

(1) On the domain $-\infty < x < +\infty$, the function has only the minimum value $\frac{4ac-b^2}{4a}$ which is obtained when $x = -\frac{b}{2a}$, and it has no maximum value.

(2) On the domain $-\infty < x \le m$, when $m \le -\frac{b}{2a}$, the function has the minimum value $y = f(m)$; when $m > -\frac{b}{2a}$, the function has the minimum value $y = f\left(-\frac{b}{2a}\right)$.

On the domain $m \le x < +\infty$, when $m > -\frac{b}{2a}$, the function has the minimum value $y = f(m)$; when $m \le -\frac{b}{2a}$, the function has the minimum value $y = f\left(-\frac{b}{2a}\right)$.

(3) On the domain $m \le x \le n$, when $n \le -\frac{b}{2a}$, the function has the maximum value $y = f(m)$ and the minimum value $y = f(n)$; when $m < -\frac{b}{2a} < n$, the function has the minimum value $y = f\left(-\frac{b}{2a}\right)$

and the maximum value is the larger value of $f(m)$ and $f(n)$; when $-\dfrac{b}{2a} \le m$, the function has the maximum value $y = f(n)$ and the minimum value $y = f(m)$.

For $a < 0$, we can follow the above discussion and get the maximum and minimum values of the quadratic function $y = ax^2 + bx + c$ on the given domain. In fact, as long as y is replaced by $-y$, we can get the corresponding conclusion.

Example 1. Given that t is a real number. If a and b are two nonnegative real roots of the quadratic equation $x^2 - 2x + t - 1 = 0$, find the minimum value of $(a^2 - 1)(b^2 - 1)$.

Solution From the given condition and the relationship between roots and coefficients, we know that

$$a \ge 0, \quad b \ge 0, \quad a + b = 2, \quad ab = t - 1.$$

Then we have $(a^2-1)(b^2-1) = (ab)^2 - a^2 - b^2 + 1 = (ab)^2 - (a+b)^2 + 2ab + 1 = (t-1)^2 - 2^2 + 2(t-1) + 1 = t^2 - 4$. Since $a \ge 0$, $b \ge 0$, we get $t - 1 \ge 0$, i.e. $t \ge 1$.

As we know, the equation has two nonnegative real roots, so $(-2)^2 - 4(t-1) \ge 0$. Solving this inequality gives $t \le 2$, thus we obtain $1 \le t \le 2$.

Hence, the minimum value of $(a^2 - 1)(b^2 - 1)$ is $1 - 4 = -3$ when $t = 1$.

Example 2. Let a and b be real numbers with $ab - 4a + 4b = 18$. Find the minimum value of $a^2 + b^2$.

Solution Let $y = a^2 + b^2$, $x = a - b$. Then

$$ab = 4a - 4b + 18 = 4x + 18.$$

Thus we have

$$y = a^2 + b^2 = (a - b)^2 + 2ab = x^2 + 2(4x + 18) = (x + 4)^2 + 20 \ge 20.$$

When $x = -4$, the minimum value of $a^2 + b^2$ is 20. At this time, $a - b = -4$, $ab = 2$, that is, $a = -2 + \sqrt{6}$, $b = 2 + \sqrt{6}$ or $a = -2 - \sqrt{6}$, $b = 2 - \sqrt{6}$.

Example 3. Consider the function $y = \sqrt{1 - x} + \sqrt{x - \frac{1}{2}}$. If the maximum value of y is a and the minimum value of y is b, find the value of $a^2 + b^2$.

Solution From the given condition, we know that $1-x \geq 0$ and $x-\frac{1}{2} \geq 0$. Thus we get $\frac{1}{2} \leq x \leq 1$.

$$y^2 = \frac{1}{2} + 2\sqrt{-x^2 + \frac{3}{2}x - \frac{1}{2}} = \frac{1}{2} + 2\sqrt{-\left(x - \frac{3}{4}\right)^2 + \frac{1}{16}}.$$

Since $\frac{1}{2} < \frac{3}{4} < 1$, the maximum value of y^2 is 1 when $x = \frac{3}{4}$. And because $y \geq 0$, we obtain $a = 1$. The minimum value of y^2 is $\frac{1}{2}$ when $x = \frac{1}{2}$ or 1, so we get $b = \frac{\sqrt{2}}{2}$. Hence, $a^2 + b^2 = \frac{3}{2}$.

Example 4. Consider the quadratic function $y = ax^2 + bx + c$. If and only if $1 < x < 3$, the graph of the function is above the straight line $y = -2x$.

(1) If the equation $ax^2 + bx + 6a + c = 0$ has two identical roots, find the expression of the quadratic function.
(2) If the maximum value of the quadratic function is a positive number, find the value range of a.

Solution (1) From the given condition, $x = 1$ and $x = 3$ are two roots of the equation $ax^2 + bx + c = -2x$, and $a < 0$, Then we have

$$\begin{cases} a \times 1^2 + b \times 1 + c = -2 \times 1, \\ a \times 3^2 + b \times 3 + c = -2 \times 3, \end{cases} \quad \text{i.e.} \quad \begin{cases} a + b + c = -2, \\ 9a + 3b + c = -6. \end{cases}$$

Solving the simultaneous equations gives $b = -4a - 2$, $c = 3a$.

Since the discriminant of the equation $ax^2 + bx + 6a + c = 0$ is

$$\Delta = b^2 - 4a(6a + c) = (-4a - 2)^2 - 4a(6a + 3a) = 0,$$

we can get $a = 1$ or $a = -\dfrac{1}{5}$. Since $a < 0$, we discard $a = 1$. Thus $a = -\dfrac{1}{5}$, and then $b = -\dfrac{6}{5}$, $c = -\dfrac{3}{5}$. Hence, the expression of the quadratic function is $y = -\dfrac{1}{5}x^2 - \dfrac{6}{5}x - \dfrac{3}{5}$.

(2) As the maximum value of the quadratic function is a positive number, it can be seen that $\dfrac{4ac - b^2}{4a} > 0$. And from $a < 0$, we get $b^2 - 4ac > 0$.

According to (1), we know that $b = -4a - 2$, $c = 3a$. Thus we get $(-4a-2)^2 - 4a \cdot (3a) > 0$, i.e. $a^2 + 4a + 1 > 0$. Solving this inequality gives $a < -2 - \sqrt{3}$ or $a > -2 + \sqrt{3}$.

Since $a < 0$, the value range of a is $a < -2 - \sqrt{3}$ or $-2 + \sqrt{3} < a < 0$.

Example 5. Given that a and b are real numbers, find the minimum value of $a^2 + ab + b^2 - a - 2b$.

Solution Let $A = a^2 + ab + b^2 - a - 2b$. Then we have the equation

$$a^2 + (b-1)a + b^2 - 2b - A = 0,$$

which can be regarded as a quadratic equation of a.

Since a is a real number, we get

$$\Delta = (b-1)^2 - 4(b^2 - 2b - A) \geq 0,$$

i.e.

$$3b^2 - 6b - 4A - 1 \leq 0. \tag{1}$$

The inequality (1) indicates that the function $y = 3b^2 - 6b - 4A - 1$ must have a non-positive value. So the ordinate of the vertex of the graph of the function must be a non-positive number, that is,

$$\frac{4 \times 3 \times (-4A - 1) - 6^2}{4 \times 3} \leq 0.$$

Solving the above inequality gives $A \geq -1$. When $A = -1$, from (1) we can get $b = 1$, and thus we know $a = 0$. Hence, when $a = 0$ and $b = 1$, the polynomial $a^2 + ab + b^2 - a - 2b$ has the minimum value -1.

Remark Notice that there is term of ab in the equation $A = a^2 + ab + b^2 - a - 2b$, we usually consider the case of supposing $a = s - t$, $b = s + t$, then s and t are both real numbers, and we have

$$A = 3s^2 + t^2 - 3s - t = 3\left(s - \frac{1}{2}\right)^2 + \left(t - \frac{1}{2}\right)^2 - 1 \geq -1.$$

The equal sign is obtained only when $s = \frac{1}{2}$ and $t = \frac{1}{2}$.

Therefore, when $a = 0$ and $b = 1$, $a^2 + ab + b^2 - a - 2b$ has the minimum value -1.

Example 6. Let a, b and c be real numbers with $c < b < a$, $a + b + c = 0$. Consider a quadratic function $y = ax^2 + bx + c$ and a linear function $y = -bx$.

(1) Prove that the graphs of these two functions have two different intersections A and B.

(2) Draw two perpendicular lines of x-axis through the points A and B in (1) respectively, with the perpendicular feet of A_1 and B_1. Find the value range of the length of line segment A_1B_1.

Proof. (1) For $\begin{cases} y = ax^2 + bx + c \\ y = -bx \end{cases}$, eliminating y gives $ax^2 + 2bx + c = 0$.

Then the discriminant is

$$\Delta = 4b^2 - 4ac = 4(-a-c)^2 - 4ac = 4(a^2 + ac + c^2) = 4\left[\left(a + \frac{c}{2}\right)^2 + \frac{3}{4}c^2\right].$$

Since $a + b + c = 0$, $a > b > c$, we can get that $a > 0$, $c < 0$, and then $\frac{3}{4}c^2 > 0$.

Hence we obtain $\Delta > 0$, that is, the graphs of these two functions have two different intersections. □

Solution (2) Let x_1 and x_2 be the roots of the equation $ax^2+2bx+c = 0$. Then

$$x_1 + x_2 = -\frac{2b}{a}, \quad x_1x_2 = \frac{c}{a}.$$

So $|A_1B_1|^2 = (x_1 - x_2)^2 = \left(\frac{\sqrt{\Delta}}{a}\right)^2$, i.e.

$$\begin{aligned} |A_1B_1|^2 &= \frac{4b^2 - 4ac}{a^2} = \frac{4(-a-c)^2 - 4ac}{a^2} = 4\left[\left(\frac{c}{a}\right)^2 + \frac{c}{a} + 1\right] \\ &= 4\left[\left(\frac{c}{a} + \frac{1}{2}\right)^2 + \frac{3}{4}\right]. \end{aligned}$$

Since $a+b+c = 0$, $a > b > c$, we get that $a > 0$, $c < 0$. So $a > -a-c > c$, Solving it gives

$$-2 < \frac{c}{a} < -\frac{1}{2}.$$

Then we have $4\left[\left(-\frac{1}{2} + \frac{1}{2}\right)^2 + \frac{3}{4}\right] < |A_1B_1|^2 < 4\left[\left(-2 + \frac{1}{2}\right)^2 + \frac{3}{4}\right]$, i.e.

$$3 < |A_1B_1|^2 < 12.$$

Hence, $\sqrt{3} < |A_1B_1| < 2\sqrt{3}$.

Example 7. Given that a, b, c and d are all positive constants, and the real numbers x and y satisfy $ax^2 + by^2 = 1$, find the minimum value of $cx + dy^2$.

Solution Let $k = cx + dy^2$. Then from $ax^2 + by^2 = 1$ we can get $y^2 = \frac{1-ax^2}{b}$.

Therefore,

$$k = cx + \frac{d - adx^2}{b} = -\frac{ad}{b}x^2 + cx + \frac{d}{b} = -\frac{ad}{b}\left(x - \frac{bc}{2ad}\right)^2 + \frac{d}{b} + \frac{bc^2}{4ad}.$$

From $ax^2 + by^2 = 1$, we obtain $ax^2 \leq 1$, i.e. $-\frac{1}{\sqrt{a}} \leq x \leq \frac{1}{\sqrt{a}}$.

As we know, a, b, c and d are all positive constants, k is a quadratic function with regard to x, of which the graph opens downwards and the axis of symmetry is $x = \frac{bc}{2ad} > 0$. So the minimum value of k may be obtained when $x = -\frac{1}{\sqrt{a}}$ or $x = \frac{1}{\sqrt{a}}$.

When $x = -\frac{1}{\sqrt{a}}$, we get $y = 0$, $k = cx = -\frac{c}{\sqrt{a}}$.

When $x = \frac{1}{\sqrt{a}}$, we get $y = 0$, $k = cx = \frac{c}{\sqrt{a}}$.

Since $-\frac{c}{\sqrt{a}} < \frac{c}{\sqrt{a}}$, we find that the minimum value of $cx + dy^2$ is $-\frac{c}{\sqrt{a}}$ when $x = -\frac{1}{\sqrt{a}}$.

Example 8. As shown in Figure 11.1,

In the plane rectangular coordinate system xOy, two sides of the rectangular $OABC$ are on the x-axis and the y-axis respectively, $OA = 10\,\text{cm}$, $OC = 6\,\text{cm}$. Let two moving points P and Q start from O and A at the same time. Given that the moving speed of the point P is $1\,\text{cm/s}$. P moves at a constant speed in the direction of OA on the line segment OA, and the point Q moves at a constant speed in the direction of AB on the line segment AB.

(1) Suppose that the moving speed of the point Q is $\frac{1}{2}\,\text{cm/s}$ and the time for motion is ts.

  (i) When the area of $\triangle CPQ$ reaches its minimum, find the coordinates of the point Q.

  (ii) When $\triangle COP$ and $\triangle PAQ$ are similar triangles, find the coordinates of the point Q.

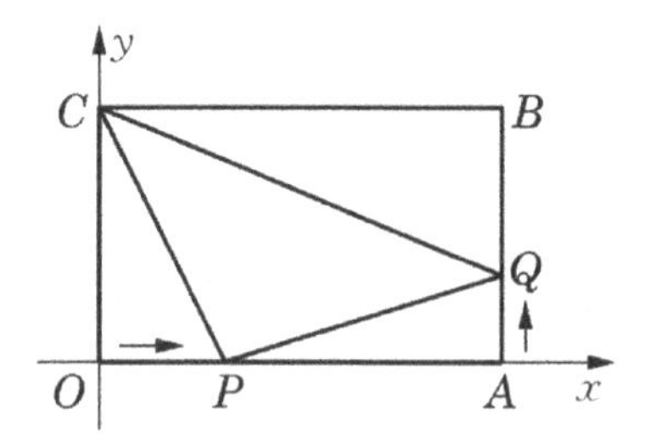

Fig. 11.1

(2) Suppose that the moving speed of the point Q is a cm/s. Whether there exists a value of a so that $\triangle OCP$ is similar to $\triangle PAQ$ and $\triangle CBQ$? If yes, find the value of a and state the coordinates of the point Q. If not, please explain the reason.

Solution

(1) (i) $S_{\triangle CPQ} = S_{\text{rectangle}\, OABC} - S_{\triangle OCP} - S_{\triangle PAQ} - S_{\triangle BCQ}$

$$= 60 - \frac{1}{2} \times 6 \times t - \frac{1}{2}(10-t) \times \frac{1}{2}t - \frac{1}{2} \times 10\left(6 - \frac{1}{2}t\right)$$

$$= \frac{1}{4}(t-6)^2 + 21 \quad (0 \le t \le 10).$$

So when $t = 6$, the minimum value of $S_{\triangle CPQ}$ is 21, and the coordinates of the point Q are $(10, 3)$.

(ii) As shown in Figure 11.2, $\triangle COP$ and $\triangle PAQ$ are similar triangles, which is equivalent to $\angle 1 = \angle 2$ or $\angle 1 = \angle 3$.

When $\angle 1 = \angle 2$, $\frac{OC}{OP} = \frac{QA}{PA}$, then we get

$$\frac{6}{t} = \frac{\frac{1}{2}t}{10-t},$$

i.e. $\frac{1}{2}t^2 + 6t - 60 = 0$. Solving the equation gives $t_1 = -6 + 2\sqrt{39}$, $t_2 = -6 - 2\sqrt{39}$ (discarded). When $\angle 1 = \angle 3$, we get $\frac{6}{t} = \frac{10-t}{\frac{1}{2}t}$. Solving the equation gives $t = 7$. Hence, when $t = -6 + 2\sqrt{39}$ or $t = 7$, that is, the coordinates of the point Q are $(10, -3 + \sqrt{39})$ or $\left(10, \frac{7}{2}\right)$, $\triangle COP$ and $\triangle PAQ$ are similar triangles.

(2) Suppose that there is a number a to make $\triangle OCP$ be similar to $\triangle PAQ$ and $\triangle CBQ$. Let the moving time of the points P and Q be ts. Then $OP = t$cm, $AQ = at$cm.

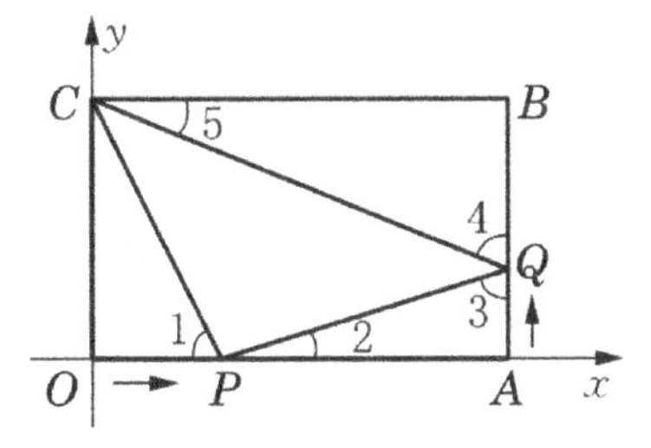

Fig. 11.2

(i) When $\angle 1 = \angle 3 = \angle 4$, we get $\dfrac{OC}{OP} = \dfrac{PA}{AQ} = \dfrac{BC}{BQ}$, $\dfrac{6}{t} = \dfrac{10-t}{at} = \dfrac{10}{6-at}$.

Solving the equations gives $t_1 = 2$, $t_2 = 18$ (discarded).

At this moment, $a = \frac{4}{3}$ and the coordinates of the point Q are $\left(10, \frac{8}{3}\right)$.

(ii) When $\angle 1 = \angle 3 = \angle 5$, $\angle CPQ = \angle CQP = 90°$, there is no such value of a.

(iii) When $\angle 1 = \angle 2 = \angle 4$, we get that $\dfrac{OC}{OP} = \dfrac{QA}{AP} = \dfrac{BC}{BQ}$, $\dfrac{6}{t} = \dfrac{at}{10-t} = \dfrac{10}{6-at}$, i.e.

$$\begin{cases} 60 - 6t = at^2, & (1) \\ 36 - 6at = 10t. & (2) \end{cases}$$

From (2) we obtian $at = \frac{36-10t}{6}$. Substituting it into (1) gives

$$60 - 6t = \frac{(36-10t)t}{6}.$$

Rearranging the equation gives $5t^2 - 36t + 180 = 0$. From $\Delta < 0$, the equation has no real roots, so there is no such value of a.

(iv) When $\angle 1 = \angle 2 = \angle 5$, we can find $\angle 1 = \angle PCB > \angle 5$ from the Figure 11.2, so there is no such value of a.

In summary, there exists a number a to make $\triangle OCP$ similar to $\triangle PAQ$ and $\triangle CBQ$. At this time, $a = \frac{4}{3}$ and the coordinates of the point Q are $\left(10, \frac{8}{3}\right)$.

Reading

Solving Problems without Decorations

To solve a problem, we should be good at grasping the essence of the problem, for which we can not apply mechanically.

Example Given that a, b and c are positive numbers and $\frac{\sqrt{2}b-2c}{a} = 1$. try to prove that $b^2 \geq 4ac$.

This problem is a very simple problem about inequality proof.

From the condition,

$$\sqrt{2}b = a + 2c. \quad (1)$$

Formula (1) is closed to the conclusion that $b^2 \geq 4ac$: b is on the left, a and c are on the right. All that remains is to derive the following conclusion from the basic inequality, that is,

$$a + 2c \geq 2\sqrt{a \cdot 2c} = 2\sqrt{2ac}.$$

Thus we have $\sqrt{2}b \geq 2\sqrt{2ac}$.

Divided two sides of the inequality by $\sqrt{2}$, then square both sides to get $b^2 \geq 4ac$.

Some people think of the discriminant of the quadratic equation $ax^2 + bx + c = 0$ as soon as they see $b^2 \geq 4ac$. So they start with the equation having real roots. Substituting $x = -\frac{\sqrt{2}}{2}$ into $ax^2 + bx + c$, we get

$$a \cdot \frac{1}{2} - \frac{\sqrt{2}}{2}b + c = \frac{a - \sqrt{2}b + 2c}{2} = 0.$$

It's such a coincidence! So it lacks in universality. If the problem is changed to "Given that $\frac{\sqrt{2}b-3c}{a} = 1$, prove that $b^2 \geq 4ac$", then the method of using the discriminant is invalid. It can be seen that the way of using discriminant is really mystifying. We should avoid spending time on those flourishes when solving problems.

(Excerpted from: Shan Zun. *Problem solving research*. Shanghai: Shanghai Education Press, 2016, 108~109.)

Exercises

1. Find the maximum and minimum values of the function $y = x^2 - 4x + 3\ (0 \leq x \leq 5)$.
2. Find the minimum value of the function $y = 3 - \sqrt{-x^2 + 6x - 5}$.
3. Suppose that real numbers x and y satisfy $x^2 - 2x - 4y = 5$. Let $t = x - 2y$, find the value range of t.
4. Let a, b, and c be the lengths of three sides of $\triangle ABC$. If the minimum value of the quadratic function $y = \left(a - \frac{b}{2}\right)x^2 - cx - a - \frac{b}{2}$ is $-\frac{8}{5}b$ when $x = 1$, then ΔABC is ( ).

   (A) an isosceles triangle
   (B) an acute triangle
   (C) an obtuse triangle
   (D) a right triangle

5. Let a be a real number. Find the maximum and minimum values of the function $y = x^2 - 2ax\ (0 \leq x \leq 1)$.
6. In $\triangle ABC$, $BC = 2$, the altitude to side BC is $AD = 1$. P is any point on the line segment BC, PE is parallel to AB and intersects AC at the point E, PF is parallel to AC and intersects AB at the point F.

(1) Let $BP = x$. Express $S_{\triangle PEF}$ in terms of x.
(2) When $S_{\triangle PEF}$ reaches its maximum value, what position of the line segment BC is the point P located?

7. There is a product that producing x tons costs $\left(\frac{1}{10}x^2 + 5x + 1000\right)$ yuan. If x tons are sold, the price per ton is $\left(a + \frac{x}{b}\right)$ yuan where a and b are real numbers. If all the products produced can be sold, and the profit reaches its maximum when producing 150 tons, with the price of 40 yuan per ton. Find the values of a and b.
8. Let a, b and c be real numbers. Consider the quadratic function $y = f(x) = ax^2 + bx + c$. Given that the axis of symmetry for the quadratic function is $2x - 3 = 0$, the reciprocal sum of the x-intercepts is 2, and the graph of the function passes through the point $(3, -3)$.

(1) Find the values of a, b and c.
(2) What is the value range of x when $y > 1$ or $y < -3$?
(3) What is the value of x when y reaches its maximum value? Find the maximum value of y.
(4) Sketch the graph of this function.

9. Let m be a real number with $m \neq -1$. Find the maximum or minimum values of the function $y = (m + 1)x^2 - 2(m + 1)x - m$.
10. Find the minimum value of function $f(x) = x|x| - 2x - |x|$ on the domain $-1 \leq x \leq \frac{3}{2}$.
11. Let b and k be real numbers where $k \neq 0$. In the rectangular coordinate system xOy, the graph of the linear function $y = kx + b + 2$ intersects with the positive semi-axis of x-axis and y-axis at the points A and B respectively such that the value of the area of $\triangle OAB$ is equal to $|OA| + |OB| + 3$. Find the minimum value of the area of $\triangle OAB$.
12. Given that the graph C_1 of the quadratic function $y = -x^2 - 3x + 4$ and the graph C_2 of the quadratic function $y = x^2 - 3x - 4$ intersect at two points A and B. The point P lies on the graph C_1 and between the points A and B; the point Q lies on the graph C_2 and also between the points A and B.

(1) Find the length of the line segment AB.
(2) When PQ is parallel to the y-axis, find the maximum value of the length of PQ.

13. Let a and b be real numbers. Given that $a^2 + b^2 = 1$, and for all real numbers x satisfying $0 \leq x \leq 1$, the inequality $a(1 - x)(1 - x - ax) - bx(b - x - bx) \geq 0$ is always true. When the product ab is the minimum, find the values of a and b.

Chapter 12

Maximum and Minimum Values of Simple Fractional Functions

We know that the simplest fractional function $y = \frac{k}{x}$ is also known as inversely proportional function. In this chapter, we will discuss the maximum and minimum values of simple fractional function $y = \dfrac{ax+d}{ax+b}$ and $y = \dfrac{dx^2+ex+f}{ax^2+bx+c}$.

Example 1. Find the maximum and minimum values of function $y = \dfrac{5x-2}{x+3}$ on the domain $0 \leq x \leq 2$.

Solution $y = \dfrac{5x-2}{x+3} = \dfrac{5(x+3)-17}{x+3} = 5 - \dfrac{17}{x+3}$.

From $0 \leq x \leq 2$, we can get

$$3 \leq x+3 \leq 5,$$

$$\frac{17}{5} \leq \frac{17}{x+3} \leq \frac{17}{3},$$

so $-\dfrac{17}{3} \leq -\dfrac{17}{x+3} \leq -\dfrac{17}{5}$.

Therefore,

$$5 - \frac{17}{3} \leq 5 - \frac{17}{x+3} \leq 5 - \frac{17}{5},$$

i.e. $-\frac{2}{3} \leq y \leq \frac{8}{5}$.

Hence, the minimum value of the function is $-\frac{2}{3}$ and the maximum value is $\frac{8}{5}$.

Remark In this example, since the function $x+3$ increases as x increases, $\frac{17}{x+3}$ decreases as x increases. Thus we know that $-\frac{17}{x+3}$ increases as x increases. Hence, the minimum value of the function $y = \frac{5x-2}{x+3}$ is obtained when $x = 0$ and the maximum value is obtained when $x = 2$.

Example 2. Find the maximum value of the function $y = \frac{x+1}{x^2+3}$.

Solution From the given condition,

$$yx^2 - x + 3y - 1 = 0.$$

Since x is a real number, the discriminant of the above quadratic equation is

$$\Delta = (-1)^2 - 4y(3y-1) \geq 0,$$

i.e. $12y^2 - 4y - 1 \leq 0$. Solving the inequality gives $-\frac{1}{6} \leq y \leq \frac{1}{2}$.

So the maximum value of the function y is $\frac{1}{2}$, which is obtained when $x = 1$.

Remark When we discuss the maximum and minimum values of the function which is in the form $y = \dfrac{dx^2 + ex + f}{ax^2 + bx + c}$, we can convert it into a quadratic equation of x, and use the discriminant of the equation to get the answer.

Example 3. Find the value range of the fraction $\dfrac{3x^2 + 6x + 5}{\frac{1}{2}x^2 + x + 1}$.

Solution Let $y = \dfrac{3x^2 + 6x + 5}{\frac{1}{2}x^2 + x + 1}$. Then we have

$$y\left(\frac{1}{2}x^2 + x + 1\right) = 3x^2 + 6x + 5,$$

i.e. $(y-6)x^2 + (2y-12)x + 2y - 10 = 0$. (1)

From the given condition, we know that the above equation of x must have real roots. If $y - 6 = 0$ i.e. $y = 6$, the equation (1) has no real roots. So $y \neq 6$. Thus the equation (1) is a quadratic equation of x and it has real roots.

Therefore, the discriminant of the equation (1) is

$$\Delta = (2y-12)^2 - 4(y-6)(2y-10) \geq 0,$$

i.e. $(y-6)(y-4) \leq 0$.

Solving the inequality gives $4 \leq y \leq 6$. Since $y \neq 6$, the value range of the fractional function is $4 \leq y < 6$.

Remark

(1) To solve the problem of the range (or the maximum and the minimum values) of the fractional function whose numerator and denominator are both quadratic trinomials, the discriminant of the equation is often used, which is the basic method to deal with this kind of problem. However, when calculating the maximum (or minimum) value with the discriminant, we should pay attention to the question whether the maximum (or minimum) value can be obtained, that is, whether there is an x corresponding to the maximum (or minimum) value.
(2) Is there any easy way to solve this problem?

Example 4. Given that the minimum value of the function $y = \frac{ax^2+bx+6}{x^2+2}$ is 2 and the maximum value is 6, find the values of real numbers a and b.

Solution By removing the denominator, the original function can be transformed into

$$(a-y)x^2+bx+(6-2y)=0. \quad (1)$$

If $y \equiv a$, that is, y is a constant, it's impossible to have a minimum of 2 and a maximum of 6. So $y \not\equiv a$, then from the given condition, we know that the equation (1) of x must have real roots. The discriminant of the function is

$$\Delta = b^2 - 4(a-y)(6-2y) \geq 0,$$

i.e. $y^2-(a+3)y+3a-\frac{b^2}{8} \leq 0$. (2)

Since the minimum value of y is 2 and the maximum value is 6, we obtain

$$(y-2)(y-6) \leq 0,$$

i.e. $y^2-8y+12 \leq 0$. (3)

Compare (2) and (3) and get

$$\begin{cases} a+3=8, \\ 3a-\frac{b^2}{8}=12. \end{cases}$$

Solving the simultaneous equations gives $a=5$, $b=\pm 2\sqrt{6}$.

Example 5. Find the maximum and minimum values of the function $y = \frac{x^4+x^2+5}{(x^2+1)^2}$.

Solution From $y = \dfrac{x^4 + x^2 + 5}{(x^2+1)^2}$, we have

$$y = \frac{5}{(x^2+1)^2} - \frac{1}{x^2+1} + 1.$$

Let $t = \dfrac{1}{x^2+1}$. Since $x^2 + 1 \geq 1$, we get $0 < \dfrac{1}{x^2+1} \leq 1$, i.e. $0 < t \leq 1$.

Thus,

$$y = 5t^2 - t + 1 = 5\left(t - \frac{1}{10}\right)^2 + \frac{19}{20}, \quad 0 < t \leq 1.$$

When $t = \dfrac{1}{10}$, the minimum value of function y is $\dfrac{19}{20}$. At this time, $\dfrac{1}{x^2+1} = \dfrac{1}{10}$, $x = \pm 3$.

When $t = 1$, the maximum value of function y is $5\left(1 - \dfrac{1}{10}\right)^2 + \dfrac{19}{20} = 5$. At this time, $\dfrac{1}{x^2+1}$ =1, $x = 0$.

Hence, the minimum value of function y is $\frac{19}{20}$ when $x = \pm 3$, and the maximum value of function y is 5 when $x = 0$.

Reading

The Early Stage of Mathematical Competition — The debate around the solution of equations

In the 16th century of Italy, it was very popular to join the "equation solving competition" mentioned. There were many mathematicians who lived by this kind of competition due to the high prize. All the questions in the competition were set to solve the cubic equations and the quartic equations. Everyone who participated in the competition had great skills and competed with each other. Cubic equations in the form of $x^3 + ax = b$ was invented by Sibion Filo (1465–1526), the mathematician of Bologna, and then passed down by his student, Flores.

In Italy, there is another mathematician, Nicolo Fontana (1499–1557), nicknamed "stuttering". It is said that he was beaten in the face by French soldiers when he was a child, resulting in a sequelae, so he was unable to speak frequently. Fontana's family was poor, so he could only learn mathematics by himself. Under his efforts, he found the solution of cubic equations in the form of $x^3 + ax = b$, just like Flores.

Fontana, who was confident in his solution, went to fight against Flores and competed in solving equations. He won at the end.

Inspired by the win of the competition, Fontana repeatedly studied and found the general solution of cubic equations in 1541.

During the same period, there is another mathematician named Cardano (1501–1576) in Milano, Italy, who is eager to win the equation solving competition. To achieve this goal, he must know the general solution of the cubic equation. Therefore, he visited Fontana, promised to "absolutely keep the secret", and finally learned the general solution.

Later, Cardano worked as a professor at the University of Padua. In his book *Ars Magna* published in 1545, he published the solution of cubic equations taught by Fontana as his own invention.

This action infuriated Fontana, and Fontana send Competition Challenge to Cardano. However, Caldano did not take part in the contest, but his student Ferrari (1522–1565) replaced.

Ferrari has a great talent in mathematics. When he was young, he had found the general solution of quartic equations. Therefore, Ferrari finally won this math competition.

In the history of mathematics, most of the discoverers of the solution of the cubic equations are recorded as Cardano, while the discoverers of the solution of the quartic equations are recorded as Ferrari.

As mentioned above, mathematicians have competed for the solution of cubic equations and quartic equations fiercely!

(Excerpted from: Jiang Teng Bang Yan. *Princess Lindera's love for Mathematics — the quadratic equation.* Taipei: Education Workshop, 2002, 90–91.)

Exercises

1. Given that $f_1(x)$ is a directly proportional function, $f_2(x)$ is an inversely proportional function, and $\frac{f_1(1)}{f_2(1)} = 2$, $f_1(2) + 4f_2(2) = 6$, find the expression of $f_1(x)$ and $f_2(x)$.
2. Given that m is a positive number less than 1, $a = 1 - \frac{1}{m}$, $b = \frac{1}{m} - 1$, $c = m - \frac{1}{m}$ and $d = \frac{1}{m} - m$, then which of the following inequalities must be true? ( )

   (A) $c < d < a < b$ (B) $b < c < d < a$
   (C) $c < a < b < d$ (D) $a < c < b < d$

3. If a, b, c and d are all integers not equal to 0, and $\frac{1}{a}+\frac{1}{b}=\frac{1}{c}$, $\frac{1}{b}+\frac{1}{c}=\frac{1}{d}$, $\frac{1}{c}+\frac{1}{d}=\frac{1}{a}$, then the minimum value of $\frac{1}{a}+\frac{1}{b}+\frac{1}{c}+\frac{1}{d}$ is ( ).

   (A) -5 (B) $-\frac{5}{2}$ (C) $-\frac{5}{6}$ (D) non-existent

4. Find the maximum and minimum values of the function $y=\frac{3x+5}{x-2}$ on $3 \leq x \leq 5$.
5. Let a and b be real numbers. Given that on the domain $0 \leq x \leq 1$, the maximum value of the function $y=\frac{ax+b}{x+1}$ is 3 and the minimum value is 1, find the values of a and b.
6. Find the maximum and minimum values of the function $y = \frac{x^2+x+2}{2x^2-x+1}$.
7. Let b and c be real numbers. If the domain of the function $y = \frac{1}{x^2-2bx+c^2}$ is all real numbers, then which of the following inequalities must meet the requirements? ( )

   (A) $b>c>0$ (B) $b>0>c$ (C) $c>0>b$ (D) $c>b>0$

8. Let x, y and z be three real numbers that are not all zero. Find the maximum value of $\frac{xy+2yz}{x^2+y^2+z^2}$.

Chapter 13

Trigonometric Functions of Acute Angle

As shown in Figure 13.1, we know that

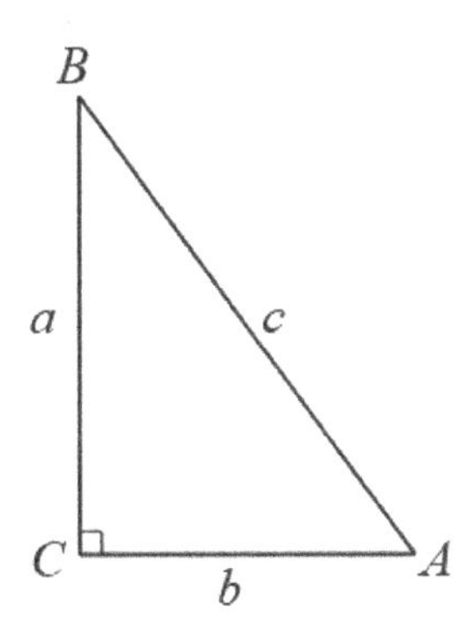

Fig. 13.1

$$\sin A = \frac{a}{c}, \quad \cos A = \frac{b}{c},$$

$$\sin B = \frac{b}{c}, \quad \cos B = \frac{a}{c},$$

$$\tan A = \frac{a}{b}, \quad \tan B = \frac{b}{a}.$$

By using the above basic relations, we can solve the problems of trigonometric functions of an acute angle. Its basic idea is to use the relation between the sides and angles of the right angled triangle to transform it into the problem to be solved.

Example 1. In $\triangle ABC$, given that both $\angle A$ and $\angle B$ are acute angles, $AC = 6$, $BC = 3\sqrt{3}$, and $\sin A = \frac{\sqrt{3}}{3}$, find the value of $\cos B$.

Solution As shown in Figure 13.2, draw the height CD of $\triangle ABC$.

Since $AC = 6$, $\sin A = \frac{\sqrt{3}}{3}$, we know that $CD = AC \cdot \sin A = 2\sqrt{3}$. Thus we have

$$DB = \sqrt{BC^2 - CD^2} = \sqrt{3(\sqrt{3})^2 - (2\sqrt{3})^2}.$$

So $\cos B = \frac{DB}{BC} = \frac{\sqrt{15}}{3\sqrt{3}} = \frac{\sqrt{5}}{3}$.

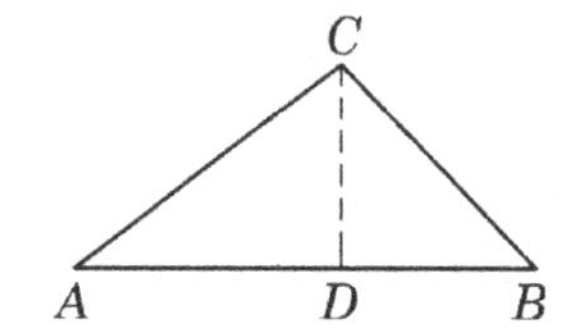

Fig. 13.2

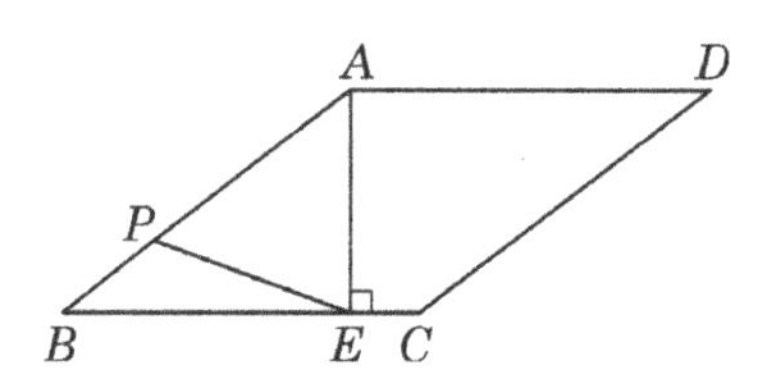

Fig. 13.3

Example 2. As shown in Figure 13.3, in a diamond $ABCD$, AE is perpendicular to BC (produced to E). If $\cos B = \frac{4}{5}$, $EC = 2$, P is a moving point on the side AB, find the minimum value of the length of the line segment PE.

Solution Let x be the length of the side of a diamond $ABCD$. Then $AB = BC = x$. And since $EC = 2$, we know that $BE = x - 2$.

Because AE is perpendicular to BC (produced to E), we obtain that in Rt$\triangle ABE$, $\cos B = \frac{x-2}{x}$. From $\cos B = \frac{4}{5}$, we get $\frac{x-2}{x} = \frac{4}{5}$. Solving the equation gives $x = 10$, that is, $AB = 10$.

Therefore, $BE = 8$, $AE = 6$, when $EP \perp AB$, PE has the minimum value.

From the formula of the area of a triangle, we have $\frac{1}{2}AB\cdot PE = \frac{1}{2}BE\cdot AE$. So the minimum value of PE is 4.8.

Example 3. As shown in Figure 13.4, α is an acute angle in $\triangle ABC$. Try to judge the relation between the value of $\sin\alpha + \cos\alpha$ and 1, and explain the reason.

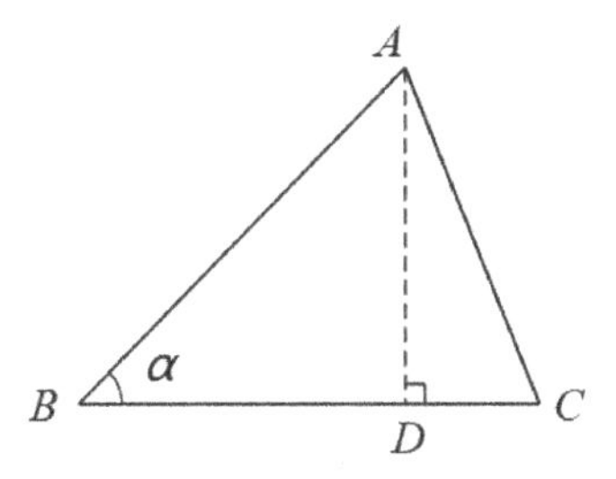

Fig. 13.4

Solution Let $\angle ABC = \alpha$. Construct the perpendicular AD from A to BC, which intersects BC at the point D, then ABD is a right triangle and $\angle ADB = 90°$.

So

$$\sin\alpha = \frac{AD}{AB}, \quad \cos\alpha = \frac{BD}{AB}.$$

Thus we have

$$\sin\alpha + \cos\alpha = \frac{AD + BD}{AB}.$$

In $\triangle ABD$, there must be an inequality $AD + BD > AB$, so we get

$$\sin\alpha + \cos\alpha > 1.$$

Example 4. Given that $\sin\alpha + \cos\alpha = \sqrt{2}$, find the value of $\sin\alpha\cos\alpha$.

Solution Square both sides of $\sin\alpha + \cos\alpha = \sqrt{2}$ and get

$$(\sin\alpha + \cos\alpha)^2 = (\sqrt{2})^2.$$

Since $\sin^2\alpha + \cos^2\alpha = 1$, we obtain that

$$1 + 2\sin\alpha\cos\alpha = 2.$$

Hence, $\sin\alpha\cos\alpha = \frac{1}{2}$.

Remark

(1) When the sum or difference between $\sin\alpha$ and $\cos\alpha$ is known, we often consider using $\sin^2\alpha + \cos^2\alpha = 1$ to convert the problem.
(2) Summarizing the process of solving this question, we know that this question is actually a familiar one for readers as follow.

Given that $a + b = \sqrt{2}$, $a^2 + b^2 = 1$, find the value of ab.

In this example, the trigonometric functions $\sin\alpha$ and $\cos\alpha$ are used instead of a and b to change the form of the problem. Therefore, it is very important to understand the essence of the problem when solving it.

Example 5. In a right triangle ABC, given that $\angle C = 90°$, $\sin A + \sin B = m$, prove that $\sin A \cdot \sin B = \frac{m^2-1}{2}$.

Proof. Since $\angle A + \angle B = 90°$, we know that $\sin B = \cos A$. Thus

$$\sin^2 A + \sin^2 B = \sin^2 A + \cos^2 A = 1.$$

Since $\sin A + \sin B = m$, we get that

$$m^2 = (\sin A + \sin B)^2 = \sin^2 A + \sin^2 B + 2\sin A \cdot \sin B,$$

i.e. $2\sin A \cdot \sin B = m^2 - (\sin^2 A + \cos^2 A) = m^2 - 1$.

So it is proved. □

Remark In $\triangle ABC$, $AC = b$, $AB = c$, $BC = a$. As shown in Figure 13.5, draw the height perpendicular to BC through the point A, and the height is h_a. Then we know that

$$\sin B = \frac{h_a}{c}, \quad \sin C = \frac{h_a}{b}.$$

Therefore, we get $\dfrac{b}{\sin B} = \dfrac{c}{\sin C}$.

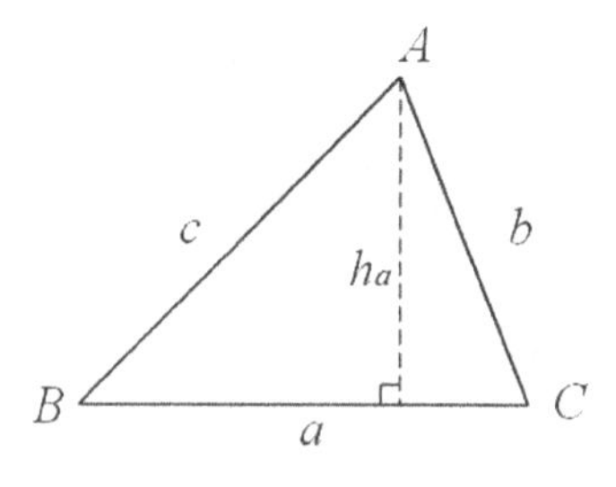

Fig. 13.5

Similarly, $\dfrac{a}{\sin A} = \dfrac{b}{\sin B}$.
For obtuse angled triangles, the same conclusion can be obtained.

Hence, we get that

$$\frac{a}{\sin A} = \frac{b}{\sin B} = \frac{c}{\sin C}.$$

This is the sine rule. The ratio is $2R$, where R is the radius of the circumscribed circle of $\triangle ABC$. The sine rule can be used to transform the sides and angles into each other.

Example 6. In $\triangle ABC$, a, b and c are the opposite sides of the angles A, B and C respectively, and $\dfrac{b+c}{10} = \dfrac{c+a}{15} = \dfrac{a+b}{11}$. Find $\sin A : \sin B : \sin C$.

Solution According to the given condition, we can convert the sides into angles.

Let $\dfrac{a}{\sin A} = \dfrac{b}{\sin B} = \dfrac{c}{\sin C} = k$. Then

$$a = k \sin A, \quad b = k \sin B, \quad c = k \sin C.$$

So the given condition turns into

$$\frac{\sin B + \sin C}{10} = \frac{\sin C + \sin A}{15} = \frac{\sin A + \sin B}{11}.$$

Let t be the ratio above. Then we have

$$\sin B + \sin C = 10t,$$

$$\sin C + \sin A = 15t,$$

$$\sin A + \sin B = 11t.$$

Thus we get $\sin A = 8t$, $\sin C = 7t$, $\sin B = 3t$,

Hence, $\sin A : \sin B : \sin C = 8 : 3 : 7$.

Example 7. Given that the sum of the lengths of two sides of the triangle is 10, the angle between these two sides is $30°$, and the area of the triangle is $\frac{25}{4}$, prove that this triangle is an isosceles triangle.

Proof. Let $a + b = 10$, $\alpha = 30°$. From the given condition, we have that

$$S_{\triangle} = \frac{1}{2} ab \sin \alpha = \frac{25}{4},$$

i.e. $\dfrac{1}{2} ab \cdot \dfrac{1}{2} = \dfrac{25}{4}$, so $ab = 25$.

From $a + b = 10$ and $ab = 25$, we know that a and b are two roots of the equation $x^2 - 10x + 25 = 0$. Since the equation has two identical roots, we get $a = b = 5$, that is, the triangle is an isosceles triangle. □

Remark We can also get $a = b$ directly from $(a-b)^2 = (a+b)^2 - 4ab = 0$.

Think about it In Example 7, if the condition "included angle is 30°" is changed to "included angle is 90°" and "$\frac{25}{4}$" is changed to "m", can you find the value range of m?

Example 8. As shown in Figure 13.6, $A(1, 0)$, $B(2, 0)$, $C(3, 0)$ and $M(0, m)$ $(m > 0)$ are four points in the plane rectangular coordinate system xOy. Given that $OP \perp AM$, $AQ \perp BM$, $BR \perp CM$, and the vertical points are P, Q and R respectively, if the three points P, Q and R are collinear, find the value of m.

Solution Let $\angle OAM = \alpha$, $\angle OBM = \beta$, $\angle OCM = \gamma$. Then $0° < \alpha$, β, $\gamma < 90°$.

From the given condition,

$$\sin\alpha = \frac{m}{\sqrt{1+m^2}}, \quad \cos\alpha = \frac{1}{\sqrt{1+m^2}},$$
$$\sin\beta = \frac{m}{\sqrt{4+m^2}}, \quad \cos\beta = \frac{2}{\sqrt{4+m^2}},$$
$$\sin\gamma = \frac{m}{\sqrt{9+m^2}}, \quad \cos\gamma = \frac{3}{\sqrt{9+m^2}}.$$

From the given condition, $OA = AB = BC = 1$, then we know that in Rt$\triangle OPA$,

$$PP_1 = PA \cdot \sin\alpha = OA \cdot \cos\alpha \cdot \sin\alpha = \sin\alpha\cos\alpha,$$
$$OP_1 = OP \cdot \sin\alpha = OA \cdot \sin\alpha \cdot \sin\alpha = \sin\alpha\sin\alpha = \sin^2\alpha.$$

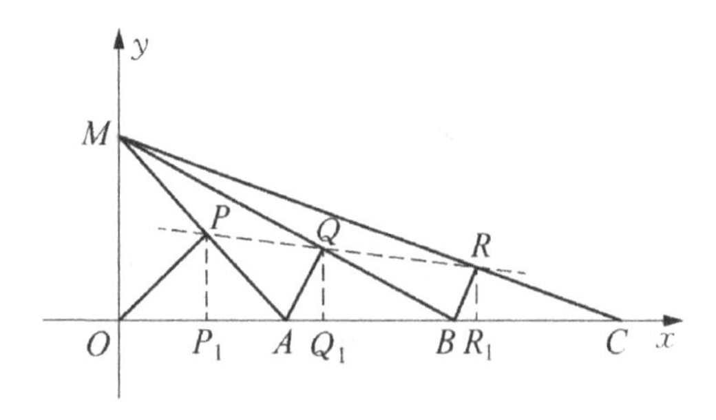

Fig. 13.6

So the coordinates of the point P are $(\sin^2\alpha, \sin\alpha\cos\alpha)$.

Similarly, in $Rt\triangle AQB$, $QQ_1 = \sin\beta\cos\beta$, $AQ_1 = \sin\beta\sin\beta = \sin^2\beta$, so the coordinates of the point Q are $(1+\sin^2\beta, \sin\beta\cos\beta)$. In Rt$\triangle BRC$, $RR_1 = \sin\gamma\cos\gamma$, $BR_1 = \sin\gamma\sin\gamma = \sin^2\gamma$, so the coordinates of the point R are $(2+\sin^2\gamma, \sin\gamma\cos\gamma)$. Since the three points P, Q and R are collinear, we obtain that

$$\frac{\sin\beta\cos\beta - \sin\alpha\cos\alpha}{1+\sin^2\beta - \sin^2\alpha} = \frac{\sin\gamma\cos\gamma - \sin\alpha\cos\alpha}{2+\sin^2\gamma - \sin^2\alpha},$$

i.e. $$\frac{\frac{2m}{4+m^2} - \frac{m}{1+m^2}}{1+\frac{m^2}{4+m^2} - \frac{m^2}{1+m^2}} = \frac{\frac{3m}{9+m^2} - \frac{m}{1+m^2}}{2+\frac{m^2}{9+m^2} - \frac{m^2}{1+m^2}}.$$

Since $m > 0$, we get

$$\frac{m^2-2}{4+2m^2+m^4} = \frac{m^2-3}{9+6m^2+m^4}.$$

Simplifying the equation gives $5m^4 - m^2 - 6 = 0$, i.e. $(m^2+1)(5m^2-6) = 0$. Solving the equation gives $m^2 = \frac{6}{5}$. And because $m > 0$, we get $m = \frac{\sqrt{30}}{5}$.

Reading

Solving Problems with Constructional Methods

Example Given that $\frac{\pi}{2} > x_1 > x_2 > 0$, prove that $\dfrac{\tan x_1}{x_1} > \dfrac{\tan x_2}{x_2}$.

Solution As shown in Figure 13.7, draw a right angled triangle OAC where $\angle A = 90°$, $\angle COA = x_1$. Draw a point B on line segment AC so

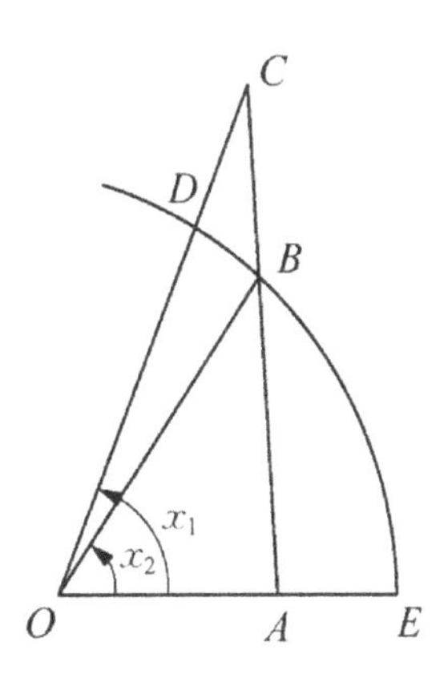

Fig. 13.7

that $\angle BOA = x_2$. Make an arc with O as the center through the point B and intersects OC and OA at D and E respectively. Then we know that

$$\frac{\tan x_1}{\tan x_2} = \frac{\triangle OAC}{\triangle OAB} = 1 + \frac{\triangle OBC}{\triangle OAB} > 1 + \frac{S_{sector\,OBD}}{S_{sector\,OEB}} = 1 + \frac{x_1 - x_2}{x_2} = \frac{x_1}{x_2}.$$

Q.E.D.

This method of proof is called "construction model". The method of proving an abstract equality or inequality with a specific figure or Example like this is often used in mathematical reasoning. In many proofs of the ancient Pythagorean theorem, the method of constructing model is mostly used.

(Excerpted from: Zhang Jingzhong and Ren hongshuo. Beijing: China children's publishing house, 2003, 232~233)

Exercises

1. If the acute angles α and β satisfy $\sin\alpha > \sin\beta$, then which of the following expressions is correct? ( ).

   (A) $\cos\alpha > \cos\beta$ (B) $\tan\alpha > \tan\beta$
   (C) $\cot\alpha > \cot\beta$ (D) None of the above is true

2. Given that A and B are both acute angles and satisfy $\sin^2 A + \cos^2 B = \frac{5}{4}t$, $\cos^2 A + \sin^2 B = \frac{3}{4}t^2$, then the sum of all possible values of real number t is ( ).

   (A) $-\frac{8}{3}$ (B) $-\frac{5}{3}$ (C) 1 (D) $\frac{11}{3}$

3. In $\triangle ABC$, $\angle A = 30°$, $AB = 4$, $BC = \frac{4}{3}\sqrt{3}$, then the size of angle B is ( ).

   (A) 30° (B) 90° (C) 30° or 60° (D) 30° or 90°

4. As shown in Figure 13.8, in $\triangle ABC$, $AB = 5$, $AC = 3$, D is the midpoint of BC, $AD = 2$, find the value of $\tan\angle BAD$.

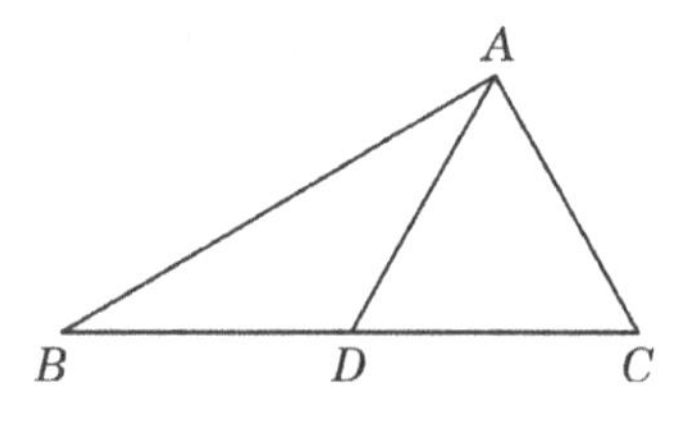

Fig. 13.8

5. As shown in Figure 13.9, in $\triangle ABC$, AD is the height of the triangle which is perpendicular to BC, $DC = 1$, $BD = 2$, $\tan B = \cos\angle DAC$, then the length of AB is ( ).

   (A) $\sqrt{5}$ (B) $\sqrt{7}$ (C) 3 (D) 7

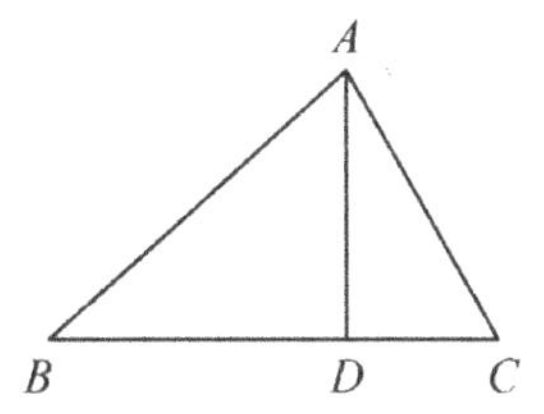

Fig. 13.9

6. In $\triangle ABC$, if $|\sin A - 1| + \left(\frac{\sqrt{3}}{2} - \cos \mathrm{B}\right)^2 = 0$, find the value of $\tan C$.
7. Given that the line segment $AB = 10$ and the distance from the point P to AB is 3, if the value of $PA \cdot PB$ reaches its minimum value, find the value of $PA + PB$.
8. Given that $y = 4\cos x \sin x + 2\cos x - 2\sin x - 1$, $0 \leq x \leq \frac{\pi}{2}$, what is the value of x if y can be a non-negative number?
9. In $\triangle ABC$, $AB = AC$, $\angle A = 40°$. P is a point on the line segment AB and $\angle ACP = 20°$. Find the value of $\frac{BC}{AP}$.
10. Consider the function $y = (\cos\theta x^2) - (4\sin\theta)x + 6$. Given that $y > 0$ holds for any real number x, and θ is an interior angle of a triangle, find the value range of $\cos\,\theta$. (Hint: Using the formula $\cos^2\theta + \sin^2\theta = 1$.

Chapter 14

Solving Right Triangles

In order to deal with some problems in triangles, we often convert them into the problems of right triangles, and use the trigonometric functions of acute angles and Pythagorean theorem to get the results.

Example 1. In $\triangle ABC$, let a, b and c be the opposite sides of angles A, B and C respectively. Given that a and b are the roots of the quadratic equation $x^2 + 4(c+2) = (c+4)x$ with one unknown x,

(1) determine the shape of $\triangle ABC$;
(2) if $\tan A = \frac{3}{4}$, find the values of a, b and c.

Solution

(1) According to the given condition, try to judge by the sides.

From the given condition and the relationship between the roots and coefficients, we have that

$$a + b = c + 4, \tag{1}$$

$$ab = 4(c+2). \tag{2}$$

So we get $a^2 + b^2 = (a+b)^2 - 2ab = (c+4)^2 - 2 \times 4(c+2) = c^2$.
Hence, $\triangle ABC$ is a right angled triangle where $\angle C = 90°$.

(2) From $\angle C = 90°$ and $\tan A = \dfrac{3}{4}$, we get that $\dfrac{a}{b} = \dfrac{3}{4}$.

If $a = 3k$, $b = 4k$ $(k > 0)$, then $c = 5k$. From the formula (1), we obtain that

$$7k = 5k + 4.$$

Solving the equation gives $k = 2$.

Hence, $a = 6$, $b = 8$, $c = 10$.

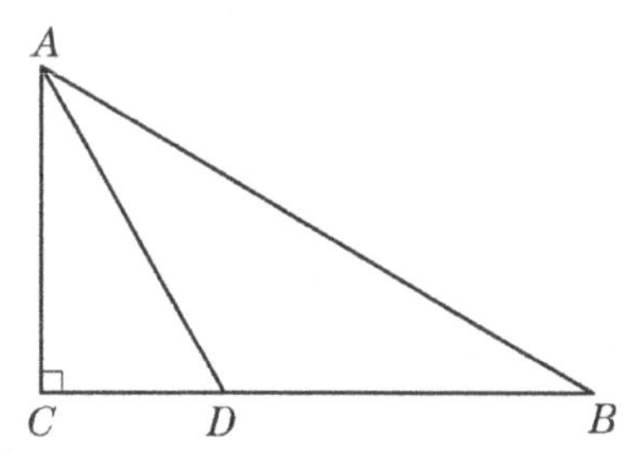

Fig. 14.1

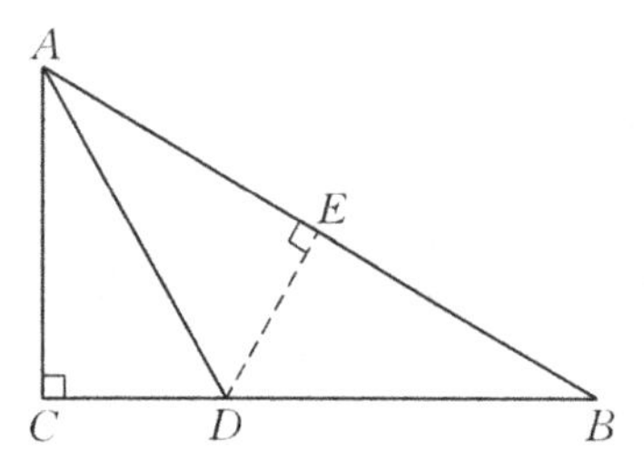

Fig. 14.2

Example 2. As shown in Figure 14.1, in right triangle ABC, $\angle C = 90°$, AD is the bisector of angle A, and $CD = \sqrt{6}$, $DB = 2\sqrt{6}$. Find the lengths of three sides of $\triangle ABC$.

Solution From the angle bisector, we think of symmetry. As shown in Figure 14.2, construct the perpendicular DE from D to AB which intersects AB at the point E. From $\angle C = 90°$, we know that $CD = DE = \sqrt{6}$.

In the right triangle BDE,

$$\sin B = \frac{DE}{DB} = \frac{\sqrt{6}}{2\sqrt{6}} = \frac{1}{2},$$

so $\angle B = 30°$.

Thus we get that

$$BC = CD + DB = 3\sqrt{6},$$

$$AC = BC \cdot \tan B$$

$$= 3\sqrt{6}\cdot\frac{\sqrt{3}}{3}$$
$$= 3\sqrt{2},$$
$$AB = \frac{\text{AC}}{\sin} = 2AC = 6\sqrt{2}.$$

Hence, the lengths of three sides of $\triangle ABC$ are $6\sqrt{2}$, $3\sqrt{2}$ and $3\sqrt{6}$.

Example 3. As shown in Figure 14.3, given that the telegraph pole AB stands upright on the ground, its shadow happens to shine on the slope CD and the ground BC. If CD makes an angle of 45° with the ground, $\angle A = 60°$, $CD = 4$m, $BC = (4\sqrt{6} - 2\sqrt{2})$m, find the length of the telegraph pole AB (to the nearest 0.1 m).

Solution As shown in Figure 14.4, extend AD to the ground at the point E, draw a line DF through the point D so that DF is perpendicular to CE (produced to F).

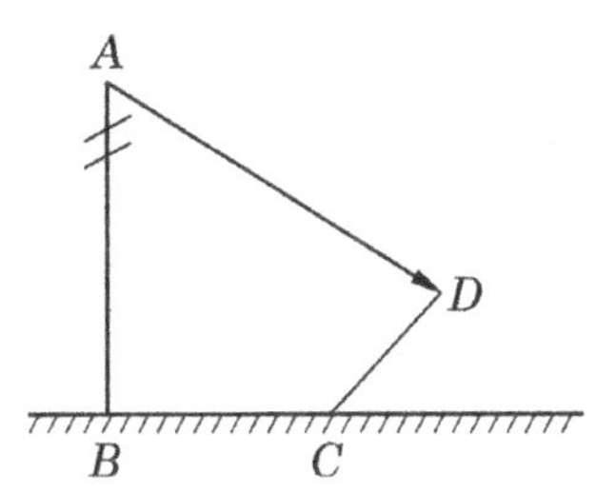

Fig. 14.3

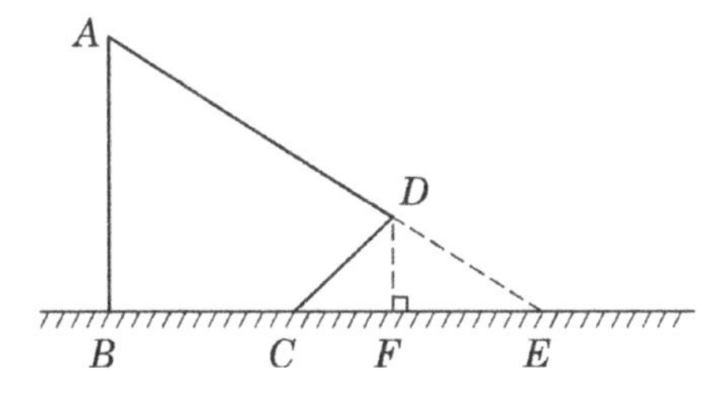

Fig. 14.4

Since $\angle DCF = 45^\circ$, $\angle A = 60^\circ$, $CD = 4$, we get that

$$\angle EDF = \angle A = 60^\circ,$$

$$CF = DF = CD \cdot \sin 45^\circ = 4 \times \frac{\sqrt{2}}{2} = 2\sqrt{2},$$

$$EF = DF \cdot \tan \angle EDF = DF \cdot \tan 60^\circ = 2\sqrt{2} \times \sqrt{3} = 2\sqrt{6}.$$

Because $\frac{AB}{BE} = \tan 30^\circ = \frac{\sqrt{3}}{3}$, we obtain

$$\begin{aligned} AB &= BE \times \frac{\sqrt{3}}{3} = (BC + CF + EF) \times \frac{\sqrt{3}}{3} \\ &= (4\sqrt{6} - 2\sqrt{2} + 2\sqrt{2} + 2\sqrt{6}) \times \frac{\sqrt{3}}{3} \\ &= 6\sqrt{2} \approx 8.5\,(\text{m}). \end{aligned}$$

Example 4. As shown in Figure 14.5, there are a large number of reefs in 42 nautical miles around an island S. Now a ship is sailing from west to east at a speed of 15 nautical miles per hour. S measured at A is in the direction 60 degrees east of north, 2 hours later, S measured at B is in the northeast direction. Whether the ship needs to change its course to avoid the risk of hitting the reefs? And explain the reason.

Solution Draw a line SM through the point S so that $SM \perp AB$ and SM intersects the extended line of AB at the point M (see Figure 14.6).

In the right triangle SMB, $\angle SBM = 45^\circ$, then $SM = BM$.

Let $SM = x$. From the given condition, in the right triangle AMS, $\angle SAM = 30^\circ$, $AB = 15 \times 2 = 30$, thus

$$\tan \angle SAM = \tan(90^\circ - 60^\circ) = \tan 30^\circ = \frac{SM}{AM} = \frac{x}{30 + x}.$$

Then we get $\frac{\sqrt{3}}{3} = \frac{x}{30 + x}$. Solving the equation gives $x = 15\sqrt{3} + 15$.

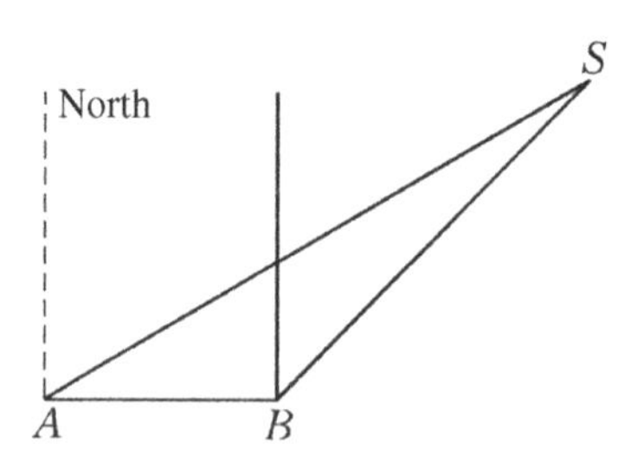

Fig. 14.5

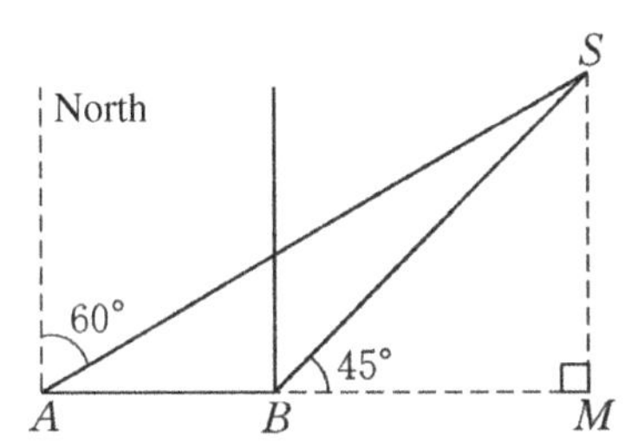

Fig. 14.6

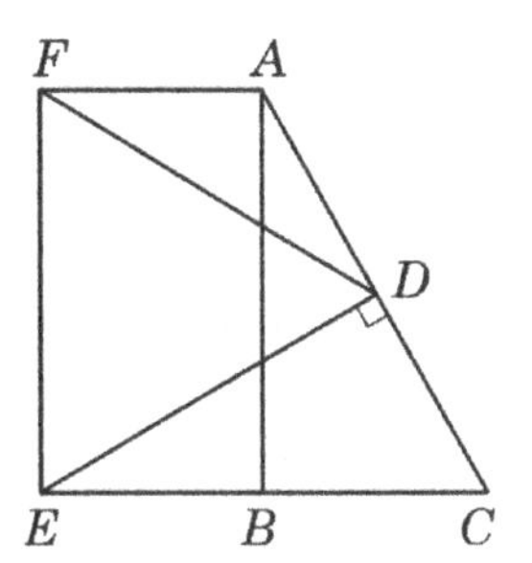

Fig. 14.7

Therefore, $SM = 15\sqrt{3} + 15 \approx 41.0$.

Since $SM < 42$, the ship needs to change the course.

Example 5. As shown in Figure 14.7, in $\triangle ABC$, given that $\angle ABC = 90°$, D is the midpoint of AC. Draw a line DE through point D so that $DE \perp AC$ and DE intersects the extended line of CB at the point E. Make a rectangle $BAFE$ with BA and BE as adjacent sides, and connect FD. If $\angle C = 60°$, $DF = \sqrt{3}$ cm, find the length of BC.

Solution As shown in Figure 14.8, draw a line DM through the point D so that $DM \parallel CB$ and DM intersects EF at M.

Since $DM \parallel CB \parallel AF$ and D is the midpoint of AC, we know that M is the midpoint of EF and $DM \perp EF$. Therefore, $DF = DE$.

From $\angle C = 60°$, we can get $\angle DEC = 30°$.

Since $DF = \sqrt{3}$cm, $DE = \sqrt{3}$ cm.

From $\frac{CD}{DE} = \tan 30°$, we obtain that $CD = \sqrt{3} \times \frac{\sqrt{3}}{3} = 1$ cm. Thus $AC = 2CD = 2$ cm.

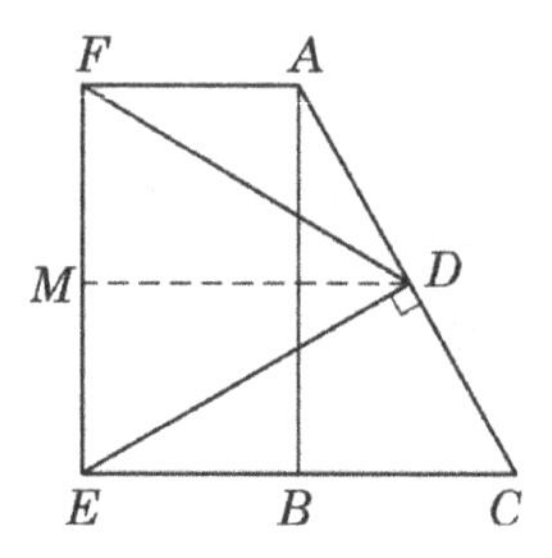

Fig. 14.8

Since $\angle C = 60°$, $\angle BAC = 30°$, we know that $\frac{BC}{AC} = \sin 30°$, then $BC = 1\,\text{cm}$.

So the length of BC is $1\,\text{cm}$.

Example 6. As shown in Figure 14.9, in $\triangle ABC$, CD is the height of $\triangle ABC$, CE is the bisector of $\angle ACB$. If $AC = 15$, $BC = 20$, $CD = 12$, find the length of CE.

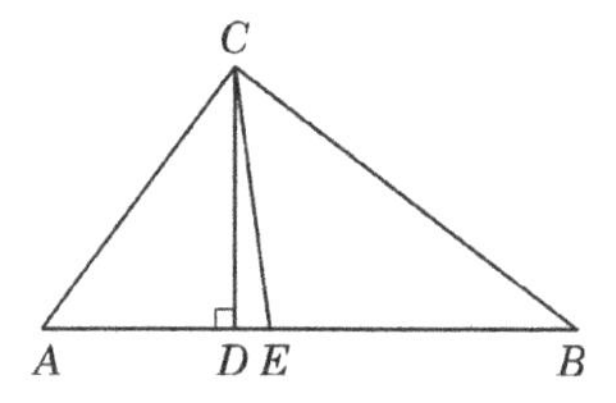

Fig. 14.9

Solution Since CE is the bisector of $\angle ACB$, the distances from the point E to sides AC and BC are equal. Thus we have

$$\frac{S_{\triangle ACE}}{S_{\triangle BCE}} = \frac{AC}{BC} = \frac{AE}{EB}.$$

Since $AC = 15$, $BC = 20$, $CD = 12$, and CD is the height of $\triangle ABC$, we obtain that

$$AD = \sqrt{AC^2 - CD^2} = 9,$$
$$BD = \sqrt{BC^2 - CD^2} = 16.$$

Therefore, $AB = AD + BD = 25$.

From $\dfrac{AC}{BC} = \dfrac{AE}{EB}$, we get $\dfrac{15}{20} = \dfrac{AE}{EB}$, so

$$AE = \frac{3}{4+3} \times AB = \frac{3}{7} \times 25 = \frac{75}{7}.$$

From $AE - AD = DE$, we get $DE = \frac{75}{7} - 9 = \frac{12}{7}$.

Hence, in the right triangle CDE,

$$CE^2 = CD^2 + DE^2 = 12^2 + \left(\frac{12}{7}\right)^2 = 12^2 \times \frac{50}{7^2},$$

i.e. $CE = \frac{60}{7}\sqrt{2}$.

Example 7. In the isosceles right triangle ABC, $AB = 1$, $\angle A = 90°$. Point E is any point on the waist AC where $AE = a$. Point F is on the base BC and $FE \perp BE$. Prove that $S_{\triangle CEF} = \frac{a(1-a)^2}{2(1+a)}$.

Proof. As shown in Figure 14.10, draw a line FD through the point F so that FD is perpendicular to AC (produced to D).

Since $\angle ABE + \angle BEA = \angle BEA + \angle DEF$, we have $\angle ABE = \angle DEF$. Then we know that

$$\triangle ABE \backsim \triangle DEF.$$

So $\frac{AB}{DE} = \frac{AE}{DF}$.

Since $AB \parallel DF$, $\angle ABC = \angle ACB$, we obtain $FD = CD$. Let $FD = x$. Then $DE = AB - AE - DC = 1 - a - x$.

Thus we have that $\frac{1}{1-a-x} = \frac{a}{x}$. Solving the equation gives $x = \frac{a(1-a)}{1+a}$.

Hence, $S_{\triangle CEF} = \frac{1}{2}EC \cdot FD = \frac{1}{2} \cdot (1-a) \cdot \frac{a(1-a)}{1+a} = \frac{a(1-a)^2}{2(1+a)}$. □

Remark

(1) When the point E is the midpoint of AC, this question is the 11th question of the National Senior School Maths Tournament in 1998.

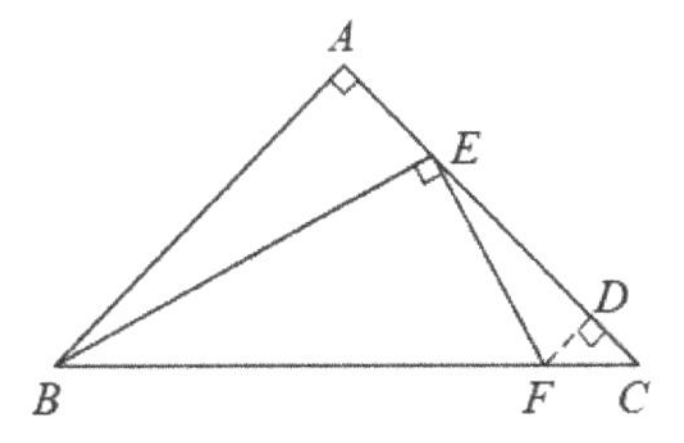

Fig. 14.10

(2) The solution of this problem can also be as follows: extend BA to the point G so that $AG = AE$, try to prove $\triangle GBE \backsim \triangle CEF$ and then get the result. (This solution is given by Mr. Zhu Dafu of Huai'an)
(3) This question can also be generalized to get Example 8.

Example 8. As shown in Figure 14.11, in the right triangle ABC, $\angle A = 90°$, $AB = a$, $AC = b$. Point E is a moving point on the side AC and the point F is on the side BC. If the point E moves from A to C and keeps $EF \perp BE$, try to find the relational expression between $S_{\triangle EFC}$ and x which is the length of AE.

Solution As shown in Figure 14.12, draw a line CD through the point C so that $CD \perp EF$ and CD intersects EF at the point D.

Then we have

$$\triangle ABE \backsim \triangle DEC.$$

So $\dfrac{AB}{ED} = \dfrac{BE}{EC} = \dfrac{AE}{CD}$.

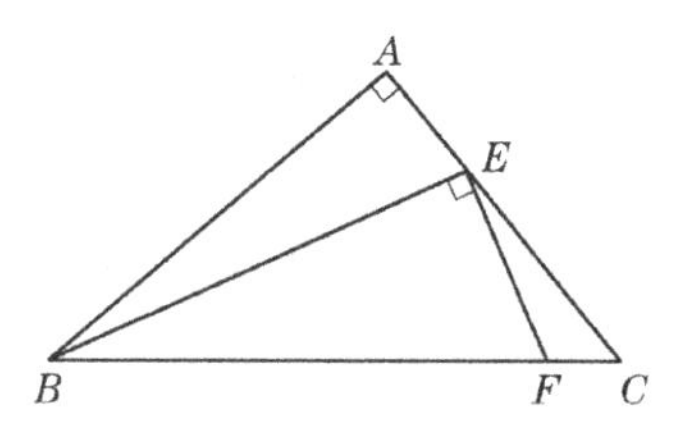

Fig. 14.11

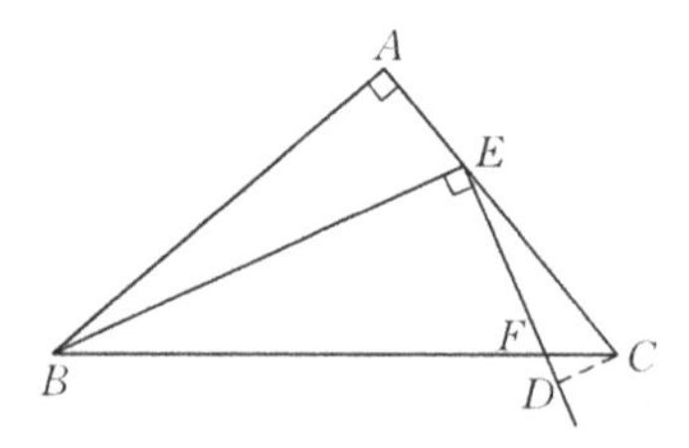

Fig. 14.12

Since $AE = x$, we get $EC = b - x$. Then the above equality can be transformed into

$$\frac{a}{DE} = \frac{\sqrt{a^2+x^2}}{b-x} = \frac{x}{CD}.$$

Thus we obtain that

$$DE = \frac{a(b-x)}{\sqrt{a^2+x^2}},$$

$$CD = \frac{(b-x)x}{\sqrt{a^2+x^2}}.$$

Since $\triangle BEF \backsim \triangle CDF$, we get $\frac{BE}{CD} = \frac{EF}{DF}$, that is,

$$\frac{BE}{BE+CD} = \frac{EF}{EF+FD} = \frac{EF}{ED}.$$

Then we obtain that $EF = \frac{\sqrt{a^2+x^2} \cdot \frac{a(b-x)}{\sqrt{a^2+x^2}}}{\sqrt{a^2+x^2} + \frac{(b-x)x}{\sqrt{a^2+x^2}}} = \frac{a(b-x)\sqrt{a^2+x^2}}{a^2+bx}$.

Hence, $S_{\triangle CEF} = \frac{1}{2}EF \cdot CD = \frac{1}{2} \cdot \frac{ax(b-x)^2}{a^2+bx}$, $0 < x < b$.

Remark In the above formula, let $a = b = 1$, which is the result of Example 7.

Example 9. Find all right triangles satisfying that the lengths of two legs are integers, and the perimeter is exactly equal to the integral times of the area.

Solution Let the two legs of a right triangle be a and b ($0 < a \leq b$). Then from the given condition, we get that

$$a + b + \sqrt{a^2+b^2} = k \cdot \frac{1}{2}ab \text{ (where } a, b \text{ and } k \text{ are all positive integers)},$$

i.e. $2\sqrt{a^2+b^2} = kab - 2(a+b)$. Simplifying the equation gives $(ka-4)(kb-4) = 8$.

Since a, b and k are all positive integers, we have $ab > 0$, then

$$\begin{cases} ka - 4 = 1, \\ kb - 4 = 8 \end{cases} \text{ or } \begin{cases} ka - 4 = 2, \\ kb - 4 = 4. \end{cases}$$

Solving the simultaneous equations gives $\begin{cases} k = 1, \\ a = 5, \\ b = 12 \end{cases}$ or $\begin{cases} k = 2, \\ a = 3, \\ b = 4 \end{cases}$ or $\begin{cases} k = 1, \\ a = 6, \\ b = 8. \end{cases}$

That is, there are three groups of solutions.

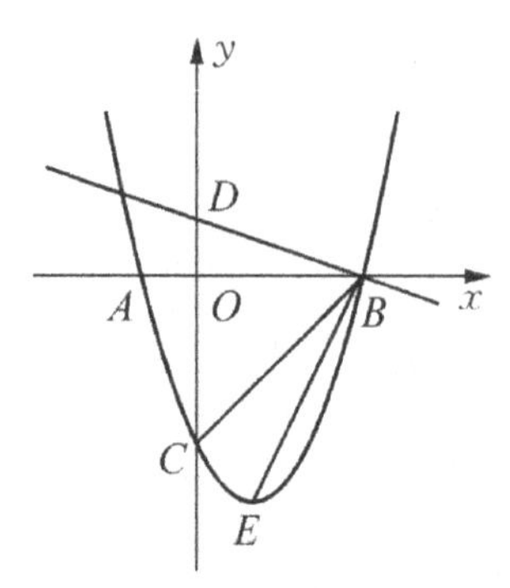

Fig. 14.13

So there are three right triangles satisfying the requirements, which are triangles with three sides of 5, 12, 13; 3, 4, 5 or 6, 8, 10.

Example 10. Let a and b be real numbers with $a \neq 0$. As shown in Figure 14.13, the vertex of the graph of quadratic function $y = ax^2 + bx - 3$ is the point E. The parabola intersects the x-axis at two points A and B, and intersects the y-axis at the point C, with $OB = OC = 3OA$. The graph of the linear function $y = -\frac{1}{3}x + 1$ intersects the y-axis at the point D. Find the size of $\angle DBC - \angle CBE$.

Solution Substituting $x = 0$ into $y = -\frac{1}{3}x + 1$ and $y = ax^2 + bx - 3$ respectively gives $D(0, 1)$, $C(0, -3)$. From $OB = OC = 3OA$, we get $B(3, 0)$, $A(-1, 0)$. So the line $y = -\frac{1}{3}x + 1$ passes through the point B.

Substituting the coordinates of the point $C(0, -3)$ into $y = a(x + 1)(x - 3)$ gives $a = 1$. So the vertex of the graph of quadratic function $y = x^2 - 2x - 3$ is the point $E(1, -4)$. From the formula of a distance between two points, we get that

$$BC = 3\sqrt{2}, \quad CE = \sqrt{2}, \quad BE = 2\sqrt{5}.$$

Therefore, $BC^2 + CE^2 = BE^2$, $\triangle BCE$ is a right triangle where $\angle BCE = 90°$.

Then we obtain $\tan \angle CBE = \frac{CE}{CB} = \frac{1}{3}$.

And because $\tan \angle DBO = \frac{OD}{OB} = \frac{1}{3}$, we get $\angle DBO = \angle CBE$.

Hence,

$$\angle DBC - \angle CBE = \angle DBC - \angle DBO = \angle OBC = 45°.$$

Exercises

1. Given that a, b and c are the lengths of three sides of a triangle, if only one group of numbers (x, y) satisfies the simultaneous equations

$$\begin{cases} x^2 - ax - y + b^2 + ac = 0, \\ ax - y + bc = 0, \end{cases}$$

determine what kind of triangle it is.

2. In $\triangle ABC$, $AB = AC$, $\angle BAC = 120^\circ$, D is the midpoint of BC, DE is perpendicular to AB (produced at E). Find the ratio of $AE : BD$.
3. As shown in Figure 14.14, in the quadrilateral $ABCD$, $\angle A = \angle BCD = 90^\circ$, $BC = CD$, E is a point on the extended line of AD. If $DE = AB = 3\,\text{cm}$, $CE = 4\sqrt{2}\,\text{cm}$, find the length of AD.

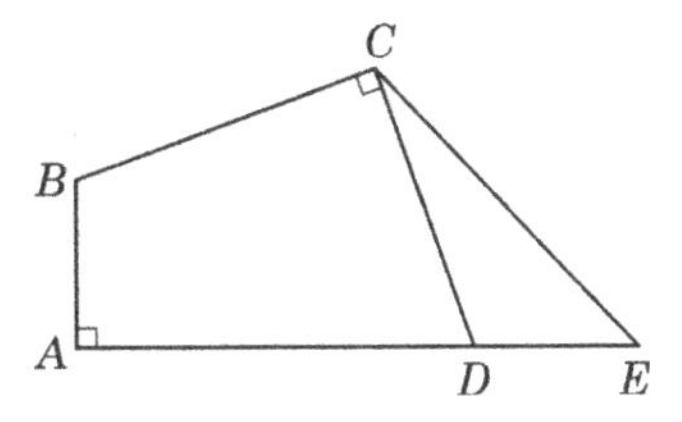

Fig. 14.14

4. As shown in Figure 14.15, let AD, BE and CF be three heights of the triangle ABC. If $AB = 6$, $BC = 5$, $EF = 3$, find the length of the line segment BE.

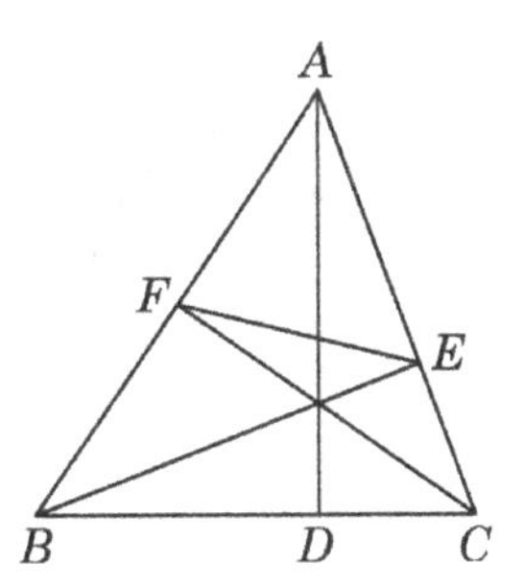

Fig. 14.15

5. In the right triangle ABC, $AB = c$, $BC = a$, $AC = b$, $\angle C = 90^\circ$, $a + c = 6$, $\tan A = \sqrt{3}$. Solve this right triangle.

6. Given that in the $\triangle ABC$, $BC = 6$, $AC = 6\sqrt{3}$, $\angle A = 30°$, find the length of AB.
7. Given that in the quadrilateral $ABCD$, $AB : BC : CD : DA = 2 : 2 : 3 : 1$, $\angle B = 90°$, find the size of $\angle DAB$.
8. As shown in Figure 14.16, there are two lighthouses A and B on the coastline. Lighthouse B is located in the south of Lighthouse A, 6 nautical miles away from A. There are two ships P and Q on the sea. Ship P is located in the point D which is in the direction 75° north by west of Lighthouse A and $4\sqrt{2}$ nautical miles away from A. Ship Q is located in the point C which is in the direction 60° north by west of Lighthouse B and 6 nautical miles away from B. Find the distance between the two ships.

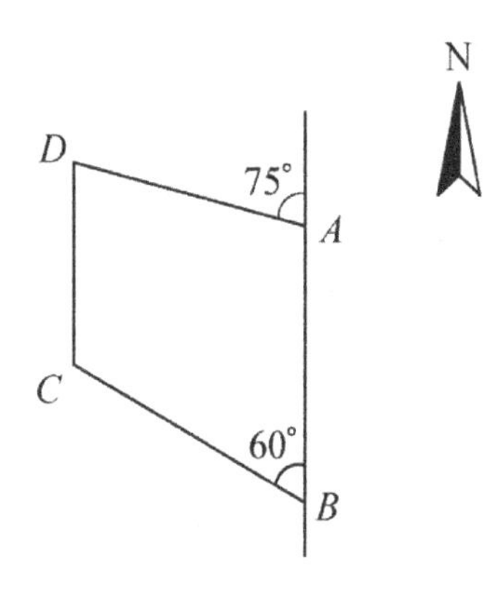

Fig. 14.16

9. Given that in the quadrilateral $ABCD$, $\angle A = 60°$, $\angle B = \angle D = 90°$, $AD = 8$, $AB = 7$, find the value of $BC + CD$.
10. Given that the perimeter of $\triangle ABC$ is 24 cm, M is the midpoint of AB, and $MC = MA = 5$ cm, find the area of $\triangle ABC$.
11. In $\triangle ABC$, $\angle C = 90°$, $\angle A = 15°$, $AB = 10$ cm, find the area of $\triangle ABC$.
12. In Rt$\triangle ABC$, $\angle C = 90°$, $BC < AC$, $BC \cdot AC = \frac{1}{4}AB^2$, find the size of $\angle A$.
13. As shown in Figure 14.17, in $\triangle ABC$, D and E are the midpoints of BC and AC respectively. Given that $\angle ACB = 90°$, $BE = 4$ cm, $AD = 7$ cm, then the length of AB is ( ).

   (A) 10 (B) $5\sqrt{3}$ (C) $2\sqrt{13}$ (D) $2\sqrt{15}$

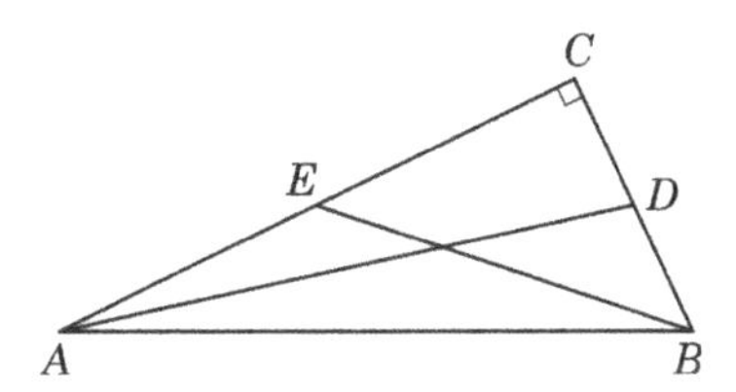

Fig. 14.17

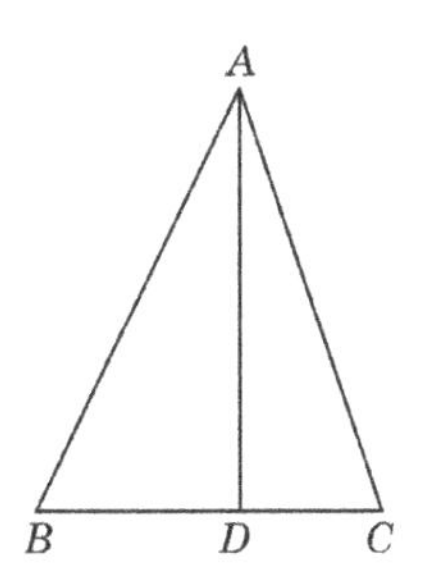

Fig. 14.18

14. As shown in Figure 14.18, in $\triangle ABC$, $\angle BAC = 45°$, AD is perpendicular to BC (produced at D). If $BD = 3$, $CD = 2$, find the area of $\triangle ABC$.
15. Given that a, b and c are all positive numbers, and the equation

$$(c + a)x^2 + 2bx + (c - a) = 0$$

has two identical real roots, whether the numbers a, b and c can be the lengths of three sides of a triangle? If so, what is the shape of this triangle?

Chapter 15

Rotations

Rotations are transformations in which a figure is rotated through an angle θ about a fixed point O in a certain direction. O is called the center of rotation and θ is known as the angle of rotation. If the point P on the figure is rotated to the point P', then these two points are called the corresponding points of rotation.

The properties of rotation are as follows:

(1) The distances from the corresponding points to the center of rotation are equal.
(2) The perpendicular bisector of the line segment connecting the corresponding points must pass through the center of rotation.
(3) The included angle of the two line segments between the corresponding points and the center of rotation is the angle of rotation.
(4) The figures before and after rotation are congruent.

Three elements of rotation are:

(1) the center of rotation,
(2) the direction of rotation,
(3) the angle of rotation.

Notice that as long as any one of the three elements is changed, the figure will be different.

Some scattered quantities and line segments can be concentrated by rotation, so that the problem can be solved. Three key points should be grasped when using rotation to deal with problems: first, the angle of rotation; second, the congruence of figures; third, the special angles and equal sides.

Example 1. As shown in Figure 15.1, press and hold one of the two mutually overlapped square paper pieces with the side length of 2, and the other is rotated clockwise through an angle about the point B. If the area of the overlapping part is $\frac{4\sqrt{3}}{3}$, find the size of the angle of rotation.

Solution As shown in Figure 15.2, join BF. It's obvious that $\text{Rt}\triangle A'FB \cong \text{Rt}\triangle CFB$. Since the area of the shaded parts in Figure 15.2 is $\frac{4\sqrt{3}}{3}$, we get $CF \cdot BC = \frac{4\sqrt{3}}{3}$. From $BC = 2$, we obtain $CF = \frac{2\sqrt{3}}{3}$ and then $BF = \sqrt{CF^2 + BC^2} = \frac{4\sqrt{3}}{3}$. Now, in $\text{Rt}\triangle BFC$, from $BF = 2CF$, we get $\angle FBC = 30°$. As $\angle A'BF = \angle FBC = 30°$, then $\angle A'BC = 60°$. Hence, $\angle CBC' = 30°$, that is, the angle of rotation is $30°$.

Think about it If the angle of rotation is known to be $30°$, how can we find the area of the overlapping part?

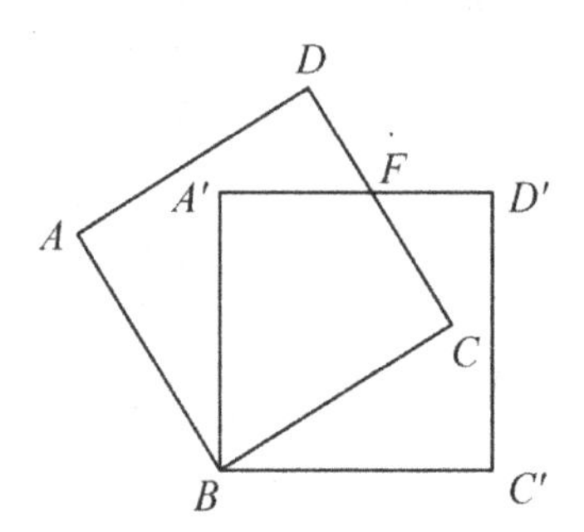

Fig. 15.1

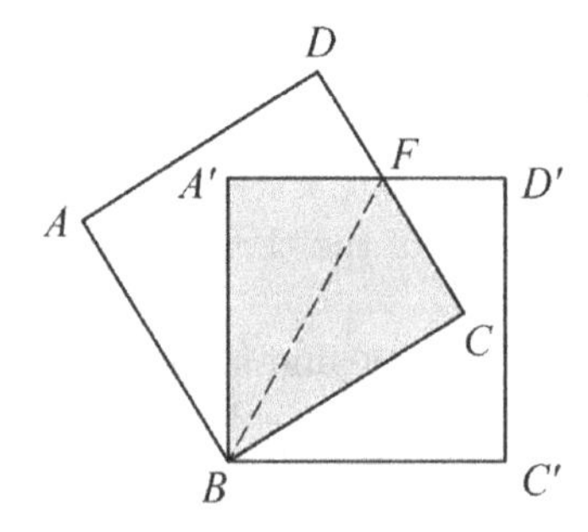

Fig. 15.2

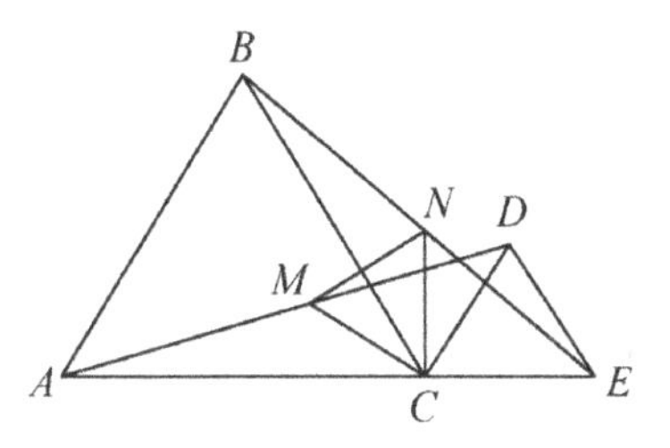

Fig. 15.3

Example 2. Given an equilateral triangle ABC, take a point E on the extended line of AC, and make an equilateral triangle CDE with CE as a side, which is on the same side of the line AE as $\triangle ABC$. Point M is the midpoint of the line segment AD, and Point N is the midpoint of the line segment BE. Prove that $\triangle CMN$ is an equilateral triangle.

Proof. As shown in Figure 15.3.

Since $\triangle ABC$ and $\triangle CDE$ are equilateral triangles, their interior angles are all $60°$.

Therefore, the line segment BE can be rotated anticlockwise through $60°$ about the point C, meanwhile, the point B falls on the point A, the point E falls on the point D, and the point N falls on the point M, from which we can know that $CN = CM$, $\angle NCM = 60°$, that is, $\triangle CNM$ is an equilateral triangle. □

Remark This problem can be solved by proving the congruence of triangles directly. Pay attention to comapre the simplicity of these two proofs.

Example 3. It is known that M and N are two points on the hypotenuse BC of the isosceles right triangle ABC, $AB = AC = 6\sqrt{2}$, $BM = 3$, $\angle MAN = 45°$, then $NC =$ ( ).

(A) 3 (B) $\frac{7}{2}$ (C) 4 (D) $\frac{9}{2}$

Solution As shown in Figure 15.4, $\triangle CAN$ is rotated clockwise through $90°$ about the point A to $\triangle ABD$. Join the point M to the point D.

Then we have that $\angle MBD = \angle ABM + \angle ABD = 45° + 45° = 90°$, $BD = CN$, $AD = AN$.

Thus,

$$\angle MAD = \angle BAC - \angle MAN = 45° = \angle MAN.$$

So $\triangle AMN \cong \triangle AMD$, then we get $MN = MD$.

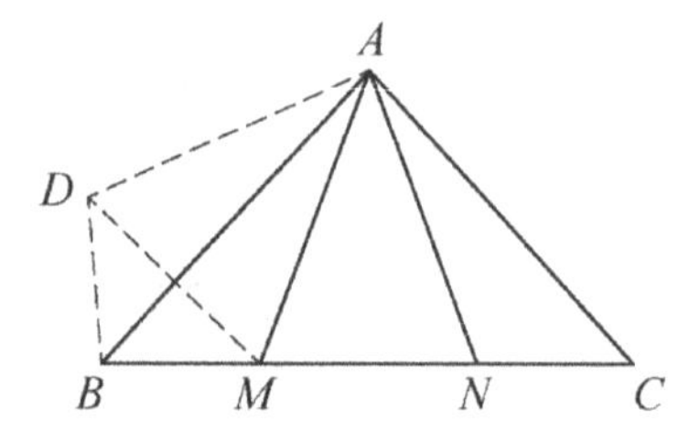

Fig. 15.4

Since $BC = \sqrt{AB^2 + AC^2} = \sqrt{2}\cdot 6\sqrt{2} = 12$, $BM = 3$, we obtain $MC = 9$, that is,

$$MN + NC = 9. \tag{1}$$

From $BD^2 + BM^2 = DM^2$,

$$NC^2 + 3^2 = MN^2. \tag{2}$$

Combine (1) and (2) to get that

$$NC = 4, \quad MN = 5.$$

So the answer is C.

Example 4. As shown in Figure 15.5, $ABCD$ and $OEFG$ are two squares with sides that are a in length, where O is the center of the square $ABCD$. The line segments OG and OE intersect CD and BC at the points H and K respectively. Prove that $S_{OKCH} = \frac{1}{4}a^2$.

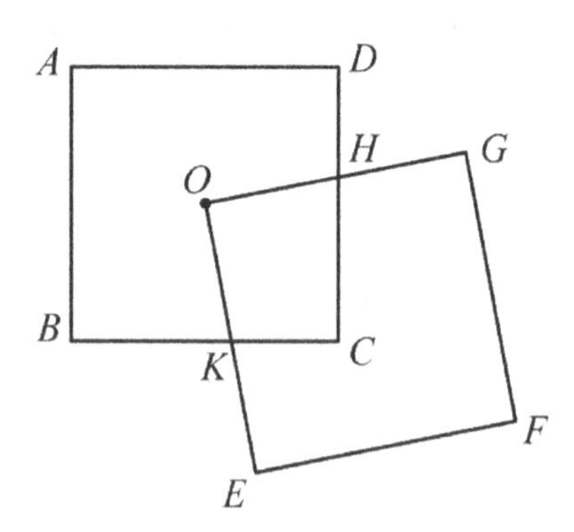

Fig. 15.5

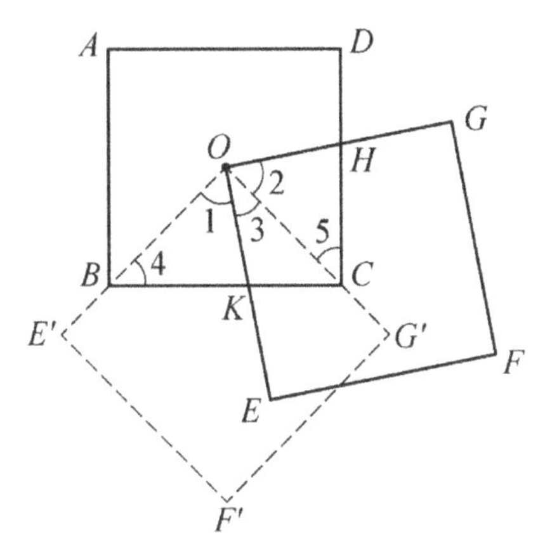

Fig. 15.6

Proof. Rotate the square $OEFG$ about the point O clockwise to the position of $OE'F'G'$ so that the point B is on the line segment OE' and the point C is on OG' (see Figure 15.6).

Because $\angle 1 + \angle 3 = 90°$, $\angle 2 + \angle 3 = 90°$, we get $\angle 1 = \angle 2$.

And from $\angle 4 = \angle 5 = 45°$, $OB = OC$, we obtain

$$\triangle OBK \cong \triangle OCH.$$

Then we know that $S_{\triangle OBK} = S_{\triangle OCH}$. Hence, $S_{\triangle OKCH} = S_{\triangle OKC} + S_{\triangle OCH} = S_{\triangle OKC} + S_{\triangle OBK} = S_{\triangle OBC} = \frac{1}{4}a^2$. □

Example 5. Operation: in $\triangle ABC$, $AC = BC = 2$, $\angle C = 90°$, place the right angle vertex of an isosceles triangle plate at the midpoint P of the hypotenuse AB, and rotate the triangle plate about the point P, where the two legs of the triangle plate intersect the ray AC and CB at two points D and E respectively. Figures 15.7, 15.8 and 15.9 are three cases of the figure obtained by rotating the triangle plate. Discuss the following questions:

(1) When the triangle plate is rotated about the point P, observe the quantitative relationship between the line segments PD and PE, and prove it with Figure 15.8.

(2) In Figure 15.7, it is obvious that $DE^2 = AD^2 + BE^2$. When the triangle plate is rotated about the point P, is this equation still true? If it is true, try to prove it with Figure 15.9; if not, please explain the reason.

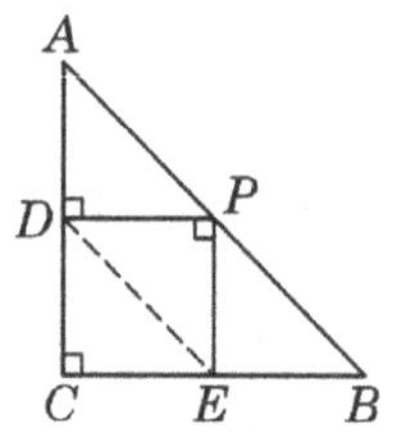

Fig. 15.7

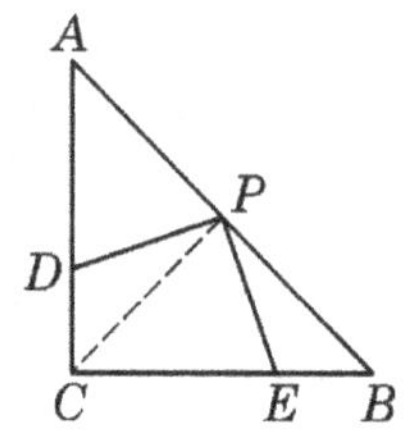

Fig. 15.8

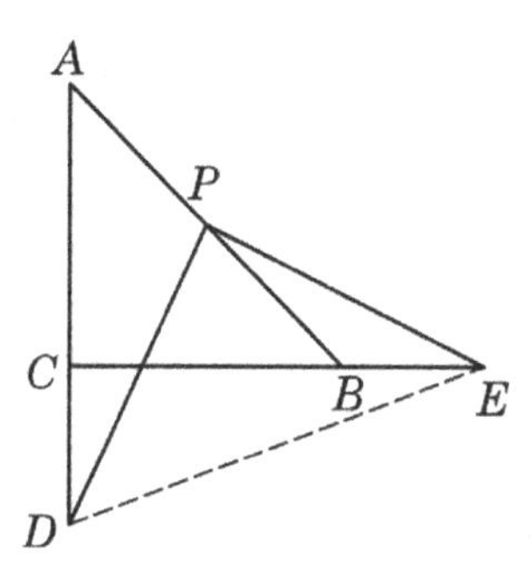

Fig. 15.9

(3) When the triangle plate is rotated about the point P, can $\triangle PBE$ become an isosceles triangle? If so, point out all the cases (i.e. find the length of CE when $\triangle PBE$ is an isosceles triangle); if not, please explain the reason.

(4) If the right angle vertex of the triangle plate is placed at the point M on the hypotenuse AB where $AM : MB = 1 : 3$, and the operation is the same as before, find the quantitative relationship between the line segments MD and ME, and prove it with Figure 15.10.

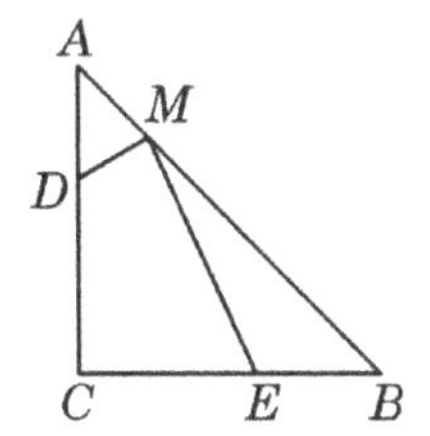

Fig. 15.10

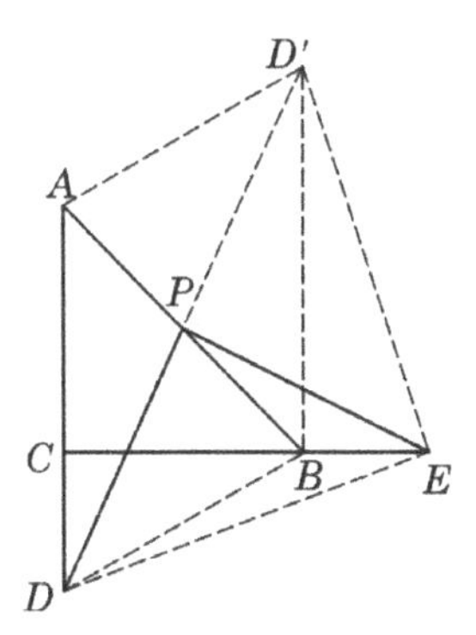

Fig. 15.11

Solution (1) $PD = PE$.

Proof. Join PC. □

Since $AC = BC$, $\angle ACB = 90°$, and P is the midpoint of AB, we know that $PC = PB$, $\angle DCP = \angle EBP = 45°$ and $\angle CPB = 90°$.

Then we get $\angle CPE + \angle EPB = 90°$. And from $\angle DPC + \angle CPE = 90°$, it is seen that $\angle DPC = \angle EPB$. So we obtain $\triangle PDC \cong \triangle PEB$. Thus, $PD = PE$.

(2) When the triangle plate is rotated about the point P, the equation $DE^2 = AD^2 + BE^2$ is still true.

Proof. As shown in Figure 15.11, use the line PE as a mirror line to reflect D to its image point D', join the line segments DE, $D'E$, $D'B$, DB and AD'.

Then we have that $DP = D'P$, $DE = D'E$. And since $AP = BP$, the quadrilateral $ADBD'$ is a parallelogram. So $AD // D'B$.

Since $AD \perp CE$, then we know $D'B \perp CE$. In Rt$\triangle D'BE$, $D'E^2 = D'B^2 + BE^2$. i.e. $DE^2 = AD^2 + BE^2$.

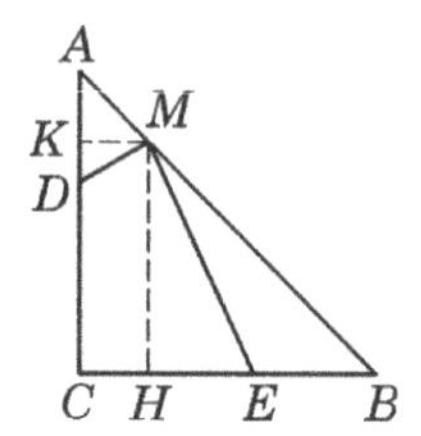

Fig. 15.12

(3) $\triangle PBE$ may become an isosceles triangle, which can be divided into the following three situations. □

Situation 1: When the point E coincides with the point C, the point D coincides with the point A, then $PE = PB$. At this time, $CE = 0$.

Situation 2: When $PE \perp BC$, we know that $PE = EB$. At this time, $CE = \frac{1}{2}BC = 1$.

Situation 3: When the point E is on the extended line of CB, we can find a point E so that $PB = BE$. Since $PB = \frac{1}{2}AB = \sqrt{2}$, at this time, $CE = BC + PB = 2 + \sqrt{2}$.

(4) As shown in Figure 15.12, draw a line MK through the point M so that MK is perpendicular to AC (produced to K), and draw another line MH which is perpendicular to CB (produced to H).

From $AC = BC$ and $\angle C = 90°$, it can be seen that $\angle A = \angle B = 45°$. So $\text{Rt}\triangle AKM \backsim \text{Rt}\triangle BHM$, then we get $\frac{MK}{MH} = \frac{AM}{BM} = \frac{1}{3}$. Since $\angle KMD + \angle DMH = 90°$, $\angle DMH + \angle HME = 90°$, we obtain $\angle KMD = \angle HME$. So $\text{Rt}\triangle MKD \backsim \text{Rt}\triangle MHE$. Therefore, $\frac{MD}{ME} = \frac{MK}{MH} = \frac{1}{3}$, that is, $ME = 3MD$.

Remark For the transformation problems, we should be good at "seek stillness in movement" and try to find the invariable relationship between quantities. The equation in the sub problem (2) can be determined by proving the congruence of triangles, but it is also ingenious to use axisymmetric transformation.

Example 6. As shown in Figure 15.13, in $\triangle ABC$, $\angle BAC = 120°$, P is a point inside $\triangle ABC$. Prove that $PA + PB + PC > AB + AC$.

Analysis PA, PB and PC are scattered in three triangles. In this case, it is necessary to consider concentrating them through rotation for easy

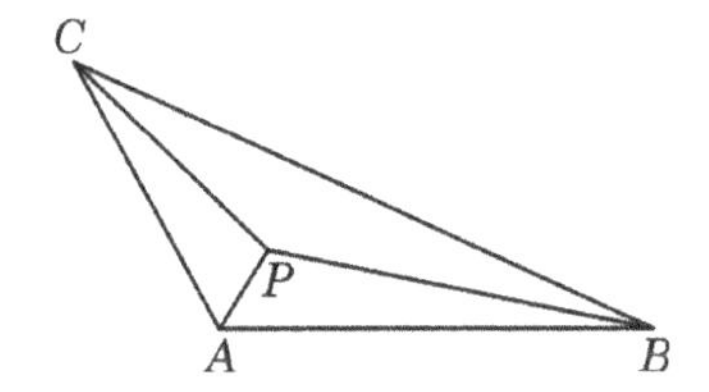

Fig. 15.13

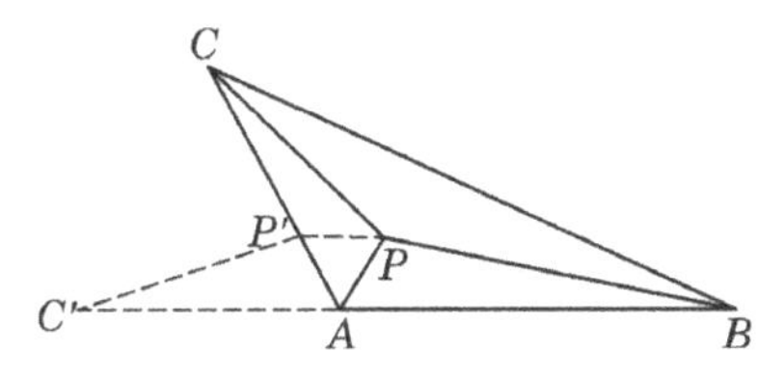

Fig. 15.14

observation, so as to find the ideas of solution. $\triangle APC$ is rotated counterclockwise through $60°$ about the point A to obtain $\triangle AP'C'$. From the properties of rotation, we can get that $\triangle APP'$ is an equilateral triangle. Then from $\angle BAC = 120°$, it can be seen that $\angle BAC' = 120° + 60° = 180°$, that is, the points B, A and C' are collinear. According to the trilateral relationship of triangles, we can draw a conclusion.

Solution As shown in Figure 15.14, $\triangle APC$ is rotated counterclockwise through $60°$ about the point A to obtain $\triangle AP'C'$. Then we get $\angle CAC' = \angle PAP' = 60°$, $AC = AC'$, $AP = AP'$, $PC = P'C'$, so $\triangle APP'$ is an equilateral triangle, $PP' = AP$.

Because $\angle BAC = 120°$, we know that $\angle BAC' = 120° + 60° = 180°$, that is, the points B, A and C' are collinear. Therefore, $BC' < BP + PP' + P'C'$, i.e. $PA + PB + PC > AB + AC$.

Reading

Change and Constant

All over the world, there are obvious or hidden movements and changes. Rapid changes are dazzling, and slow changes are imperceptible. However, just like some examples above, there are often relatively unchanged things in the process of change.

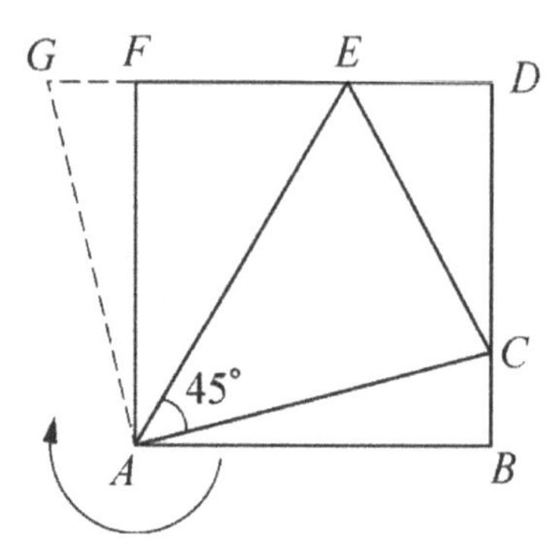

Fig. 15.15

Mathematicians often focus on the unchangeable things in the change. It is these unchangeable things that link different scenes in the change, help us understand the essence of the change process, and help us solve various problems.

Draw two points C and E on two sides of the square $ABDF$ so that $\angle CAE = 45°$, try to prove that

$$BC + EF = CE.$$

This is not an easy problem. If you think about rotation, you will turn $\triangle BAC$ about the point A to the position of $\triangle FAG$ (see Figure 15.15), and suddenly you will find that the problem becomes clear. In the rotation, the lengths do not change, so $AB = AF$, $AC = AG$, and the angles do not change, so $\angle BAC = \angle FAG$. Thus, it can be proved that $\triangle AGE \cong \triangle ACE$, then

$$BC + EF = GF + FE = GE = CE.$$

The problem was solved.

(Excerpted from: Zhang Jingzhong. *Mathematician's vision.* Beijing: China children's publishing house, 1990. $48 \sim 49$)

Exercises

1. As shown in Figure 15.16, in the right angled trapezoid $ABCD$, $\angle BCD = 90°$, $AD \parallel BC$, $BC = CD$. E is a point inside the trapezoid,

and $\angle BEC = 90°$. Rotate $\triangle BEC$ through $90°$ about the point C so that BC coincides with DC, then get $\triangle DFC$. Join EF which intersects CD at the point M. If $BC = 5$, $CF = 3$, then the value of $DM : MC$ is ( ).

(A) $5:3$ (B) $3:5$ (C) $4:3$ (D) $3:4$

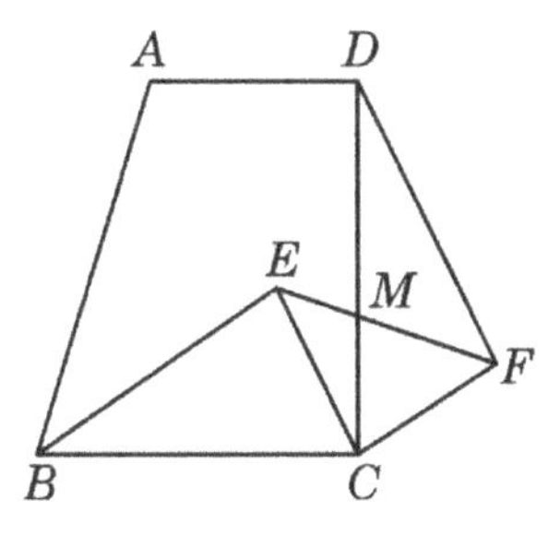

Fig. 15.16

2. As shown in Figure 15.17, given that Rt$\triangle ABC \cong$ Rt$\triangle DEC$, $\angle E = 30°$, D is the midpoint of AB, and $AC = 1$. If $\triangle DEC$ is rotated clockwise about the point D to make ED and CD intersect the leg BC of Rt$\triangle ABC$ at the points M and N respectively, when $\triangle DMN$ is an equilateral triangle, the length of AM is ( ).

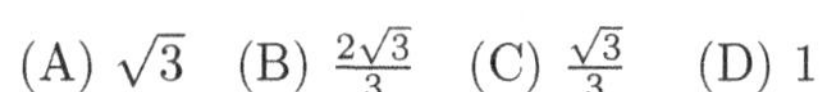
(A) $\sqrt{3}$ (B) $\frac{2\sqrt{3}}{3}$ (C) $\frac{\sqrt{3}}{3}$ (D) 1

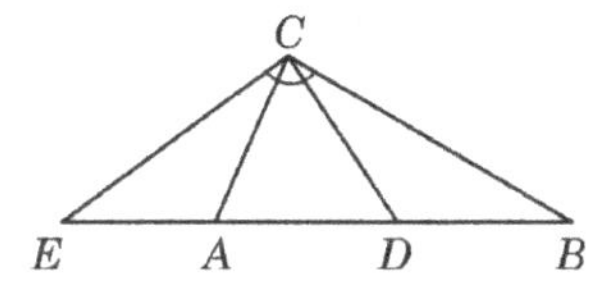

Fig. 15.17

3. As shown in Figure 15.18, the length of the side of an equilateral triangular wood board ABC is 1. Now flip the board along the horizontal line (rotated about a point). Then the path length from the beginning to the end of the point A in figure is ( ).

(A) 4 (B) 2π (C) $\frac{2\pi}{3}$ (D) $\frac{4\pi}{3}$

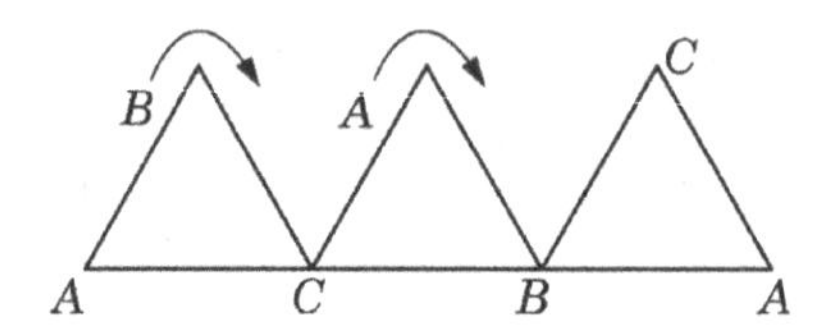

Fig. 15.18

4. As shown in Figure 15.19, the length of the leg of the isosceles right angle triangle ABC is 5 cm. Rotate $\triangle ABC$ anticlockwise through $15°$ about the point A to get $\triangle AB'C'$, then the area of shadow part in the figure is ____cm^2.

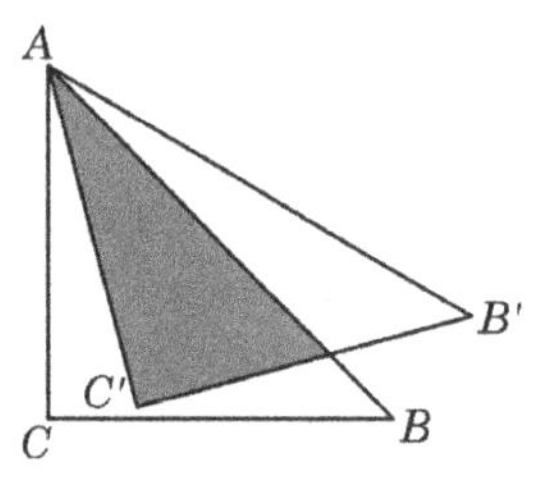

Fig. 15.19

5. As shown in Figure 15.20, P is a point inside the equilateral triangle ABC, $\angle APB = 100°$, $\angle BPC = 120°$, and $\angle CPA = 140°$.

   (1) Prove that a triangle can be formed with PA, PB and PC as its sides.
   (2) Find the degrees of the interior angles of the triangle formed.

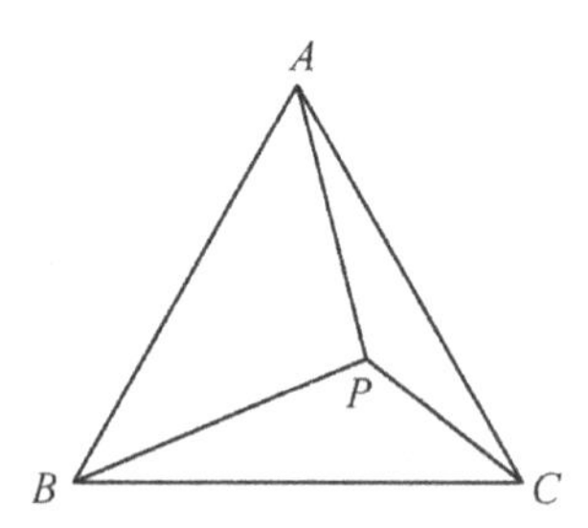

Fig. 15.20

6. In $\triangle ABC$, $AB = AC$, D is a point inside the triangle, and $\angle ADC > \angle ADB$. Prove that $DB > DC$.
7. As shown in Figure 15.21, in the pentagon $ABCDE$, $\angle B = \angle AED = 90°$, $AB = CD = AE = BC + DE = 1$, find the area of the pentagon.

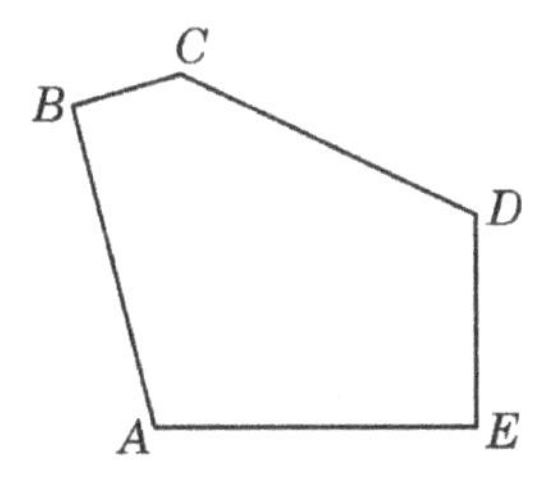

Fig. 15.21

8. When reviewing the knowledge of "congruent triangle", the teacher assigned an assignment as follow: as shown in Figure 15.22(1), in $\triangle ABC$, $AB = AC$, P is any point inside $\triangle ABC$, rotate AP clockwise about the point A to the position of AQ so that $\angle QAP = \angle BAC$, join BQ and CP, then $BQ = CP$.

   Xiao Liang is a brainy student. He proved that $\triangle ABQ \cong \triangle ACP$ by analyzing Figure 15.22(1), and then got $BQ = CP$. After that, he moved the point P out of the isosceles triangle ABC while the conditions in the original question remains unchanged, he found that "$BQ = CP$" is still true. Please prove it with Figure 15.22(2).

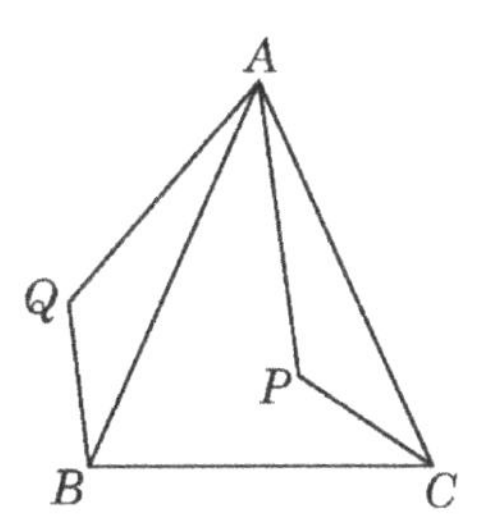

Fig. 15.22(1)

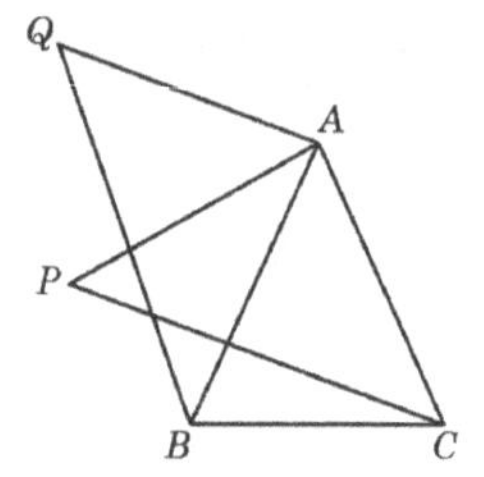

Fig. 15.22(2)

9. As shown in Figure 15.23, The right angled triangle plate ABC with the angle of $30°$ ($\angle B = 30°$) is rotated counterclockwise through an angle $\alpha(\alpha < 90°)$ about the right angle vertex C, and then reflected to $\triangle A'B'C$ by using the opposite side of $\angle A$ as a mirror line. The line segments AB and $B'C$ intersect at the point M, $A'B'$ and BC intersect at the point N, $A'B'$ and AB intersect at the point E.

   (1) Prove that $\triangle ACM \cong \triangle A'CN$.
   (2) When $\alpha = 30°$, find the quantitative relationship between ME and MB' and explain it.

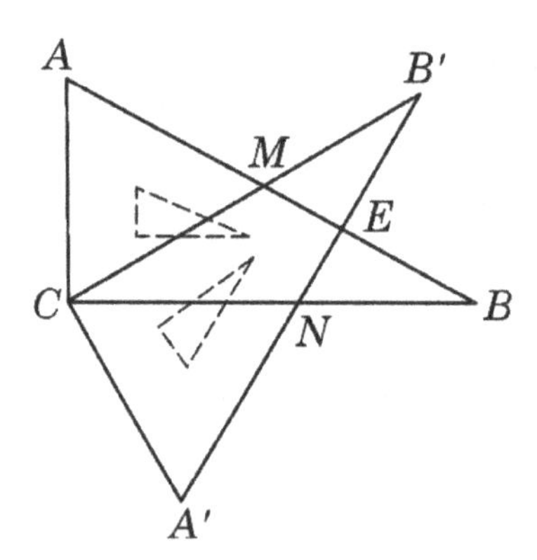

Fig. 15.23

10. As shown in Figure 15.24, quadrilateral $ABDE$ and $ACFG$ are both squares, try to prove that $BG = EC$.

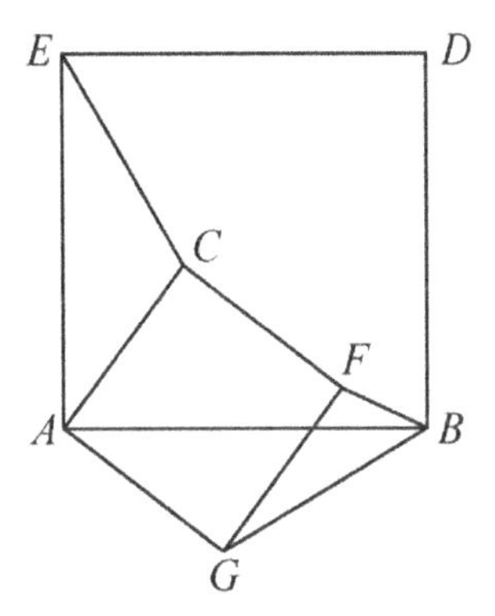

Fig. 15.24

11. As shown in Figure 15.25, $\triangle ABC$ is an isosceles right triangle, $AB = AC$, D is the midpoint of the hypotenuse BC, E and F are the points on the sides of AB and AC respectively, and $DE \perp DF$. If $BE = 12$, $CF = 5$, find: (1) the sum of the areas of $\triangle BDE$ and $\triangle DCF$; (2) the area of $\triangle DEF$.

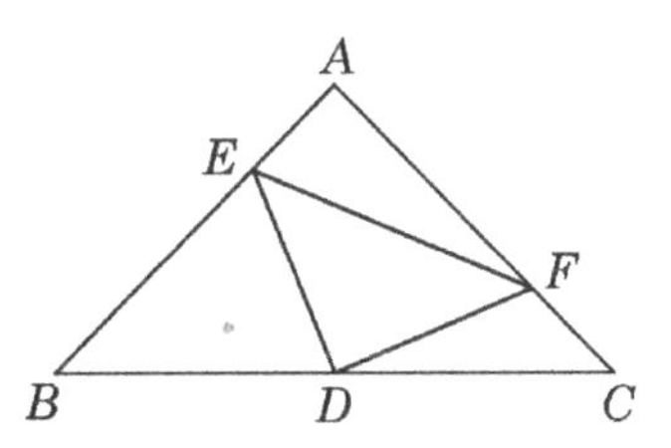

Fig. 15.25

12. As shown in Figure 15.26, $\triangle ABC$ is an isosceles right triangle, $\angle C = 90°$, O is a point inside $\triangle ABC$ and the distance from the point O to each side of $\triangle ABC$ is equal to 1. Rotate $\triangle ABC$ clockwise through 45° about the point O to obtain $A'B'C'$, and the common part of these two triangles is the polygon $KLMNPQ$.
    (1) Prove that $\triangle AKL$, $\triangle BMN$ and $\triangle CPQ$ are all isosceles right triangles.
    (2) Find the area of the common part of $\triangle ABC$ and $\triangle A'B'C'$.

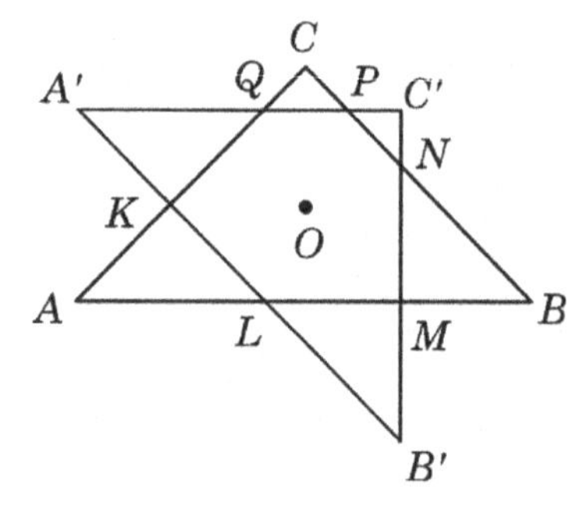

Fig. 15.26

Chapter 16

Basic Properties of Circles

Circle is an important content in junior high school mathematics, and its basic properties have strong applicability. Therefore, it is also an important content in mathematics competition.

Example 1. As shown in Figure 16.1, the acute triangle ABC is inscribed in a circle O, $\angle ABC = 60°$, $\angle BAC = 36°$. Construct the perpendicular OE from O to AB which intersects $\overset{\frown}{AB}$ at the point E. Join EC, find the size of $\angle OEC$.

Solution Since $OE \perp AB$, it can be seen that E is the midpoint of the minor arc $\overset{\frown}{AB}$, then

$$\angle BCE = \angle ECA = \frac{1}{2}\angle BCA.$$

From $\angle ABC = 60°$ and $\angle BAC = 36°$, we know that

$$\angle BCA = 180° - 60° - 36° = 84°.$$

Thus, $\angle BCE = 42°$.

Because $\angle OEC + 90° = \angle ABC + \angle BCE$, we get

$$\angle OEC = 12°.$$

Example 2. As shown in Figure 16.2, in the rectangular coordinate system xOy, $\odot C$ passes through the origin and intersects the coordinate axis at two points A and D respectively. Given that $\angle OBA = 30°$, and the coordinates of the point D are (0, 2), find the coordinates of the points A and C.

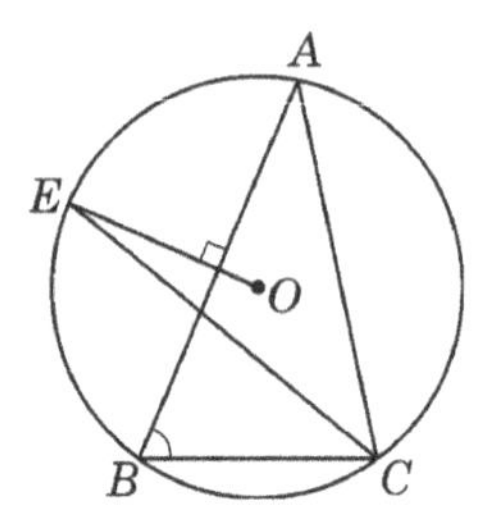

Fig. 16.1

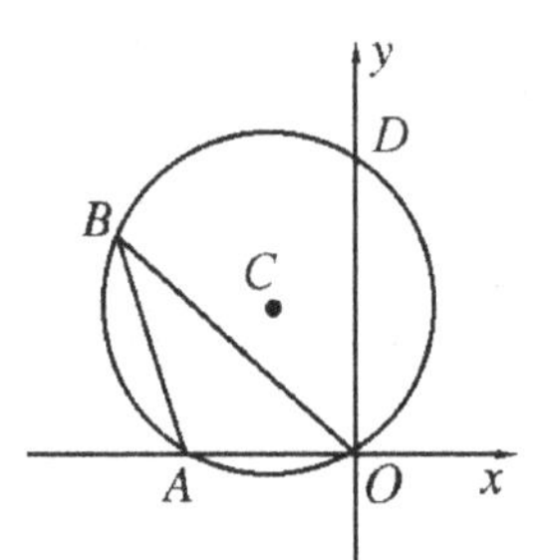

Fig. 16.2

Solution Join AD.

Because $\angle AOD = 90°$, AD is the diameter of $\odot C$. Since C is the center of $\odot C$, we know $AC = CD = OC$.

Because $\angle ABO$ and $\angle ADO$ are opposite angles of the same minor arc, we get

$$\angle ADO = \angle ABO = 30°.$$

Thus, $OA = OD \cdot \tan 30° = 2 \times \frac{\sqrt{3}}{3} = \frac{2\sqrt{3}}{3}$. So the coordinates of the points A are $\left(-\frac{2\sqrt{3}}{3}, 0\right)$.

Because C is the midpoint of AD, the coordinates of point C are (0,1) according to the midpoint coordinate formula. Hence, the coordinates of the points A and C are $\left(-\frac{2\sqrt{3}}{3}, 0\right)$ and $\left(-\frac{\sqrt{3}}{3}, 1\right)$ respectively.

Example 3. As shown in Figure 16.3, in $\triangle ABC$, the semicircle with the diameter of BC intersects the sides AB and AC at the points D and E

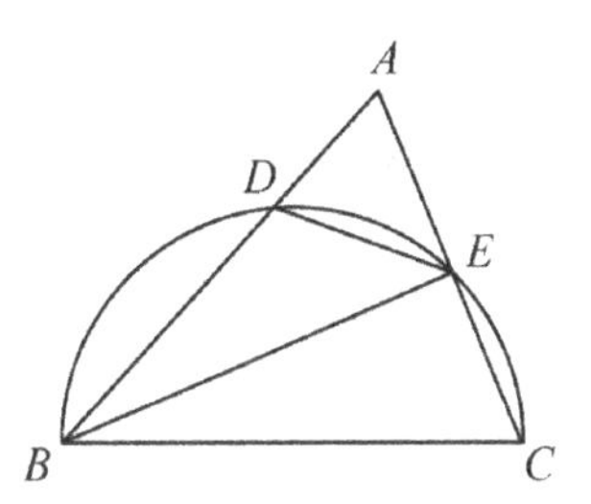

Fig. 16.3

respectively. If $DE = EC = 4$, $BC - BD = \frac{16}{5}$, then the value of $\sqrt{\frac{BD-AD}{BC}}$ is ( ).

(A) $\dfrac{3}{5}$ (B) $\dfrac{2}{3}$ (C) $\dfrac{4}{5}$ (D) $\dfrac{5}{6}$

Solution Since BC is the diameter of the semicircle, we know $\angle BEC = 90°$.

From $DE = EC$, we get $\angle ABE = \angle CBE$, then $AB = BC$, $AE = EC = DE$, so

$$AD = AB - BD = BC - BD = \frac{16}{5}.$$

From $AE = DE = EC = 4$, we obtain $AC = 8$. Since $\angle AED = \angle ABC$, $AE = ED$, $AB = BC$, it can be seen that $\triangle AED \backsim \triangle ABC$. Then we have $\frac{AE}{AB} = \frac{AD}{AC} = \frac{2}{5}$, so $AB = BC = 10$. Therefore, $BD = AB - AD = 10 - \frac{16}{5} = \frac{34}{5}$, then we get $\sqrt{\frac{BD-AD}{BC}} = \frac{3}{5}$. So the answer is A.

Example 4. Make a semicircle with the diameter of AB and the center of O. Point C is a point on the semicircle, and $OC^2 = AC \cdot BC$. Find the size of $\angle CAB$.

Solution From the given condition, there are two possible positions for the point C as shown in Figures 16.4(1), 16.4(2), and $\angle ACB = 90°$.

As shown in Figure 16.4(1), make a line segment CD through C so that $CD \perp AB$ at the point D, then

$$S_{\triangle ABC} = \frac{1}{2} AB \cdot CD = \frac{1}{2} AC \cdot BC.$$

That is, $AB \cdot CD = AC \cdot BC = OC^2$.

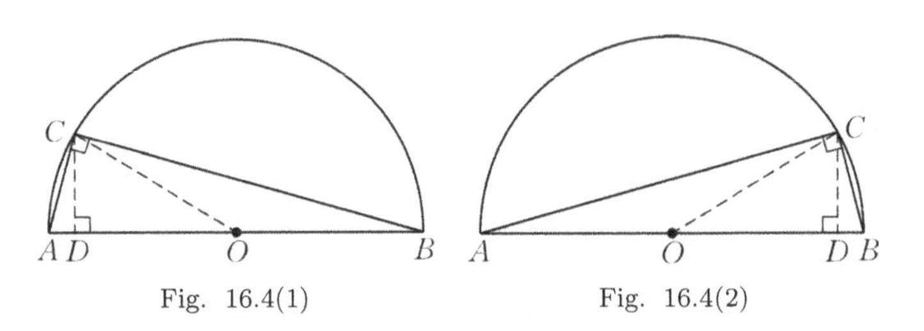

Fig. 16.4(1) Fig. 16.4(2)

And since $AB = 2OC$, we get

$$CD = \frac{1}{2}OC.$$

Thereby it can be deduced that $\angle AOC = 30°$, then

$$\angle CAB = \frac{1}{2}(180° - 30°) = 75°.$$

As shown in Figure 16.4(2), make a line segment CD through C so that $CD \perp AB$ at the point D. According to symmetry, we get

$$\angle CAB = 90° - 75° = 15°.$$

Hence, $\angle CAB = 15°$ or $75°$.

Example 5. Given that AB is a chord of circle O with radius 1, and $AB = a < 1$. Take AB as one side to make an equilateral triangle ABC in the circle O. D is a point on circle O which is different from the point A, and $DB = AB = a$. The extended line of DC intersects the circle O at the point E, find the length of AE.

Solution 1 As shown in Figure 16.5, join OE, OA and OB. Let $\angle BDC = \alpha$. Then

$$\begin{aligned}\angle ECA &= 180° - \angle ACB - \angle BCD \\ &= 180° - 60° - \angle BDC = 120° - \alpha.\end{aligned}$$

Since four points A, B, D and E coexist with a circle, we know that $\angle EAC + \angle CAB + \angle BDC = 180°$, then

$$\angle EAC = 180° - 60° - \alpha = 120° - \alpha.$$

Thus, $\angle ECA = \angle EAC$.

Since $\angle ABO = \frac{1}{2}\angle ABD = \frac{1}{2}(60° + 180° - 2\alpha) = 120° - \alpha$, we get $\angle ABO = \angle ACE$. Then $\triangle ACE \cong \triangle ABO$.

Hence, $AE = OA = 1$.

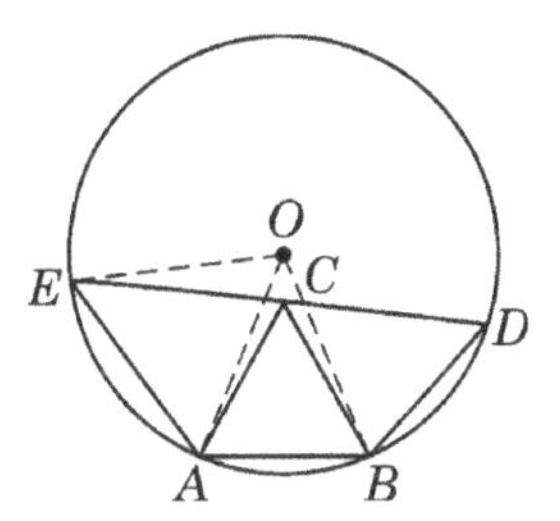

Fig. 16.5

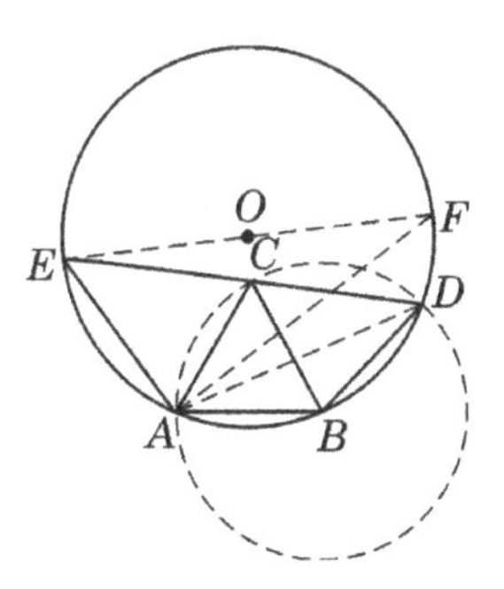

Fig. 16.6

Solution 2 As shown in Figure 16.6, draw a diameter EF and join AF. Make $\odot B$ with the center of B and the radius of AB, and join DA.

Since $AB = BC = BD$, the points A, C and D all lie on $\odot B$. Thus we get $\angle F = \angle EDA = \frac{1}{2}\angle CBA = \frac{1}{2} \times 60^\circ = 30^\circ$. Hence, $AE = EF \times \sin \angle F = 2 \times \sin 30^\circ = 1$.

Example 6. As shown in Figure 16.7, the acute triangle ABC is inscribed in a circle O, $AD \perp BC$, $BE \perp AC$, and the perpendicular feet are D and E respectively.

(1) If $\angle ACB = 60^\circ$, find the size of $\angle ABO$.
(2) Prove that the areas of $\triangle OBD$ and $\triangle OAE$ are equal.

(1) **Solution** Because $\angle ACB$ is the inscribed angle and $\angle AOB$ is the central angle intercepting the same arc AB, we can get

$$\angle AOB = 2\angle ACB = 120^\circ.$$

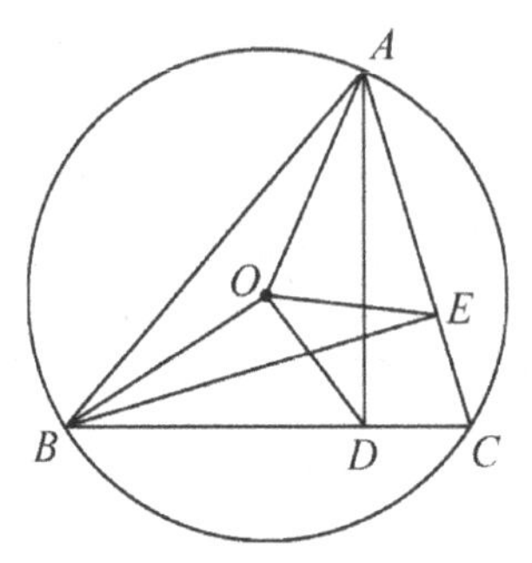

Fig. 16.7

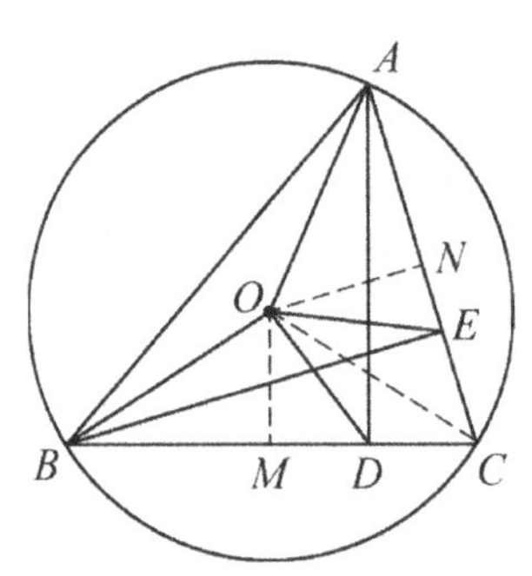

Fig. 16.8

Since $OA = OB$, it can be seen that

$$\angle ABO = \frac{1}{2}(180° - \angle AOB) = 30°.$$

(2) **Proof** As shown in Figure 16.8, draw the line segments OM and ON so that $OM \perp BC$, $ON \perp AC$, and the perpendicular feet are M and N respectively, join OC.

In $\triangle OBC$, since $OB = OC$, it can be known that OM is the bisector of the angle $\angle BOC$, then

$$\angle BOM = \frac{1}{2}\angle BOC. \tag{1}$$

Because the inscribed angle and the central angle intercepting the minor arc BC are $\angle BOC$ and $\angle BAC$ respectively, we get

$$\angle BAC = \frac{1}{2}\angle BOC. \tag{2}$$

From (1) and (2), it can be deduced that $\angle BOM = \angle BAC$.

In Rt$\triangle OMB$ and Rt$\triangle AEB$,

$$\angle BOM = \angle BAE, \quad \angle OMB = \angle AEB = 90^\circ,$$

so Rt$\triangle OMB \backsim$ Rt$\triangle AEB$. Therefore, $\frac{OB}{AB} = \frac{OM}{AE}$. (3)

In the same way, from Rt$\triangle ANO \backsim$ Rt$\triangle ADB$, it can be deduced that

$$\frac{OA}{AB} = \frac{ON}{BD}. \tag{4}$$

From (3), (4) and $OA = OB$, we get $\frac{OM}{AE} = \frac{ON}{BD}$. Therefore, $OM \cdot BD = ON \cdot AE$. So $\frac{1}{2}OM \cdot BD = \frac{1}{2}ON \cdot AE$, i.e. $S_{\triangle OBD} = S_{\triangle OAE}$. That is, the areas of $\triangle OBD$ and $\triangle OAE$ are equal.

Example 7. As shown in Figure 16.9, AB, AC and AD are three chords in a circle, point E is on line segment AD, and $AB = AC = AE$. Please explain the reasons for each of the following expressions:

(1) $\angle CAD = 2\angle DBE$;
(2) $AD^2 - AB^2 = BD \cdot DC$.

Solution (1) As shown in Figure 16.10, join BC.

Since $AB = AC = AE$, we obtain that

$$\angle 5 = \angle 2, \quad \angle 2 + \angle 3 = \angle 6.$$

From $\angle 4 + \angle 5 = \angle 6 = \angle 2 + \angle 3$, it can be known that $\angle 4 = \angle 3$.

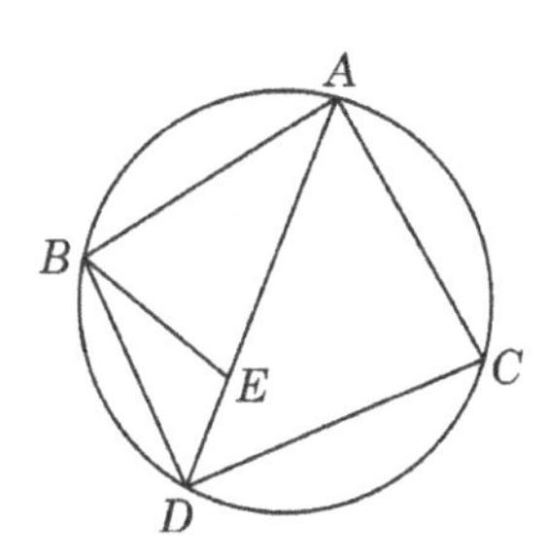

Fig. 16.9

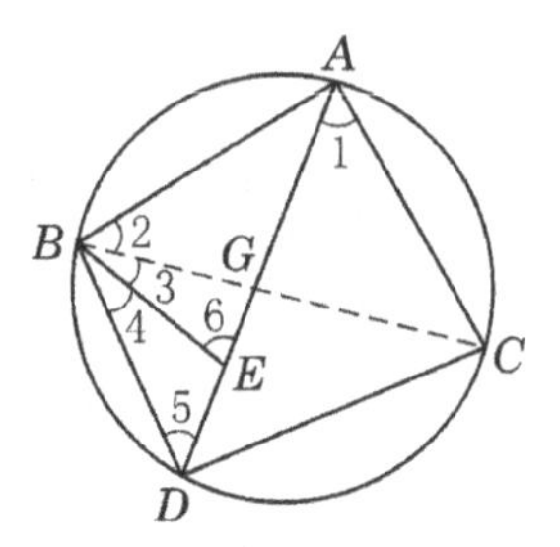

Fig. 16.10

Since $\angle 1 = \angle 4 + \angle 3$, we get $\angle 1 = 2\angle 4$, that is, $\angle CAD = 2\angle DBE$.

(2) Let BC and AD intersect at the point G.

Because $\angle 2 = \angle 5$, $\angle BAG = \angle DAB$, we get $\triangle BAG \backsim \triangle DAB$, then $AB^2 = AG \cdot AD$. Therefore, $AD^2 - AB^2 = AD^2 - AG \cdot AD = AD \cdot DG$. Since $\angle 5 = \angle ADC$, $\angle DBG = \angle 1$, we get $\triangle BDG \backsim \triangle ADC$, then $\frac{DB}{AD} = \frac{DG}{DC}$, that is, $AD \cdot DG = BD \cdot DC$. Hence, $AD^2 - AB^2 = BD \cdot DC$.

Example 8. Given that the points A, B, C and D are on the circle O in turn, $\overset{\frown}{AB} = \overset{\frown}{BD}$, $BM \perp AC$ and the perpendicular feet is M, prove that $AM = DC + CM$.

Solution As shown in Figure 16.11, extend DC so that $CN = MC$, then

$$\angle BCN = \angle CBD + \angle BDC = \angle BAD.$$

Since $\angle ACB = \angle ADB$ (the inscribed angles subtended by the same arc are equal), and $\overset{\frown}{AB} = \overset{\frown}{BD}$, we can get $\angle ACB = \angle BAD$, $AB = BD$. Then we know $\angle BCN = \angle BCM$. Therefore, $\triangle BCN \cong \triangle BCM$, then

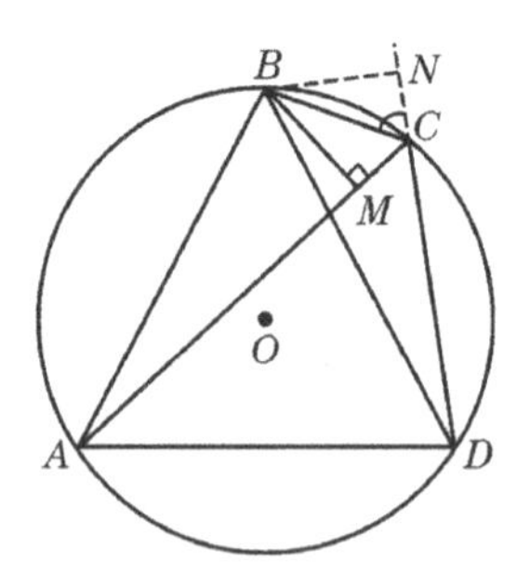

Fig. 16.11

$\angle BNC = \angle BMC = 90°$. Since $\angle BAM = \angle BDN$, we obtain $\triangle BAM \cong \triangle BDN$. Hence, $AM = DN = CD + CM$.

Remark This example is the famous Archimedes' Theorem of the Broken Chord.

Example 9. As shown in Figure 16.12, the radius of sector OMN is 1, and the central angle is $90°$. Point B is a moving point on $\overset{\frown}{MN}$, $BA \perp OM$ at the point A and $BC \perp ON$ at the point C. Points D, E, F and G are the midpoints of the line segments OA, AB, BC and CO respectively. GF and CE intersect at the point P, DE and AG intersect at the point Q.

(1) Prove that the quadrilateral $EPGQ$ is a parallelogram.
(2) Find the length of OA if the quadrilateral $EPGQ$ is a rectangle.
(3) Join PQ, and try to explain that $3PQ^2 + OA^2$ is a fixed value.

Solution (1) As shown in Figure 16.13, join OB.

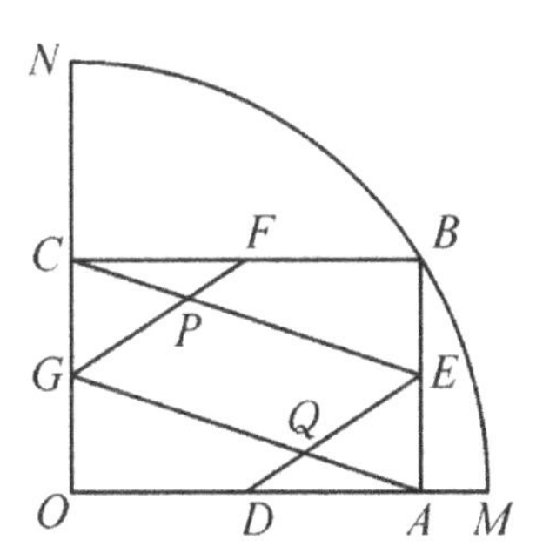

Fig. 16.12

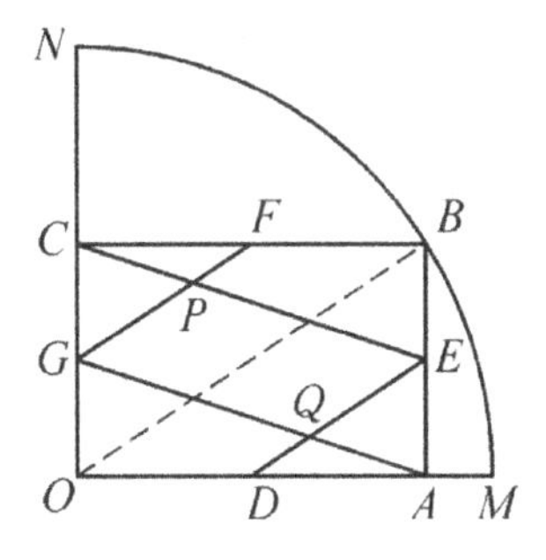

Fig. 16.13

Since $\angle AOC = 90°$, $BA \perp OM$, $BC \perp ON$, we know that the quadrilateral $OABC$ is a rectangle, then $AB \parallel OC$, $AB = OC$.

Since E and G are the midpoints of AB and CO respectively, it can be seen that $AE \parallel GC$, $AE = GC$. So the quadrilateral $AECG$ is a parallelogram, then $CE \parallel AG$.

Since D, E, F and G are the midpoints of OA, AB, BC and CO respectively, we obtain that $GF \parallel OB$, $DE \parallel OB$, then $PG \parallel EQ$. Hence, the quadrilateral $EPGQ$ is a parallelogram.

(2) As shown in Figure 16.14, when $\angle CED = 90°$. $\square EPGQ$ is a rectangle. At this time, $\angle AED + \angle CEB = 90°$. Since $\angle DAE = \angle EBC = 90°$, we have $\angle AED = \angle BCE$. So $\triangle AED \backsim \triangle BCE$, and then $\frac{AD}{BE} = \frac{AE}{BC}$. Let $OA = x$, $AB = y$. Then $\frac{x}{2} : \frac{y}{2} = \frac{y}{2} : x$, that is, $y^2 = 2x^2$. From $OA^2 + AB^2 = OB^2$, i.e. $x^2 + y^2 = 1^2$, we get $x^2 + 2x^2 = 1$, then $x = \frac{\sqrt{3}}{3}$. Hence, when the length of OA is $\frac{\sqrt{3}}{3}$, the quadrilateral $EPGQ$ is a rectangle.

(3) As shown in Figure 16.15, join GE which intersect PQ at the point O', then $O'P = O'Q$, $O'G = O'E$. Make a parallel line of OC through the point P that intersect BC and GE at the points B' and A'.

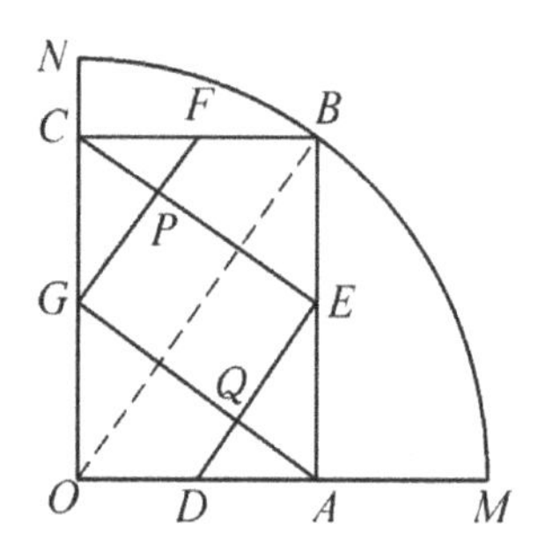

Fig. 16.14

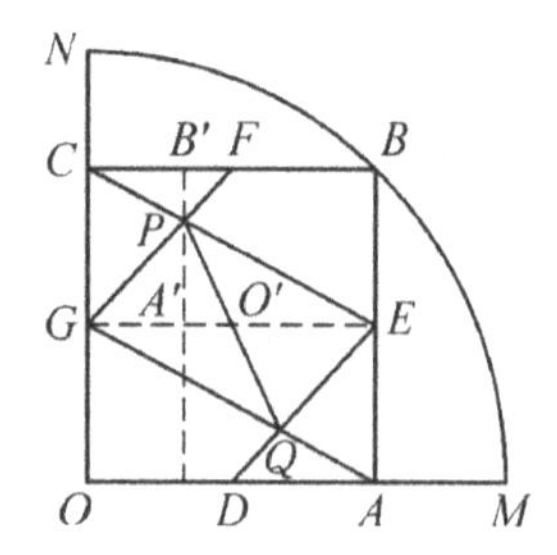

Fig. 16.15

From $\triangle PCF \backsim \triangle PEG$, it can be known that

$$\frac{PG}{PF} = \frac{PE}{PC} = \frac{GE}{FC} = \frac{2}{1}.$$

Then we have that $PA' = \frac{2}{3}A'B' = \frac{1}{3}AB$, $GA' = \frac{1}{3}GE = \frac{1}{3}OA$. So

$$A'O' = \frac{1}{2}GE - GA' = \frac{1}{6}OA.$$

In Rt$\triangle PA'O'$, $PO'^2 = PA'^2 + A'O'^2$, that is,

$$\frac{PQ^2}{4} = \frac{AB^2}{9} + \frac{OA^2}{36}.$$

And since $AB^2 + OA^2 = 1$, we get

$$3PQ^2 = AB^2 + \frac{1}{3}.$$

Hence, $OA^2 + 3PQ^2 = OA^2 + \left(AB^2 + \frac{1}{3}\right) = \frac{4}{3}$, which is a fixed value.

Exercises

1. Given that in $\triangle ABC$, $AB = AC = 4\sqrt{3}$, the height $AD = 4$, then the radius of circumscribed circle of $\triangle ABC$ is ( ).
2. As shown in Figure 16.16, the side AB of the rectangular $ABCD$ passes through the center of a circle O. Points E and F are the intersection points of the sides AB, DC and the circle O respectively. If $AE = 3$, $AD = 4$, $DF = 5$, find the diameter of the circle O.

   (A) 3 (B) 4 (C) 5 (D) 6

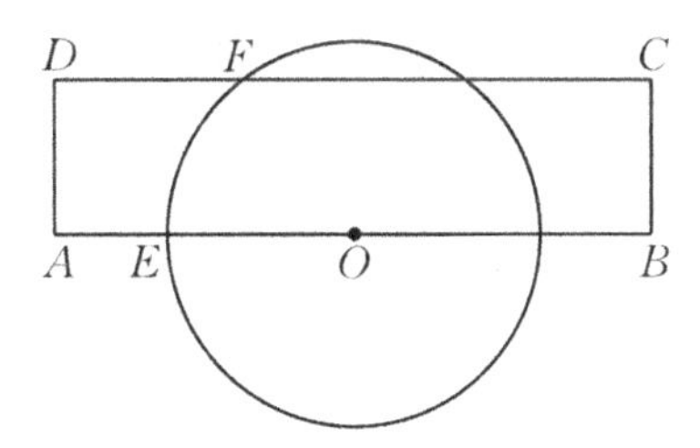

Fig. 16.16

3. The diameter of a circle O is $AB = 20$, the chord CD intersects AB at the point G, $AG > BG$, $CD = 16$, $AE \perp CD$ at the point E, $BF \perp CD$ at the point F. Find the value of $AE - BF$.

4. Fold $\overset{\frown}{BC}$ along the chord BC and $\overset{\frown}{BC}$ intersects AB at the point D. If $AD = 4$, $DB = 5$, then the length of BC is ( ).

   (A) $3\sqrt{7}$ (B) 8 (C) $\sqrt{65}$ (D) $2\sqrt{15}$

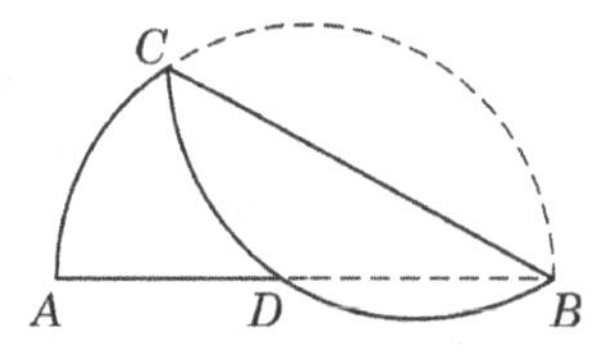

Fig. 16.17

5. As shown in Figure 16.18, AB is the diameter of the circle, CD is the chord parallel to AB, AC and BD intersect at the point E, and $\angle AED = \alpha$. Find the ratio of the area of $\triangle CDE$ to the area of $\triangle ABE$.

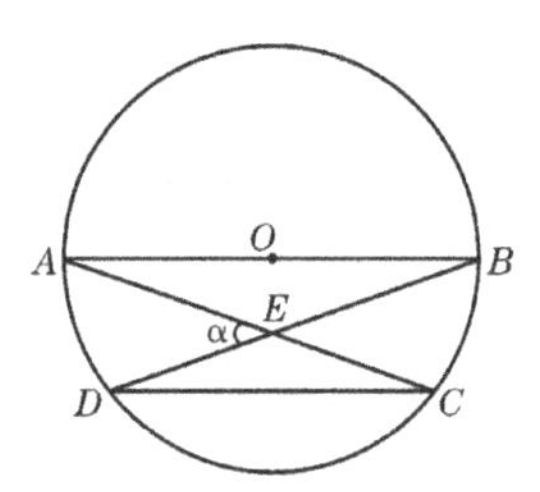

Fig. 16.18

6. As shown in Figure 16.19, given that $\triangle ABC$ is inscribed in circle O with the diameter of d. Let $BC = a$, $AC = b$, find the height CD of $\triangle ABC$.

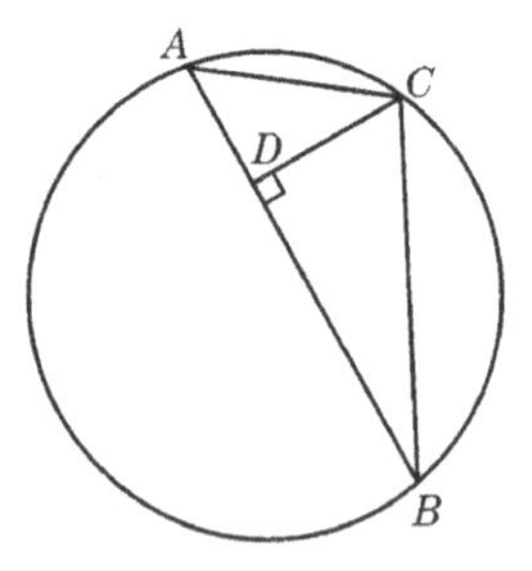

Fig. 16.19

7. Given that the points C and D are on the semicircle circumference with the diameter of AB, AD is the angular bisector of $\angle BAC$, $AB = 20$, $AD = 4\sqrt{15}$, then the length of AC is ______.
8. As shown in Figure 16.20, $\triangle ABC$ is inscribed in a circle O, and $AB = AC$, the diameter AD intersects BC at the point E, F is the midpoint of OE. If $BD \parallel CF$, $BC = 2\sqrt{5}$, then the length of the line segment CD is ______.

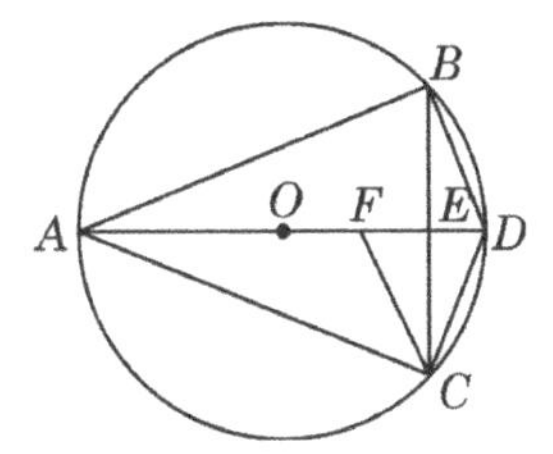

Fig. 16.20

9. As shown in Figure 16.21, in Rt$\triangle ACB$, $AC = BC$, the center of $\widehat{DEF}$ is A. If the areas of two shadow parts in the figure are equal, find the value of $AD : DB$.

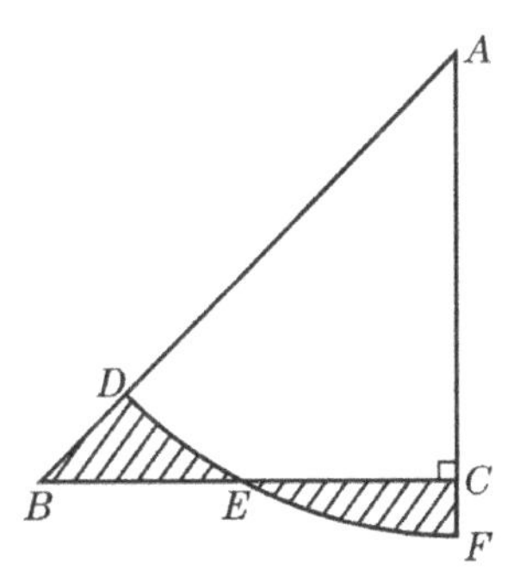

Fig. 16.21

10. As shown in Figure 16.22, the side lengths of rectangular $ABCD$ are $AB = 1$ and $BC = \sqrt{3}$, $\triangle ADE$ is an equilateral triangle. If the circle with radius R can cover the pentagon $ABCDE$ (that is, every vertex of the pentagon $ABCDE$ is in the circle or on the circle), find the minimum value of R.

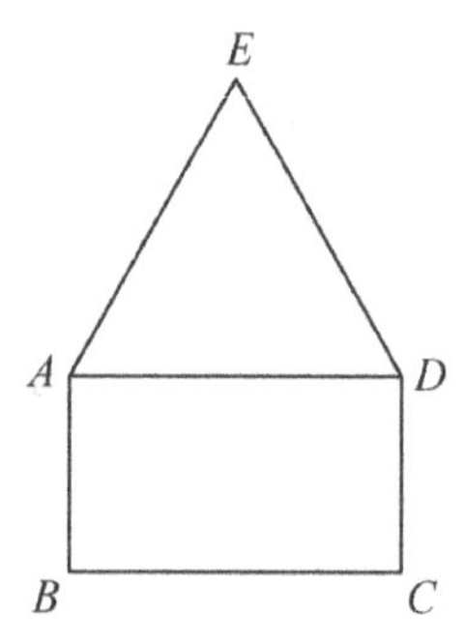

Fig. 16.22

11. Divide the line segment AD into three equal parts in turn at the points B and C. P is any point on the circumference with the diameter of BC. Find the value of $\tan\angle APB \cdot \tan\angle CPD$.
12. As shown in Figure 16.23, let the center of the semicircle with radius 1 be O, the diameter be AB, and C, D be the two points on the semicircle. If the degree of $\overset{\frown}{AC}$ is 96°, the degree of $\overset{\frown}{BD}$ is 36°, and the moving point P is on the diameter AB, find the minimum value of $CP + PD$.

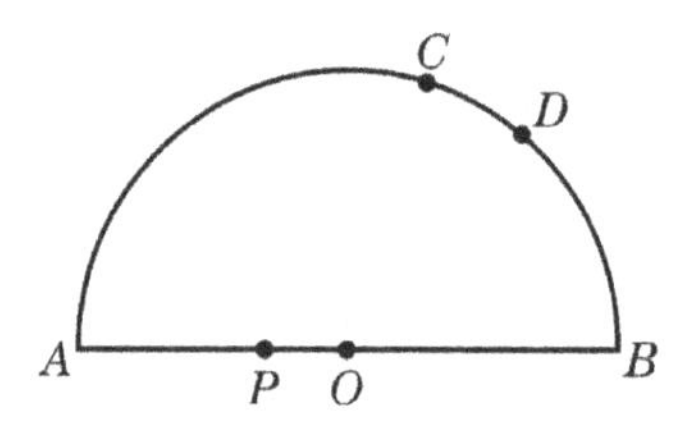

Fig. 16.23

Chapter 17

Positional Relation Between Line and Circle

There are three kinds of positional relations between a line and a circle: the line intersects the circle, the line is tangent to the circle, and the line separates from the circle.

Consider a line l and a circle O. If the radius of the circle is R and the distance between the center O and the line l is d, then we have the following conclusions.

$$\text{The line } l \text{ intersects the circle } O \Leftrightarrow d < R.$$
$$\text{The line } l \text{ is tangent to the circle } O \Leftrightarrow d = R.$$
$$\text{The line } l \text{ separates from the circle } O \Leftrightarrow d > R.$$

The line that intersects circle O in two points is called the secant of circle, and the line that is tangent to circle O is called the tangent of circle.

If the line l is tangent to the circle O and the point of tangency is T, then $l \perp OT$.

There are only two tangents from an external point P, and the tangent segments are equal in length. OP bisects the angle between the two tangents. The angle of osculation is equal to the inscribed angle intercepting the same arc.

Example 1. As shown in Figure 17.1, E and F are two points on the circumcircle of the square $ABCD$, and $\angle EBF = 45°$, AD and the extended line of BF intersect at the point P. Prove the following conclusions:

(1) $EC \parallel BP$;
(2) $BP \cdot BE = \sqrt{2}AB^2$.

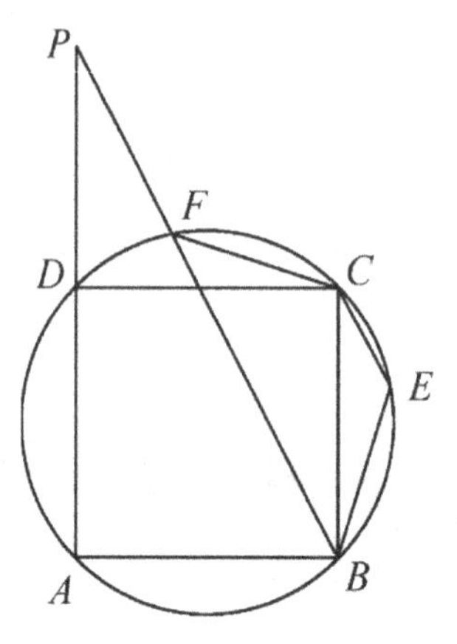

Fig. 17.1

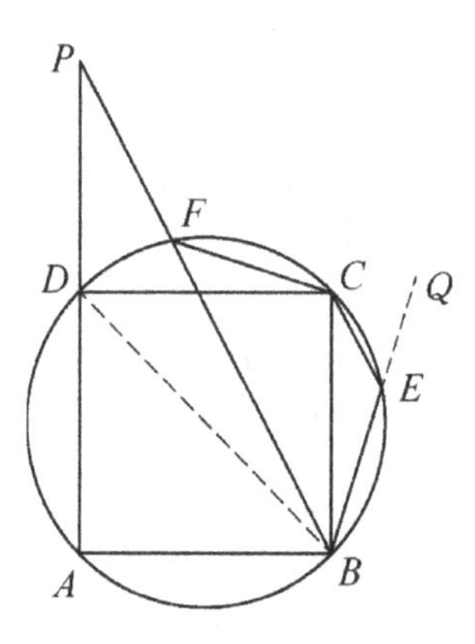

Fig. 17.2

Proof. (1) As shown in Figure 17.2, extend BE to the point Q.

Because the quadrilateral $BFCE$ is a cyclic quadrilateral, we get $\angle QEC = \angle BFC$.

Because the quadrilateral $ABCD$ is a square, we get $\angle BFC = 45°$, then $\angle QEC = 45°$.

Since $\angle EBF = 45°$, it can be known that $\angle QEC = \angle EBF$. Then $EC \parallel BP$.

(2) Join BD.

From (1), we know $EC \parallel BP$, so $\angle BCE = \angle CBF$.

Because the quadrilateral $ABCD$ is a square, we get $BC \parallel AP$, then $\angle CBF = \angle BPD$.

Thus, $\angle BCE = \angle BPD$. Since $\angle CBD = 45°$, $\angle EBF = 45°$, we get $\angle CBE = \angle PBD$. Hence, $\triangle CBE \backsim \triangle PBD$, then we obtain $\frac{BE}{BD} = \frac{BC}{BP}$.

That is, $BP \cdot BE = BD \cdot BC = \sqrt{2}AB^2$. □

Example 2. As shown in Figure 17.3, PA and PB are tangent to the circle O and the points of tangency are A and B respectively, PC satisfies

$$AB \cdot PB - AC \cdot PC = AB \cdot PC - AC \cdot PB,$$

and $AP \perp PC$, $\angle PAB = 2\angle BPC$, find the size of $\angle ACB$.

Solution From $AB \cdot PB - AC \cdot PC = AB \cdot PC - AC \cdot PB$, we get

$$(AB + AC)(PB - PC) = 0,$$

i.e. $PB = PC$.

Since PA and PB are two tangents of the circle O, we have $PA = PB$. Thus, $PA = PB = PC$. Thereby it can be deduced that the three points A, B and C are on the circle with the center of P and the radius of PA.

In the circle P, let $\angle ACB = \alpha$, then $\angle APB = 2\alpha$, $\angle BPC = 90° - 2\alpha$.

So $\angle PAB = 2\angle BPC = 180° - 4\alpha$.

And since $\angle PAB = \angle PBA = \frac{1}{2}(180° - \angle APB) = 90° - \alpha$, we get

$$180° - 4\alpha = 90° - \alpha.$$

Solving it gives $\alpha = 30°$, that is, $\angle ACB = 30°$.

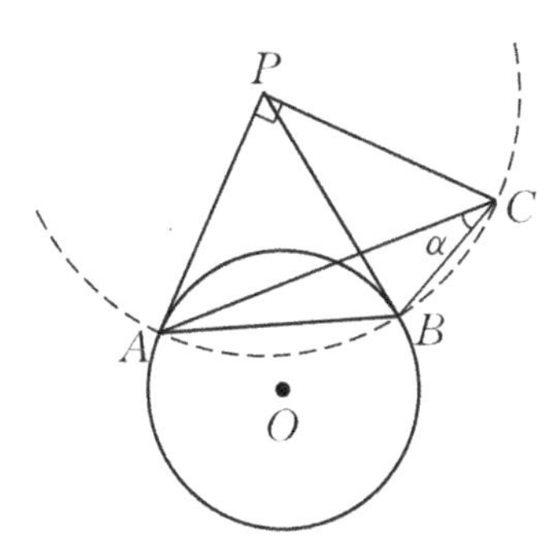

Fig. 17.3

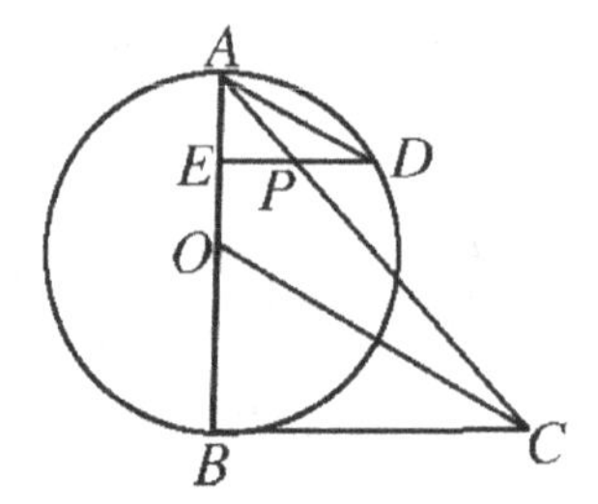

Fig. 17.4

Remark To deal with the geometric problem of tangent (secant) of circle, the basic idea is to determine the quantitative relation of line segments or angles by the positional relation; on the contrary, determine the positional relation of a line and a circle by the quantitative relation. In the process of solving such a problem, the figure is often transformed into a right triangle and similar triangles, and the result of the problem is obtained by using Pythagorean theorem and some basic properties of similar triangles.

Example 3. As shown in Figure 17.4, AB is the diameter of $\odot O$, BC is the tangent of $\odot O$, and OC is parallel to the chord AD. Draw a line segment DE so that $DE \perp AB$ at the point E, and join AC that intersects DE at the point P. Whether EP and PD are equal or not? Prove your conclusion.

Solution Because AB is the diameter of $\odot O$ and BC is the tangent, we have $AB \perp BC$.

From $\text{Rt}\triangle AEP \backsim \text{Rt}\triangle ABC$,

$$\frac{EP}{BC} = \frac{AE}{AB}. \tag{1}$$

Since $AD \parallel OC$, we get $\angle DAE = \angle COB$.

Thus, $\text{Rt}\triangle AED \backsim \text{Rt}\triangle OBC$, then we obtain

$$\frac{ED}{BC} = \frac{AE}{OB} = \frac{AE}{\frac{1}{2}AB} = \frac{2AE}{AB}. \tag{2}$$

From (1) and (2), it can be reduced that $ED = 2EP$. Hence, $DP = PE$.

Example 4. As shown in Figure 17.5, the hypotenuse AB of the right triangle ABC is tangent to the circle O at the point D, the leg AC intersects the circle O at the point E, and $DE \parallel BC$. If $AE = 2\sqrt{2}$, $AC = 3\sqrt{2}$, $BC = 6$, find the radius of $\odot O$.

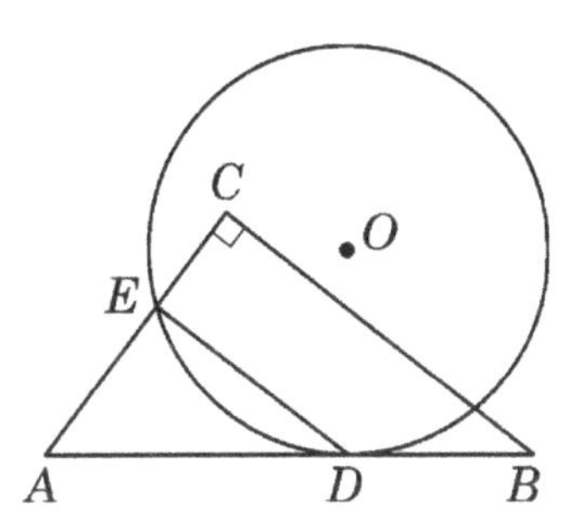

Fig. 17.5

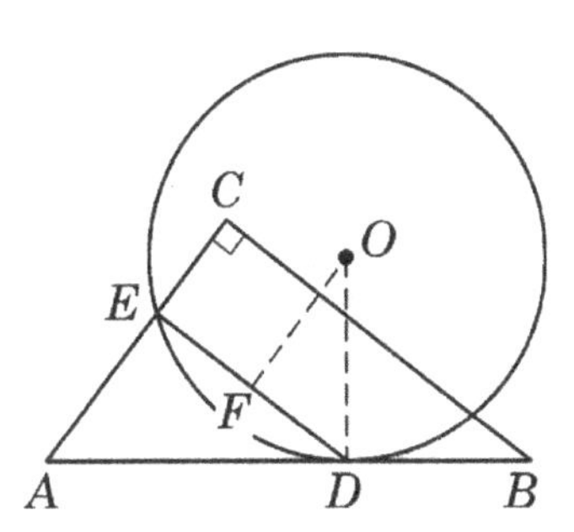

Fig. 17.6

Solution From the given condition,

$$AB = \sqrt{AC^2 + BC^2} = \sqrt{(3\sqrt{2})^2 + 6^2} = 3\sqrt{6}.$$

Since $DE \parallel BC$, we know that $\triangle AED \backsim \triangle ACB$. Then $\frac{ED}{CB} = \frac{AE}{AC}$. So $ED = 6 \times \frac{2\sqrt{2}}{3\sqrt{2}} = 4$.

As shown in Figure 17.6, draw a line segment OF so that $OF \perp DE$ at the point F, and join OD, then $FD = \frac{1}{2}ED = 2$.

Because AB is the tangent of $\odot O$, we get that $OD \perp AB$, i.e. $\angle ODA = 90°$. Since $DE \parallel BC$, we have $\angle ADE = \angle ABC$. From $\angle ODA = \angle ODF + \angle ADE$, it can be seen that $\angle ODF + \angle ABC = 90°$. And from $\angle ABC + \angle BAC = 90°$, we get $\angle BAC = \angle ODF$.

Because $\angle ACB = \angle DFO = 90°$ and $\angle BAC = \angle ODF$, it can be deduced that

$$\triangle OFD \backsim \triangle BCA.$$

Then we get $\frac{OD}{BA} = \frac{FD}{CA}$. Hence, the radius of $\odot O$ is $OD = 3\sqrt{6} \times \frac{2}{3\sqrt{2}} = 2\sqrt{3}$.

Example 5. As shown in Figure 17.7, in $\triangle ABC$, $AB = AC = \sqrt{5}$, $BC = 2$. The circle with the diameter of AB intersects the sides AC and BC at the points D and E respectively. Find the area of $\triangle CDE$.

Solution As shown in Figure 17.8, join AE, BD and DE, draw a line segment DF so that $DF \perp EC$ at the point F.

Because AB is the diameter of $\odot O$, we have

$$\angle ADB = \angle AEB = 90^\circ.$$

And since $AB = AC$, it can be known that

$$CE = \frac{1}{2}BC = 1, AE = \sqrt{AC^2 - CE^2} = 2.$$

Because $\frac{1}{2}BC \cdot AE = \frac{1}{2}AC \cdot BD$, we get $BD = \frac{4}{5}\sqrt{5}$.

In $\triangle ABD$, $AD = \sqrt{AB^2 - BD^2} = \frac{3}{5}\sqrt{5}$, then

$$CD = AC - AD = \frac{2}{5}\sqrt{5}.$$

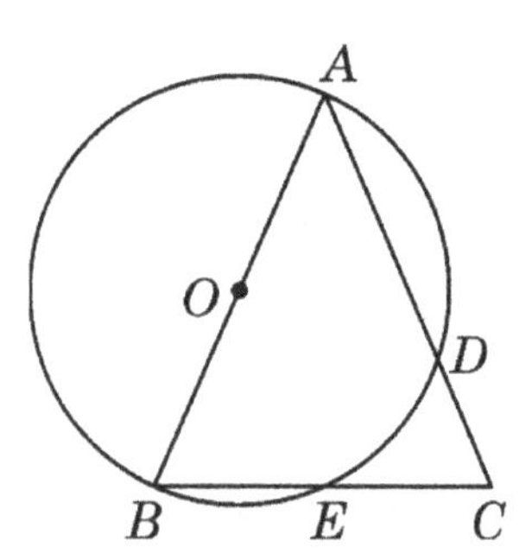

Fig. 17.7

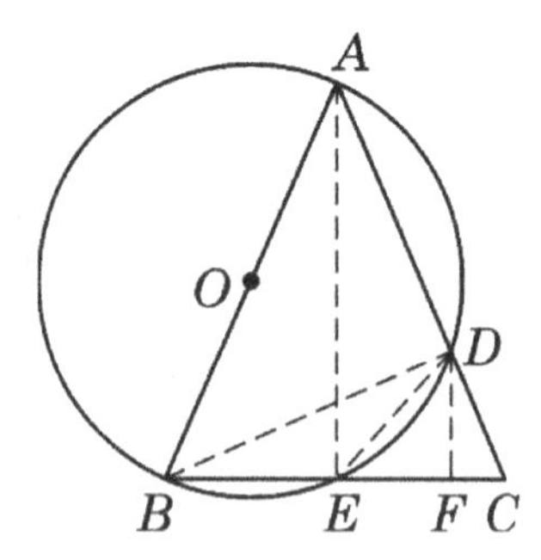

Fig. 17.8

Since $\triangle CDF \backsim \triangle CAE$, we get that $\frac{CD}{CA} = \frac{DF}{AE}$. Then $DF = \frac{4}{5}$ is obtained. Hence, the area of $\triangle CDE$ is $\frac{1}{2}CE \cdot DF = \frac{2}{5}$.

Remark In the above solution, after getting $CD = \frac{2\sqrt{5}}{5}$, we can obtain the result directly by using the area relation, that is,

$$S_{\triangle CDE} = \frac{1}{2}S_{\triangle BDC} = \frac{1}{2} \times \frac{1}{2} \times BD \times CD = \frac{1}{2} \times \frac{1}{2} \times \frac{4}{5}\sqrt{5} \times \frac{2}{5}\sqrt{5} = \frac{2}{5}.$$

Example 6. As shown in Figure 17.9, in $\odot O$, the chord CD is perpendicular to the diameter AB. M is the midpoint of OC, the extended line of AM intersects $\odot O$ at the point E, DE and BC intersect at the point N. Prove that $BN = CN$.

Proof. As shown in Figure 17.10, join AC and BD. Because the chord CD is perpendicular to the diameter AB, we know that $BC = BD$. Then

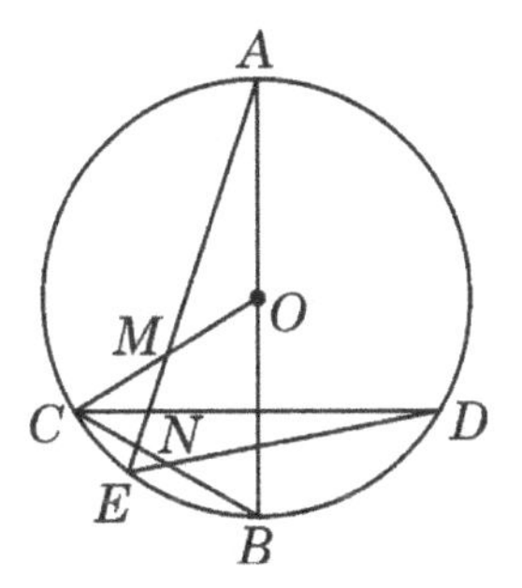

Fig. 17.9

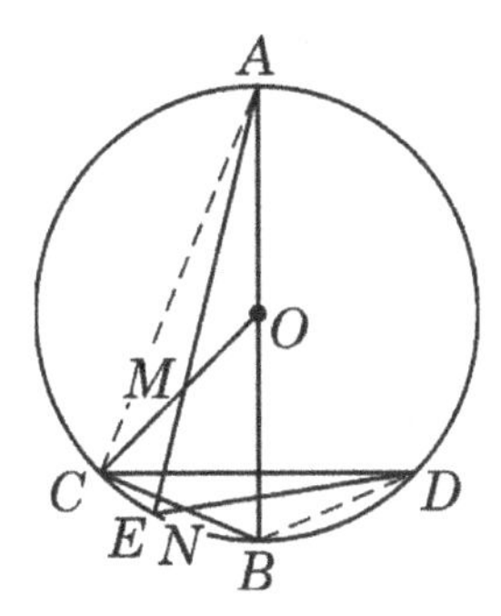

Fig. 17.10

$\angle BCD = \angle BDC$. Since $OA = OC$, $\angle OCA = \angle OAC$ is obtained. As $\angle BDC = \angle OAC$, then $\angle BCD = \angle OCA$. □

Therefore, $\triangle BCD \backsim \triangle OCA$, then we get

$$\frac{CB}{CO} = \frac{CD}{CA}.$$

In $\triangle CDN$ and $\triangle CAM$, because

$$\angle DCN = \angle ACM,$$

$$\angle CDN = \angle CAM,$$

it can be obtained that $\triangle CDN \backsim \triangle CAM$.

Thus we have $\frac{CN}{CM} = \frac{CD}{CA} = \frac{CB}{CO} = \frac{CB}{2CM}$, then $CN = \frac{1}{2}CB$. Hence, $BN = CN$.

Example 7. As shown in Figure 17.11, AB is the diameter of the semicircle, $AC \perp AB$, $AC = AB$. For any point D on the semicircle, construct the perpendicular DE to CD, which intersects AB at the point E. And construct the perpendicular BF to AB, which intersects the extended line of AD at the point F.

(1) Let $\overset{\frown}{AD}$ be an arc of $x°$. In order to make the point E on the extended line of BA, find the value range of x.
(2) No matter where the point D is on the semicircle, there must be two line segments equal to each other besides $AB = AC$. State these two equal line segments and prove your conclusion.

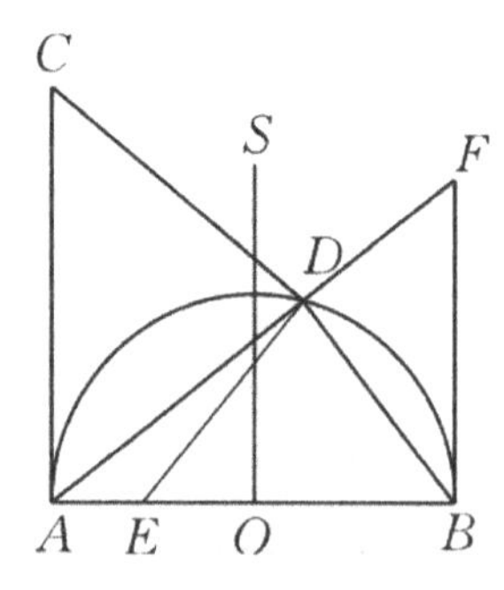

Fig. 17.11

Solution

(1) When the point E moves closer to the point A from the right, $\triangle ADB$ will become an isosceles right triangle, that is, the point D is the intersection of OS and the circle O, where $OS \perp AB$. So when the point E moves from the right to the point A, $\overset{\frown}{AD}$ is an arc of $90°$, that is, $x = 90$.

When the point E leaves the point A on the extended line of BA, the farther away it leaves from the point A, the closer point D is to the point A, and the closer x is to 0. Hence, the value range of x is $0 < x < 90$.

(2) From the given condition, $\angle CDE = 90°$, $\angle CAB = \angle EBF = 90°$, $\angle ADB = 90°$.

Since AC is the tangent of the circle, we know that $\angle CAD = \angle ABD$. And because $\angle DEB = 180° - \angle AED = 180° - (360° - 180° - \angle C) = \angle C$, we get $\triangle ACD \backsim \triangle BED$, then

$$\frac{AD}{BD} = \frac{AC}{BE}. \tag{1}$$

Because $\angle ABD = \angle BFD$, we get that $\triangle ABD \backsim \triangle AFB$, then

$$\frac{AD}{BD} = \frac{AB}{BF}. \tag{2}$$

Since $AB = AC$, from (1) and (2), $BE = BF$ is obtained.

Example 8. There are four points inside the circle, and the distance between any two points is greater than the radius of the circle. Prove that we can always find two diameters that are perpendicular to each other, which divide the circle into four parts to make the interior of each part have and only have one point.

Proof. Let these four points be A, B, C and D in the order of a quadrilateral. Since the distance between any two points is greater than the radius, the center of the circle is not any of these four points, and any two points are not on the same radius.

As shown in Figure 17.12, join AB. □

In $\triangle AOB$, since AO and BO are both less than the radius, and AB is greater than the radius, we know that $\angle AOB$ is the greatest angle in $\triangle AOB$, then $\angle AOB > 60°$.

Similarly, $\angle BOC > 60°$, $\angle COD > 60°$ and $\angle DOA > 60°$.

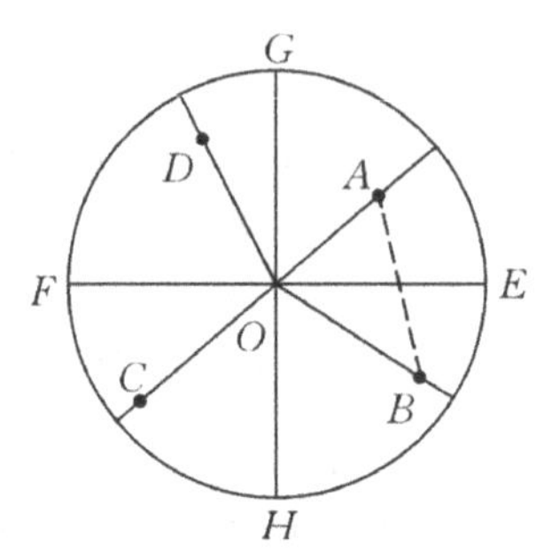

Fig. 17.12

(i) If one of these four angles is greater than $90°$, might as well let it be $\angle COD$. Draw the bisector OE of $\angle AOB$ and extend it to be the diameter EF in reverse. Construct the other diameter GH which satisfies $GH \perp EF$. At this time, GH and EF divide the circle into four parts, the point A is inside GOE and the point B is inside EOH.

Consider the positions of C and D. Because $\angle AOB > 60°$, then $\angle EOB > 30°$, it can be deduced that $\angle BOH < 60°$. And since $\angle BOC > 60°$, the point C is not inside EOH.

Similarly, the point D is not inside EOG. However, the central angle of each part is just $90°$ and $\angle COD > 90°$, so the two points C and D cannot be in the same part. Therefore, the point C is inside HOF and the point D is inside FOG.

(ii) If these four angles are not more than $90°$, then they are all $90°$. Therefore, four points A, B, C and D are on two mutually perpendicular diameters respectively. Construct two diameters that bisect the angle between these two diameters, which are what we desire.

To sum up, the proposition holds.

Exercises

1. Given that there is only one common point between the line segment PQ and the circle O, then which of the following situations can the two endpoints P and Q of this segment only be? ( )

   (A) One point is inside the circle O and the other point is outside the circle O, or PQ is the tangent of the circle while one of P and Q is the point of tangency

(B) One of the two points P and Q must be outside the circle O
(C) Case (A) or case (B)
(D) At most one point is inside the circle

2. As shown in Figure 17.13, perpendicular ruler ABC is tangent to the $\odot O$ at the point D and contacts $\odot O$ at the point A. It's measured that $AB = a$, $BD = b$, find the radius of $\odot O$.

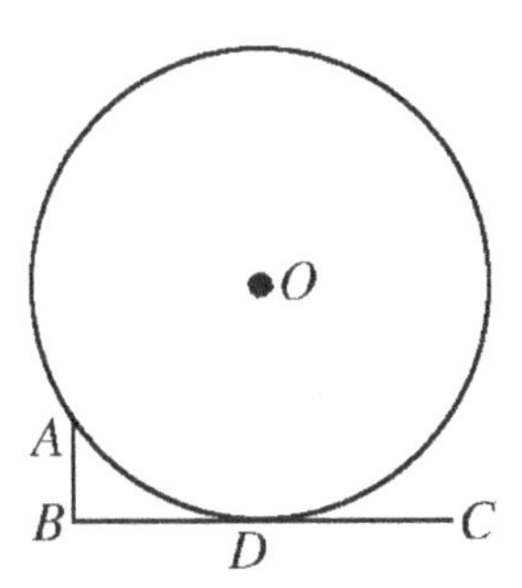

Fig. 17.13

3. As shown in Figure 17.14, in the two concentric circles with the center O, the chord AB of the big circle is tangent to the small circle at the point C, and $AB = 8\,\text{cm}$. Find the area of the shaded part in the figure.

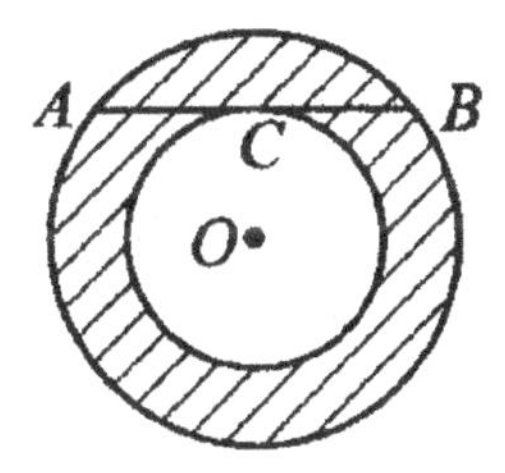

Fig. 17.14

4. As shown in Figure 17.15, the chords AC and BD intersect at the point E, $\overset{\frown}{AB} = \overset{\frown}{BC} = \overset{\frown}{CD}$, $\angle BEC = 130°$, then the size of $\angle ACD$ is ( ).

(A) 115° (B) 130° (C) 48° (D) 105°

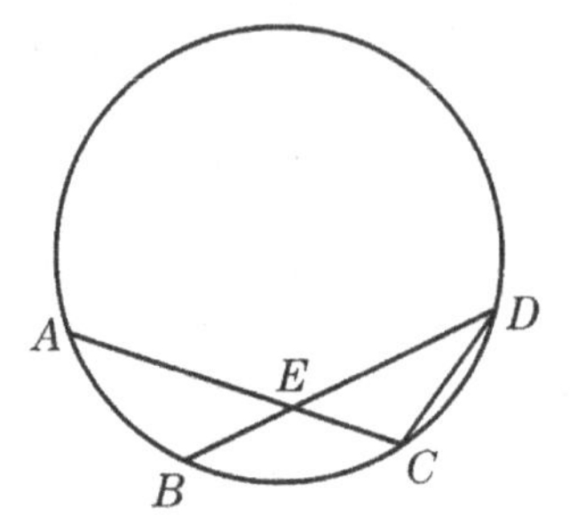

Fig. 17.15

5. Definition: the minimum distance between the point A and any point on $\odot O$ is called the distance between the point A and $\odot O$. As shown in Figure 17.16, there is a rectangle $ABCD$, where $AB = 14\,\text{cm}$, $BC = 12\,\text{cm}$. The sides AB, BC and CD of the rectangle are tangent to $\odot K$ at the points E, F and G respectively, then the distance between the point A and $\odot K$ is ( ).

   (A) 4cm (B) 8cm (C) 10cm (D) 12cm

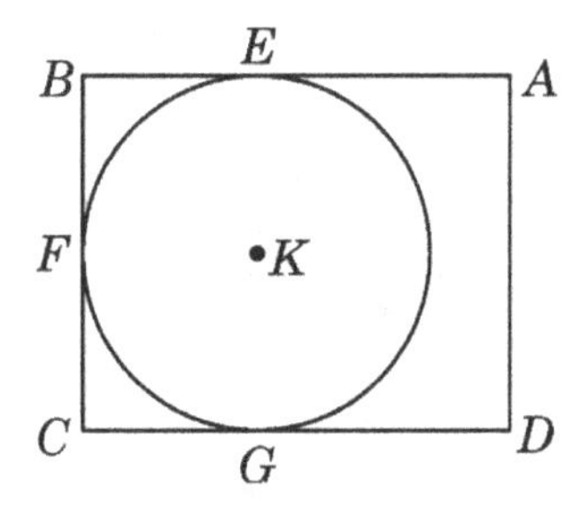

Fig. 17.16

6. As shown in Figure 17.17, the lines AB and AC are the tangent to $\odot O$ at the points B and C respectively, P is a point on the circle, the distances from P to AB and AC are $4\,\text{cm}$ and $6\,\text{cm}$ respectively. Find the distance from P to BC.

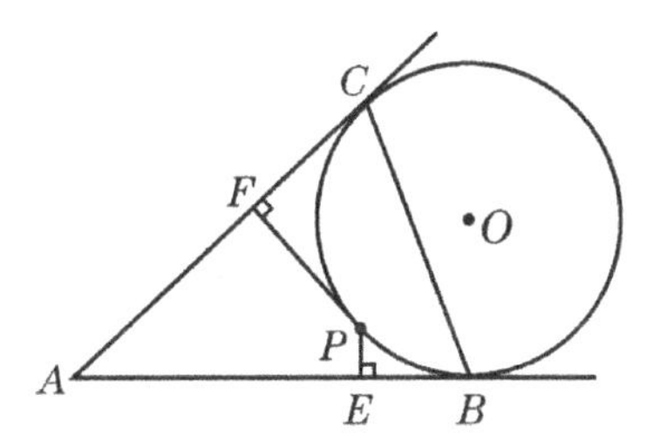

Fig. 17.17

7. Circle C is tangent to three different lines l, m and n respectively. Consider the following three cases:

   (1) the intersection angles of l and m, m and n, n and l are all $60°$;
   (2) $l \parallel m \parallel n$;
   (3) there is another circle C' which is also tangent to the lines l, m and n.

   The case that can hold is ( ).
   (A) only (1) (B) only (1) and (3)
   (C) only (1) and (2) (D) all three

8. As shown in Figure 17.18, the height of an equilateral triangle ABC is equal to the radius of $\odot O$, $\odot O$ rolls on AB and the point of tangency is T, $\odot O$ intersects AC and BC at the points M and N, then $\overset{\frown}{MTN}$ ( ).

   (A) changes from $0°$ to $30°$
   (B) changes from $30°$ to $60°$
   (C) changes from $60°$ to $90°$
   (D) keeps $30°$ unchanged
   (E) keeps $60°$ unchanged

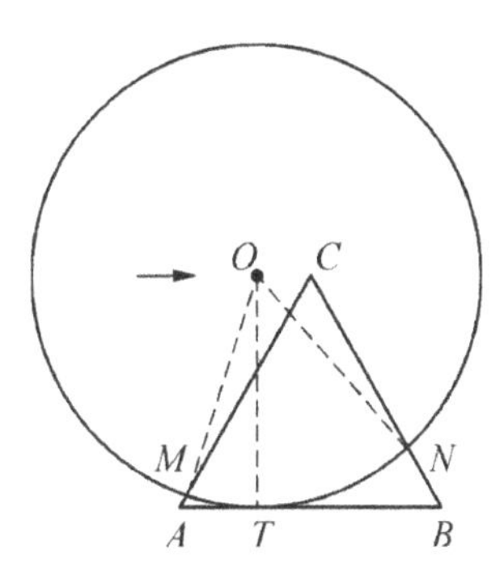

Fig. 17.18

9. As shown in Figure 17.19, the line AB and $\odot O$ intersect at the two points A and B, the point O is on AB, the point C is on $\odot O$, and $\angle AOC = 40°$. Point E is a moving point on the line AB which does not coincide with the point O, and the line EC intersect $\odot O$ at another point D, then how many points E are there that make $DE = DO$?

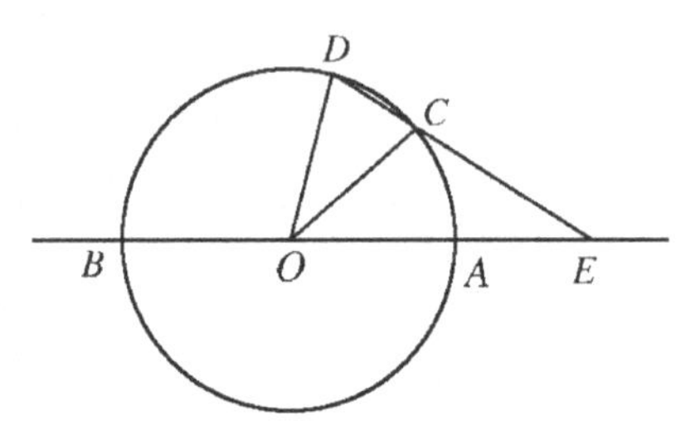

Fig. 17.19

10. As shown in Figure 17.20, in acute triangle ABC, $\angle A = 30°$. Construct a circle with the diameter of BC which intersects AB and AC at the points D and E respectively, and join DE to divide $\triangle ABC$ into $\triangle ADE$ and quadrilateral $DBCE$. Let their areas be S_1 and S_2. Find $S_1 : S_2$.

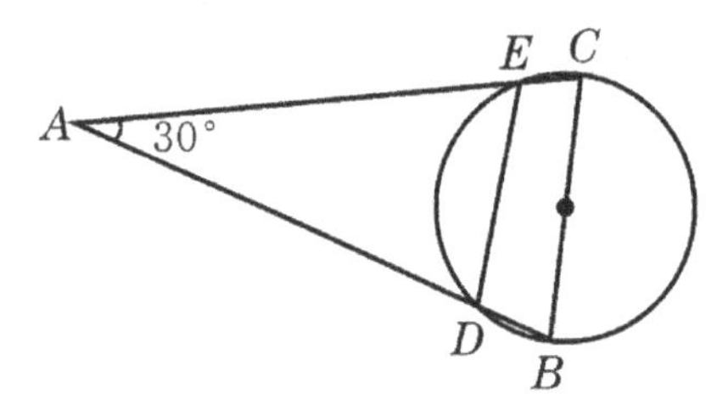

Fig. 17.20

11. Extend the radius OA of $\odot O$ to the point B so that $OA = AB$, DT is a tangent of the circle and the point of tangency is T, C is the projection of B on DT. Prove that $\angle ACB = \frac{1}{3}\angle CAO$.

12. If the diagonals of the inscribed quadrilateral of a circle are perpendicular to each other, prove the following propositions.
    (1) The straight line passing through the intersection of diagonals and bisecting one side must be perpendicular to the opposite side of this side.
    (2) The distance from the center of the circumscribed circle to one side is equal to half of the opposite side.

Chapter 18

Positional Relation of Two Circles

The positional relations of two circles can be as follow: two circles intersect at two points, the two circles are tangent to each other (internally or externally), one circle lies completely outside the other, and one circle lies completely inside the other.

Let two circles be $\odot O_1$ and $\odot O_2$, the radii be R_1 and R_2 where $R_1 \geq R_2$, and the distance between O_1 and O_2 be d, then we have the following conclusions.

$$d > R_1 + R_2 \Leftrightarrow \text{one circle lies completely outside the other;}$$
$$d = R_1 + R_2 \Leftrightarrow \text{the two circles touch externally;}$$
$$d = R_1 - R_2 \Leftrightarrow \text{the two circles touch internally (where } R_1 \neq R_2\text{);}$$
$$R_1 - R_2 < d < R_1 + R_2 \Leftrightarrow \text{the two circles intersect at two points;}$$
$$d < R_1 - R_2 \Leftrightarrow \text{one circle lies completely inside the other (where } R_1 \neq R_2\text{).}$$

Example 1. Given that the radii of the two circles are 5 and 6 respectively, the distance x between the centers of these two circles satisfies the system of inequalities

$$\begin{cases} \dfrac{x+5}{2} \leq x - \dfrac{1}{2}, \\ 3x + 14 > 8x - 36. \end{cases}$$

Try to determine the positional relation of the two circles.

Solution Determine the quantitative relationship between x, 5 and 6.

Solving the system of inequalities

$$\begin{cases} \frac{x+5}{2} \leq x - \frac{1}{2}, \\ 3x + 14 > 8x - 36, \end{cases}$$

gives $6 \leq x < 10$. So $6 - 5 < x < 5 + 6$. Hence, the two circles meeting the given conditions intersect in two points.

Example 2. As shown in Figure 18.1, $\odot O_1$ and $\odot O_2$ intersect at two points A and B, O_1 is on the circumference of $\odot O_2$, and the chord AC of $\odot O_1$ intersects $\odot O_2$ at the point D. Prove that line segment O_1D is perpendicular to BC.

Proof Join O_1C, O_1B and AB, as shown in Figure 18.2, it can be seen that $O_1B = O_1C$. We can try to prove that O_1D is the bisector of $\angle BO_1C$.

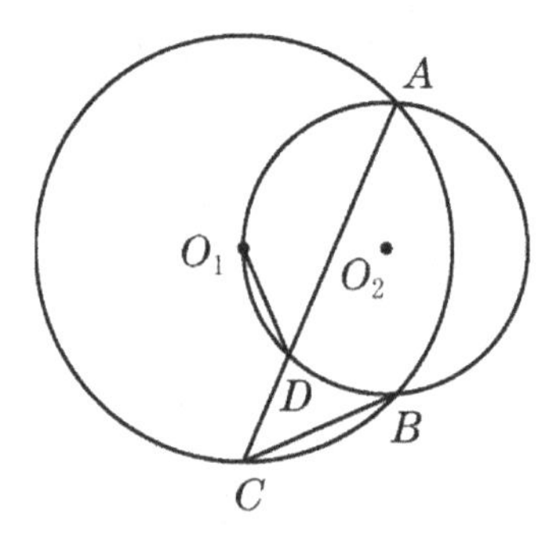

Fig. 18.1

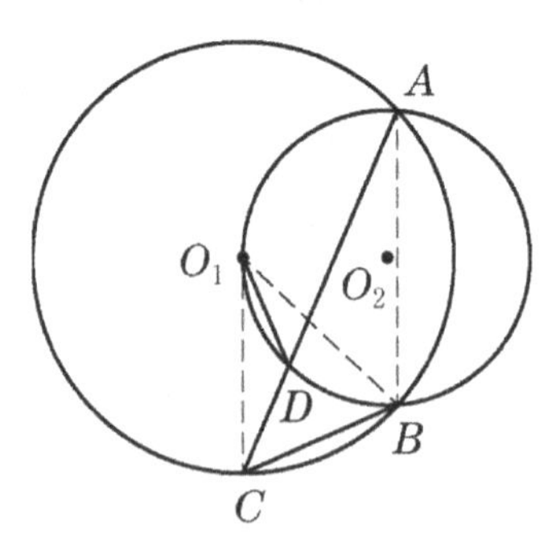

Fig. 18.2

In $\odot O_2$, $\angle CAB = \angle DO_1B$, and in $\odot O_1$, $\angle CAB = \frac{1}{2}\angle CO_1B$, then we know that $\angle DO_1B = \frac{1}{2}\angle CO_1B$, that is, O_1D is the bisector of $\angle CO_1B$.

Since $O_1C = O_1B$, $\triangle CO_1B$ is an isosceles triangle, and $\angle CO_1B$ is the vertex angle, we obtain $O_1D \perp BC$. □

Example 3. As shown in Figure 18.3, $\angle CAB = \angle ABD = 90°$, $AB = AC + BD$, AD intersects BC at the point P, construct $\odot P$ which is tangent to AB. State the positional relation between $\odot O$ with the diameter of AB and $\odot P$ and prove your conclusion.

Analysis Find the quantitative relationship between OP and the radii of two circles.

Solution As shown in Figure 18.4, let AB be tangent to $\odot P$ at the point E. Join PE and PO, then $PE \perp AB$. Suppose that $AC = a$, $BD = b$ $(a \leq b)$, then $AB = a + b$.

Let the radius of $\odot O$ be R, and the radius of $\odot P$ be r.

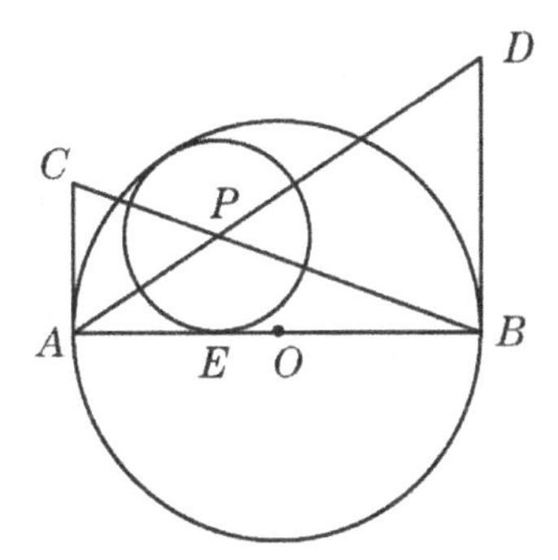

Fig. 18.3

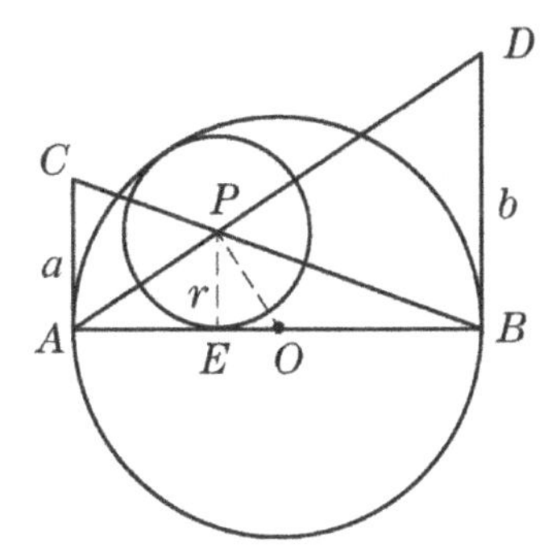

Fig. 18.4

From $AB = AC + BD$, we have $R = \frac{1}{2}(a+b)$. From $\dfrac{r}{a} + \dfrac{r}{b} = \dfrac{EB}{AB} + \dfrac{AE}{AB} = 1$, we get $r = \dfrac{ab}{a+b}$. And from $\dfrac{AE}{r} = \dfrac{AB}{b}$, it can be deduced that $AE = r \cdot \dfrac{AB}{b} = \dfrac{ab}{a+b} \cdot \dfrac{a+b}{b} = a$, then

$$OE = \frac{1}{2}(a+b) - a = \frac{1}{2}(b-a).$$

In right triangle OPE,

$$OP^2 = r^2 + OE^2 = \left(\frac{ab}{a+b}\right)^2 + \frac{1}{4}(b-a)^2 = \frac{(a^2+b^2)^2}{4(a+b)^2},$$

so

$$OP = \frac{a^2+b^2}{2(a+b)} = \frac{(a+b)^2 - 2ab}{2(a+b)} = \frac{1}{2}(a+b) - \frac{ab}{a+b} = R - r.$$

Hence, $\odot O$ and $\odot P$ touch internally.

Example 4. As shown in Figure 18.5, $\odot M$ and $\odot O$ are tangent externally at the point C, the radii of $\odot M$ and $\odot O$ are r and R respectively. The line TPQ is tangent to $\odot M$ at the point T, and intersects $\odot O$ at the points P and Q. Find the value of $\dfrac{CQ - CP}{PQ}$.

Solution As shown in Figure 18.6, extend PC and QC, which intersect $\odot M$ at the points D and E respectively. Join MO, MD and OP.

Since three points M, C and O are collinear, we can get that isosceles triangle MCD is similar to isosceles triangle OCP. Thus we have $\frac{CD}{CP} = \frac{r}{R}$, then

$$\frac{PD}{CP} = \frac{R+r}{R},$$
$$\frac{PD \cdot PC}{PC^2} = \frac{R+r}{R}.$$

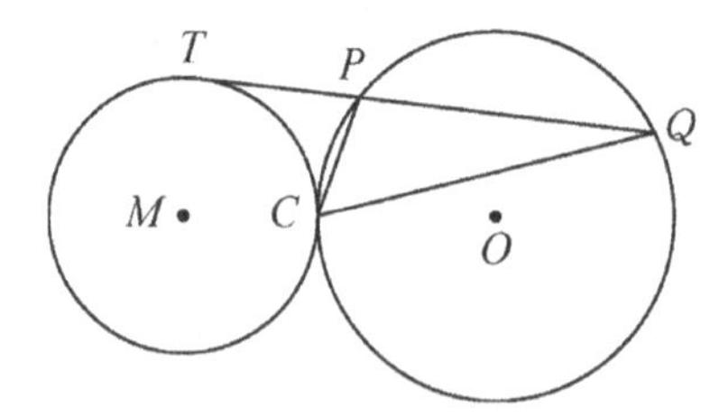

Fig. 18.5

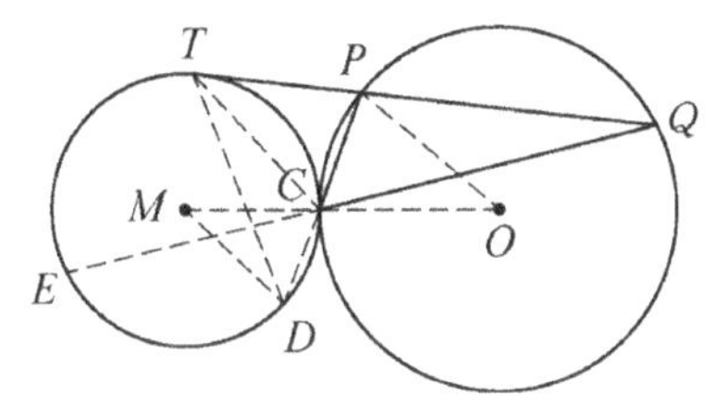

Fig. 18.6

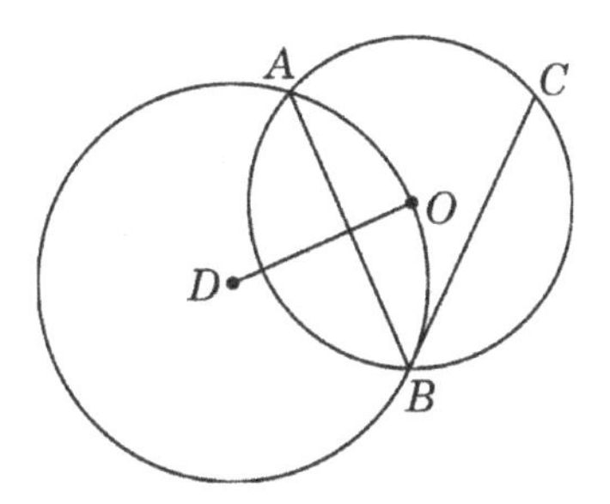

Fig. 18.7

Join TD and TC. It's easy to know that $\triangle PDT \backsim \triangle PTC$, then $\frac{PT}{PD} = \frac{PC}{PT}$, that is, $PD \cdot PC = PT^2$.

Therefore, $\frac{PT^2}{PC^2} = \frac{R+r}{R}$, and then

$$\frac{PT}{PC} = \sqrt{\frac{R+r}{R}}. \tag{1}$$

In the same way, $\frac{QT}{QC} = \sqrt{\frac{R+r}{R}}$. (2)

From (1) and (2), we can get $\frac{QT-PT}{QC-PC} = \sqrt{\frac{R+r}{R}}$, then $\frac{PQ}{QC-PC} = \sqrt{\frac{R+r}{R}}$. That is, $\frac{QC-PC}{PQ} = \sqrt{\frac{R}{R+r}}$.

Example 5. As shown in Figure 18.7, the $\odot O$ and the $\odot D$ intersect at two points A and B, BC is the tangent of the circle D, the point C is on the circle O, and $AB = BC$.

(1) Prove that the point O is on the circumference of the circle D.
(2) Let the area of $\triangle ABC$ be S, find the minimum value of the radius r of the circle D.

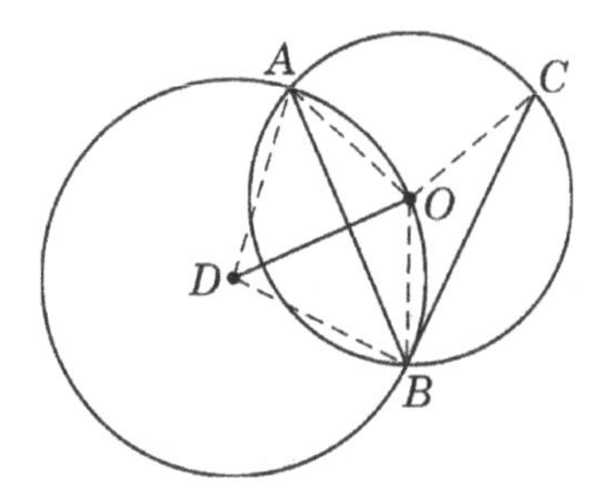

Fig. 18.8

Solution

(1) As shown in Figure 18.8, join OA, OB and OC, then $OA = OB = OC$. And since $AB = BC$, we can obtain that isosceles triangles $\triangle ABO \cong \triangle CBO$, $\angle ABO = \angle CBO$.

Because BC is the tangent of the circle D, the length of the minor arc clamped by the angle of the osculation $\angle ABC$ is twice the length of the minor arc clamped by $\angle OBC$, that is, the ray BO passes through the midpoint of the arc AB. And since $OA = OB$, the point O is the midpoint of the arc AB, that is, O is on the circle D.

(2) Join AD and BD, then $2r = AD + BD \geq AB$, so

$$4r^2 \geq AB^2 = AB \cdot BC.$$

From $AB \cdot BC \geq 2S$, we know that $4r^2 \geq 2S$, i.e. $r \geq \frac{\sqrt{2S}}{2}$, and the equality holds if and only if AB is the diameter of circle D. Hence, the minimum value of r is $\frac{\sqrt{2S}}{2}$.

Example 6. Given that AB is the diameter of the semicircle O, and P is any point on the diameter AB. Take the point A as the center and AP as the radius to construct $\odot A$, which intersect the semicircle O at the point C. Take the point B as the center and BP as the radius to construct $\odot B$, which intersects the semicircle O at the point D. And the midpoint of the line segment CD is M. Prove that MP is tangent to $\odot A$ and $\odot B$ respectively.

Proof As shown in Figure 18.9, join AC, AD, BC and BD. Draw the perpendicular lines of AB through the points C and D respectively, and the perpendicular feet are E and F respectively, then $CE \parallel DF$. □

Because AB is the diameter of $\odot O$, we have $\angle ACB = \angle ADB = 90°$.

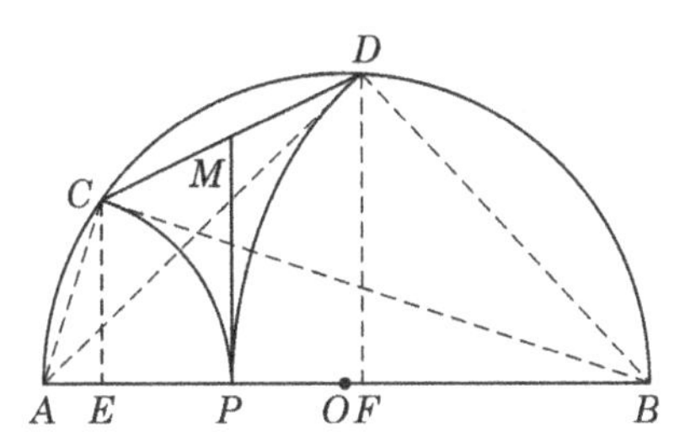

Fig. 18.9

In Rt$\triangle ABC$ and Rt$\triangle ABD$, from the right triangle altitude theorem,

$$PA^2 = AC^2 = AE \cdot AB, \quad PB^2 = BD^2 = BF \cdot AB.$$

Subtracting the two equalities gives $PA^2 - PB^2 = AB(AE - BF)$.

And from $PA^2 - PB^2 = (PA + PB)(PA - PB) = AB(PA - PB)$, we get that

$$AE - BF = PA - PB,$$

i.e. $PA - AE = PB - BF$.

So $PE = PF$, that is, the point P is the midpoint of EF. Thus MP is neutrality line of right angle trapezoid $CDFE$, then $MP \perp AB$. Hence, MP is tangent to $\odot A$ and $\odot B$ respectively.

Example 7. As shown in Figure 18.10, two circles are internally tangent at the point P, the chord AD of the big circle separates from the small circle, PA and PD intersect the small circle at the points E and F, and the line EF intersects the big circle at the points B and C. Prove that $\angle APB = \angle CPD$.

Proof As shown in Figure 18.11, draw the common tangent GH of two circles through the point P.

Since GH is tangent to two circles at the point P, we know that

$$\angle GPA = \angle EFP, \angle GPB = \angle BCP.$$

From $\angle EFP = \angle CPD + \angle BCP$, we get

$$\angle GPA - \angle GPB = \angle EFP - \angle BCP = \angle CPD.$$

So $\angle APB = \angle CPD$. □

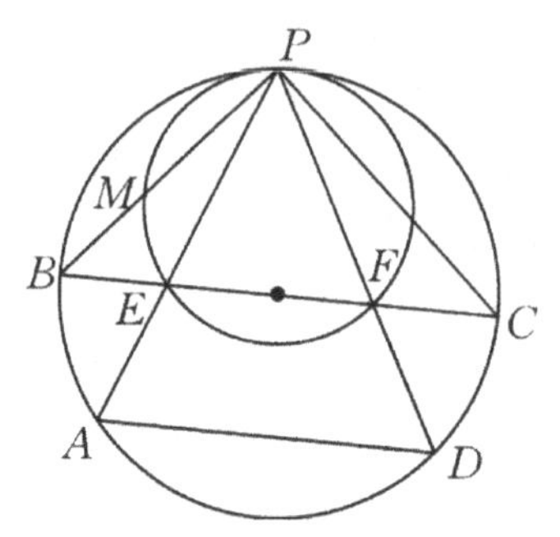

Fig. 18.10

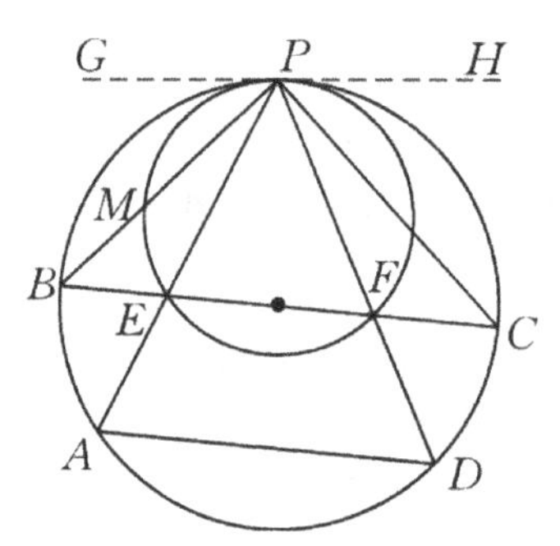

Fig. 18.11

Remark What conclusion can you draw by changing "two circles are tangent internally" into "one circle lies completely outside the other" or "one circle lies completely inside the other"?

(1) When "two circles are tangent internally" in Example 7 is changed into "one circle lies completely outside the other", we can get the following conclusion.

As shown in Figure 18.12, $\odot O_1$ and $\odot O_2$ lie completely outside each other, a line intersects $\odot O_1$ at the points A and D, and intersects $\odot O_2$ at the points B and C. The line O_1O_2 intersects $\odot O_1$ at the point E, and intersects $\odot O_2$ at the point F. The extended lines of AE and BF intersect at the point P, and the extended lines of DE and CF intersect at the point Q. Prove that $\angle APB + \angle CQD = 180°$.

Proof Join the line segments O_1D and O_2C, draw a line segment O_1M so that $O_1M \perp AD$ at the point M, and draw a line segment O_2N

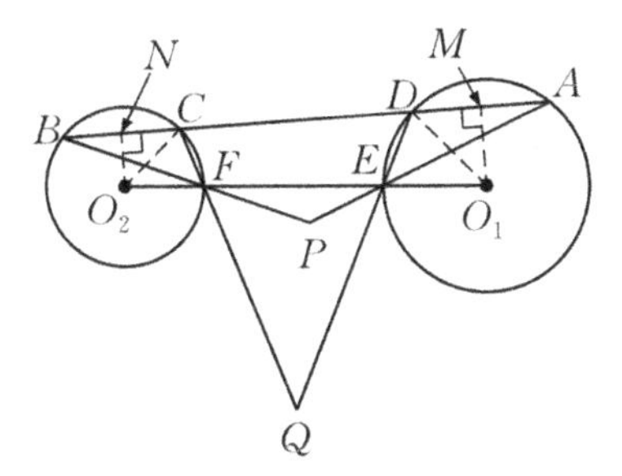

Fig. 18.12

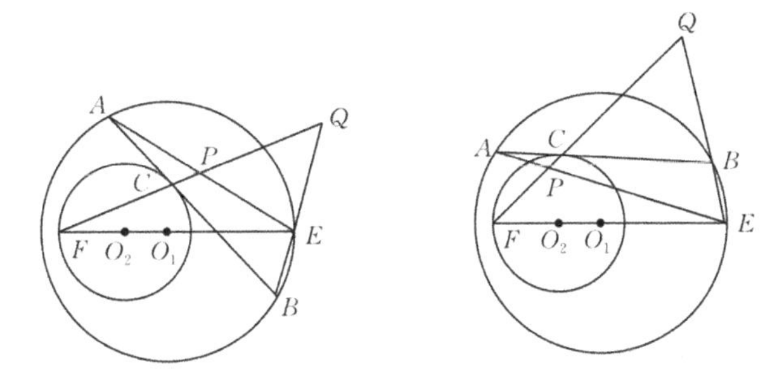

Fig. 18.13

so that $O_2N \perp BC$ at the point N. Then

$$\begin{aligned}
&\angle APB + \angle CQD \\
&\quad = (180^\circ - \angle A - \angle B) + (180^\circ - \angle DCQ - \angle CDQ) \\
&\quad = 360^\circ - 2\angle A - 2\angle B - \angle AED - \angle BFC \\
&\quad = 360^\circ - (\angle DO_1E + \angle DO_1M + \angle CO_2F + \angle CO_2N) \\
&\quad = 360^\circ - 180^\circ = 180^\circ.
\end{aligned}$$

(2) When "two circles are tangent internally" in Example 7 is changed into "one circle lies completely inside the other", we can get the following conclusion.

As shown in Figures 18.13(1) and 18.13(2), $\odot O_2$ lies completely inside $\odot O_1$, the chord AB of $\odot O_1$ is tangent to $\odot O_2$ at the point C. Line O_1O_2 intersects $\odot O_1$ at the point E, and intersects $\odot O_2$ at the point F. AE and BE intersect FC at the points P and Q respectively. Prove that $\angle APC$ and $\angle BQC$ are equal or complementary.

It is easy for readers to prove the above conclusion.

Example 8. Let P be a point on the plane where the equilateral triangle ABC is located, and make $\triangle ABP$, $\triangle PBC$ and $\triangle PCA$ all be isosceles triangles. How many points P satisfy the conditions?

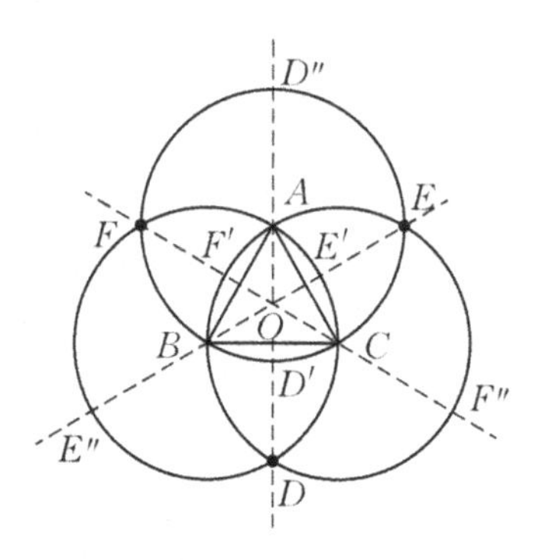

Fig. 18.14

Solution As shown in Figure 18.14, draw the perpendicular bisectors of AB, BC and CA. Construct the $\odot A$, the $\odot B$ and the $\odot C$, of which all the radii are a (a is the side length of the equilateral triangle).

Except for three points A, B and C, there are the 10 intersection points of the three perpendicular bisectors and the three circles, written as O, D, E, F, D', E', F', D'', E'' and F''.

It is easy to verify that all the 10 points meet the requirements, so there are 10 points P that meet the conditions.

Remark Generalize the Example 8, we have the following conclusion:

On a plane, a point P and any two vertices of an n regular polygon form an isosceles triangle, and the number of the points P is denoted as S_n. Then we know that $S_3 = 10$, $S_4 = 9$, $S_5 = 6$, and when $n \geq 6$, $S_n = 1$.

Exercises

1. If two circles with the radii of 13 and 15 intersect at two points and the distance between the centers of these two circles is 12, the length of the common chord of the two circles is ( ).

   (A) $3\sqrt{11}$ (B) $\frac{65}{6}$ (C) $4\sqrt{6}$ (D) 10 (E) none of the above

2. As shown in Figure 18.15, the relationship between the arc length l_1 of the large semicircle and the sum l_2 of the arc lengths of the n small semicircles is ( ).

   (A) $l_1 = l_2$ (B) $l_1 = \pi l_2$ (C) $l_1 = nl_2$ (D) $l_1 = \frac{1}{n}l_2$

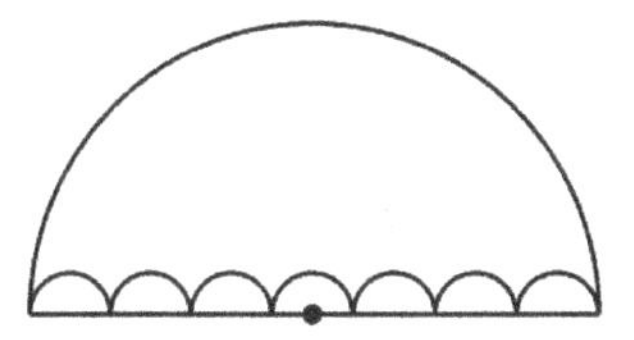

Fig. 18.15

3. As shown in Figure 18.16, A is an intersection of $\odot O_1$ and $\odot O_2$, P is the midpoint of O_1O_2. The line MN passing through the point A is perpendicular to PA, and it intersects $\odot O_1$ and $\odot O_2$ at the points M and N respectively. Prove that $AM = AN$.

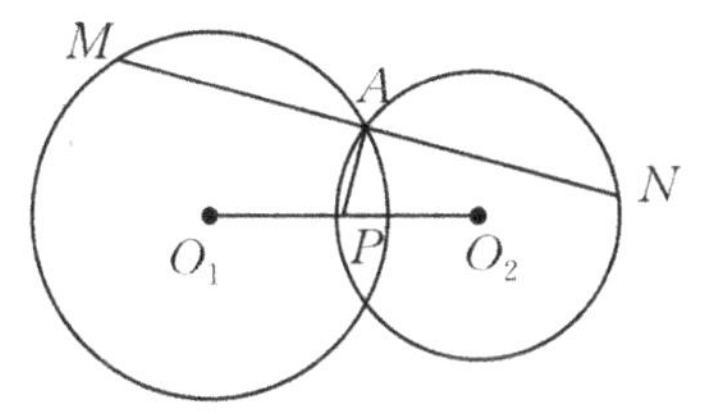

Fig. 18.16

4. As shown in Figure 18.17, the two circles intersect at the points P and Q. The secant AD of the big circle intersects the small circle at the points B and C. Prove that $\angle APB + \angle CQD = 180°$.

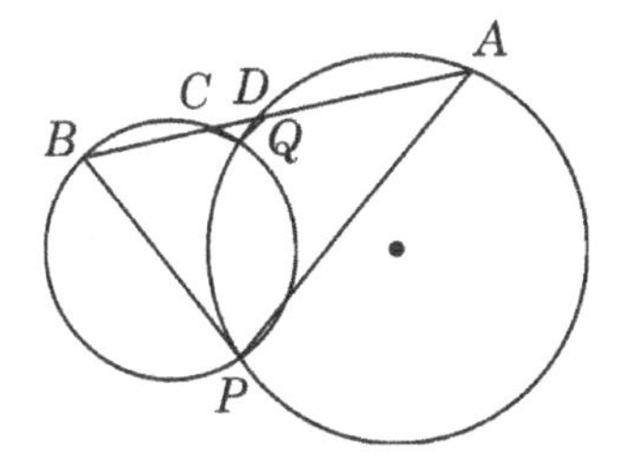

Fig. 18.17

5. Given that r_1 and r_2 are the radii of the two circles respectively, the distance between the centers of these two circles is $d = 5$, the numerical value r_1, r_2 and $(r_1 - r_2)$ are exactly the three roots of the equation $x^3 - 6x^2 + 11x - 6 = 0$. Which of the following is the positional relation of these two circles? ( ).

   (A) One circle lies completely outside the other.
   (B) The two circles touch externally.
   (C) The two circles intersect at two points.
   (D) The two circles touch internally.
   (E) Not sure.

6. As shown in Figure 18.18, $\odot O_1$ and $\odot O_2$ intersect at the two points A and B, the point C is on $\odot O_1$ and outside $\odot O_2$. The extended lines of CA and CB intersect $\odot O_2$ at the points D and E respectively. The radius of $\odot O_1$ is 5, $AC = 8$, $AD = 12$, $DE = 14$, and $S_{\triangle CDE} = 112$, then the radius of $\odot O_1$ is ( ).

   (A) 5 (B) 4 (C) 6 (D) $\frac{11}{2}$

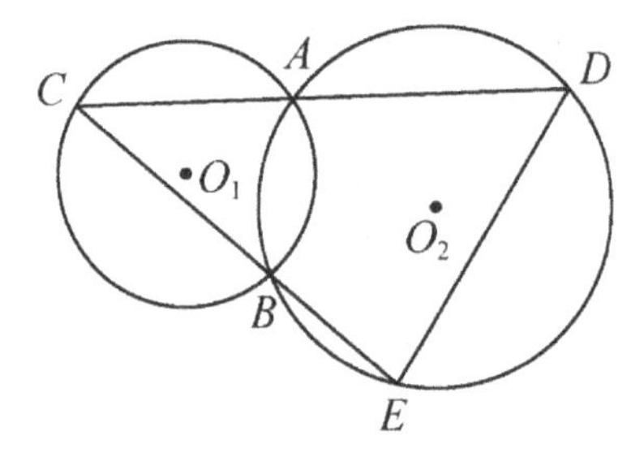

Fig. 18.18

7. As shown in Figure 18.19, let $AB = 2a$ be a fixed length. Draw a circle O with the diameter of AB. Take a moving point C on the upper semicircle, and take point C as the center to draw a circle C which is tangent to AB at the point D. Circle C cut off a part of $\triangle ABC$, and the area of remaining part is denoted as S, find the maximum value of S.

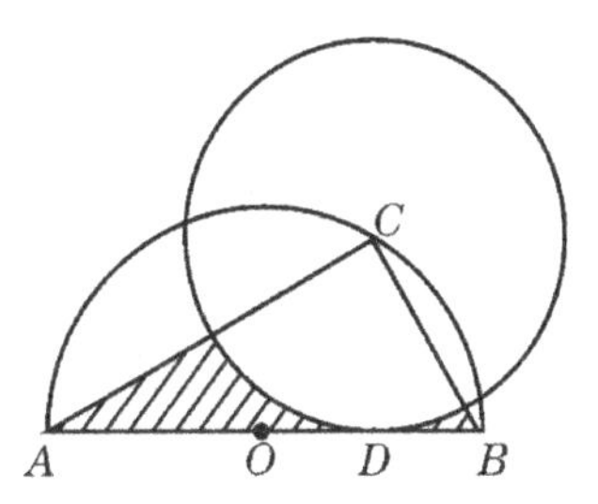

Fig. 18.19

8. As shown in Figure 18.20, $\odot O_1$ and $\odot O_2$ intersect at the two points A and B, the points O_1 and O_2 are on two sides of AB, and AC is the diameter of $\odot O_1$. Extend CB to the point D where CB intersects $\odot O_2$. E is a point on $\overset{\frown}{BC}$, extend EB to the point F where EB intersects $\odot O_2$. M and N are the midpoints of CD and EF respectively, $AC = 2CE$, then $\angle AMN =$ ______.

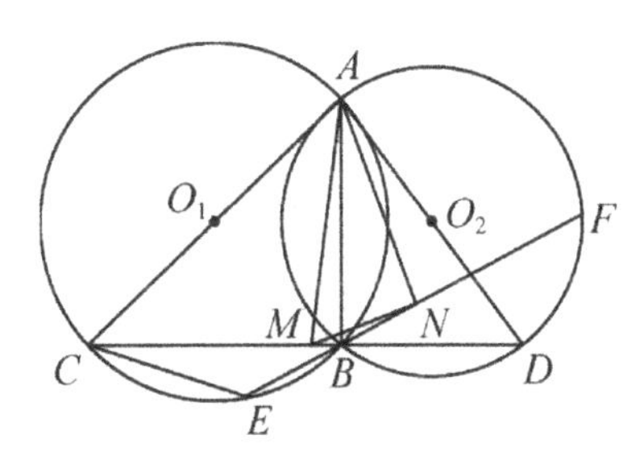

Fig. 18.20

9. Prove that among the secants drawn from one intersection of the two circles to the two circles, the secant parallel to the line of centers is the longest.
10. As shown in Figure 18.21, point H is the orthocenter of $\triangle ABC$, $\odot O_1$ with the diameter of AB and the circumscribed circle $\odot O_2$ of $\triangle BCH$ intersect at the point D. Extend AD to the point P where AD intersects CH. Prove that P is the midpoint of CH.

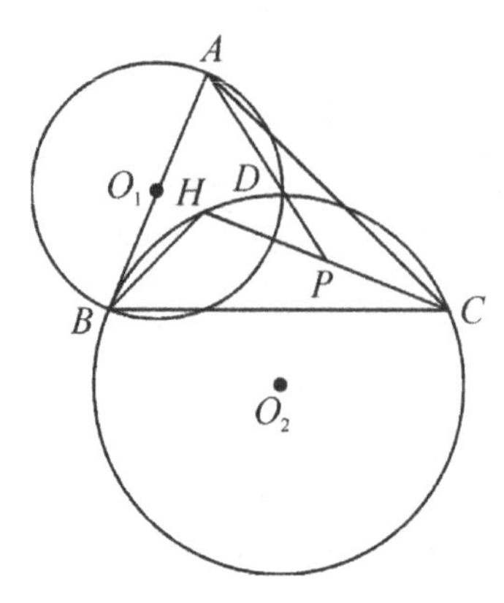

Fig. 18.21

11. As shown in Figure 18.22, in $\triangle ABC$, $\angle BAC = 90°$, AD is the height perpendicular to BC. The inscribed circles of Rt$\triangle ABC$, Rt$\triangle ADB$ and Rt$\triangle ADC$ are tangent to the hypotenuse of each triangle at the points L, M and N respectively. Prove that $AM \cdot AB + CN \cdot AC = CL \cdot BC$.

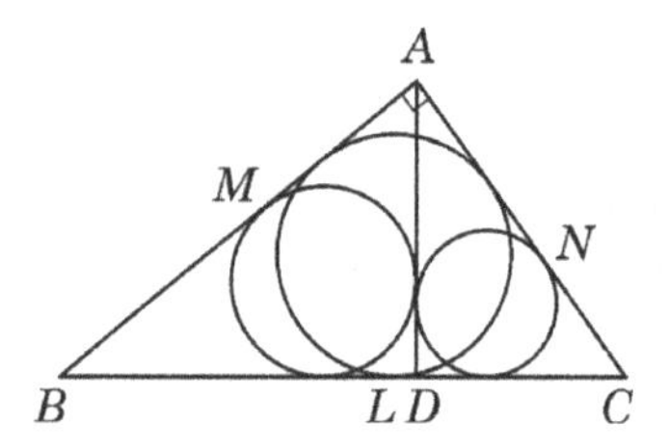

Fig. 18.22

Chapter 19

Power of A Point Theorem

In order to deal with the problem of a proportion line segment in a circle, the power of a point theorem is usually used. Power of a point theorem is one of important theorems in plane geometry.

Segments of chords theorem, segments of secants theorem and segments of secants and tangents theorem are collectively called power of a point theorem.

1. Segments of Chords Theorem
If two chords AB and CD intersect at the point P in the interior of the circle (as shown in Figure 19.1(1)), then $PA \cdot PB = PC \cdot PD$.

2. Segments of Secants Theorem
If two secant segments PAB and PCD share the same endpoint P outside the circle (as shown in Figure 19.1(2)), then $PA \cdot PB = PC \cdot PD$.

3. Segments of Secants and Tangents Theorem
If a secant segment PAB and a tangent segment PC share an endpoint P outside the circle (as shown in Figure 19.1(3)), then $PA \cdot PB = PC^2$.

In fact, the segments of secants and tangents theorem can be regarded as the limit case of the segments of secants theorem. Therefore, the above conclusions can be combined as follows:

If two intersecting lines with an intersection P intersect the circle O at the points A, B and C, D, then $PA \cdot PB = PC \cdot PD$, where P, A, B and P, C, D are collinear respectively.

Example 1. As shown in Figure 19.2, in $\square ABCD$, the circle passing through the points A, B and C intersects AD at the point E, and CD is tangent to the circle. If $AB = 4$, $BE = 5$, find the length of DE.

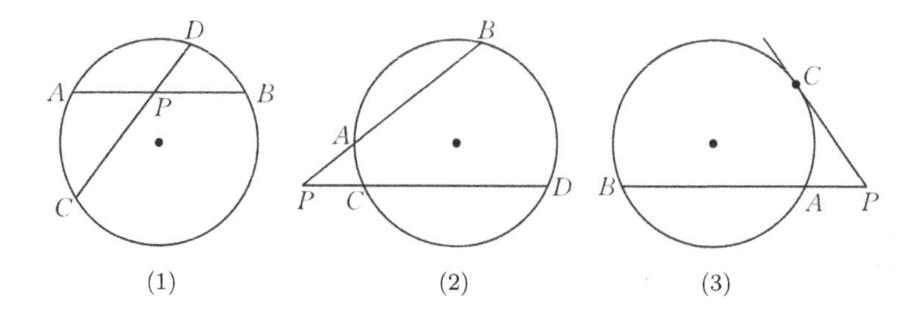

Fig. 19.1

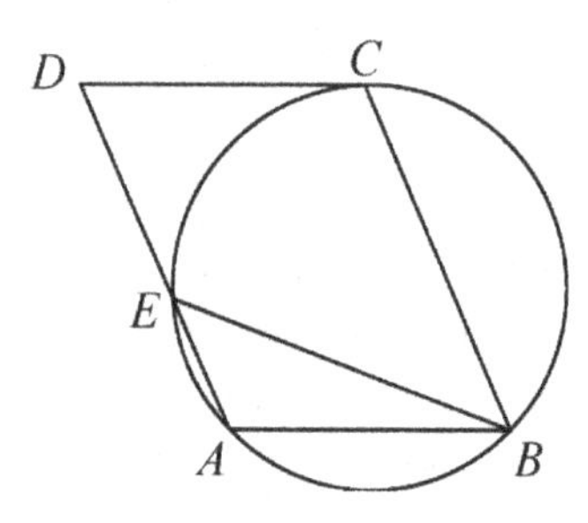

Fig. 19.2

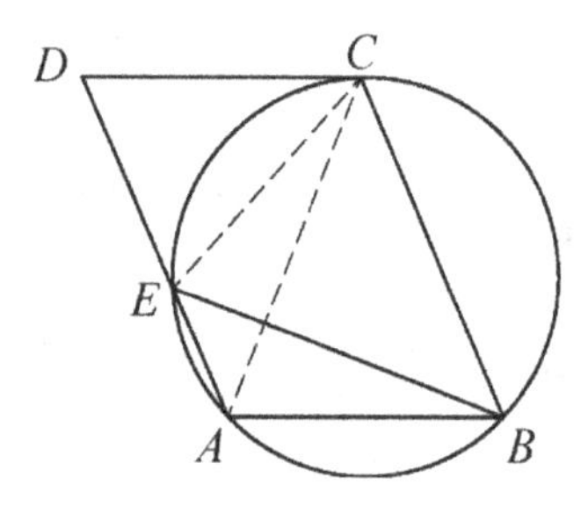

Fig. 19.3

Solution As shown in Figure 19.3, join AC and CE. From $AE \parallel BC$, we know that the quadrilateral $ABCE$ is an isosceles trapezoid, so

$$AC = BE = 5.$$

Since $DC \parallel AB$, DC is tangent to the circle, we get

$$\angle BAC = \angle ACD = \angle ABC.$$

Thus, $AC = BC = AD = 5$.

From the segments of secants and tangents theorem, we know that $DC^2 = AD \cdot DE$.

And since $DC = AB = 4$, $AD = 5$, we obtain

$$DE = \frac{DC^2}{AD} = \frac{16}{5}.$$

Example 2. As shown in Figure 19.4, $\odot O$ and $\odot O_1$ are externally tangent to each other at the point P. An external common tangent is tangent to the two circles at the points A and C respectively. AB is the diameter of $\odot O$. Prove that the length of the tangent from the point B to $\odot O_1$ is equal to AB.

Analysis Transform the relationship between line segments by using the secants and tangents theorem.

Solution As shown in Figure 19.5, join AP, BP and PC, construct the common tangent PM through the point P which intersects AC at the point M, then $MP = MA = MC$. So $\triangle APC$ is a right triangle, and $\angle APC = 90°$.

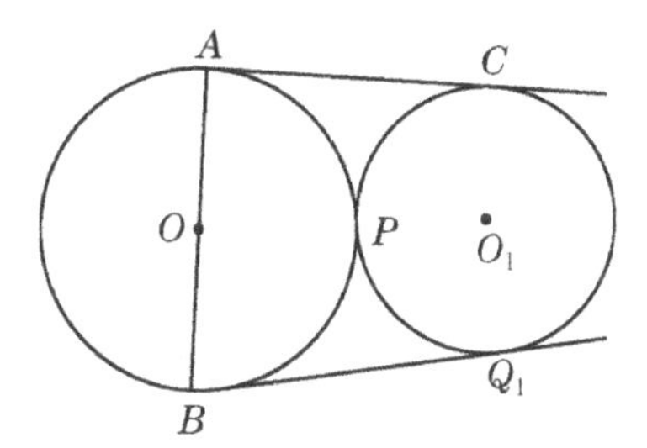

Fig. 19.4

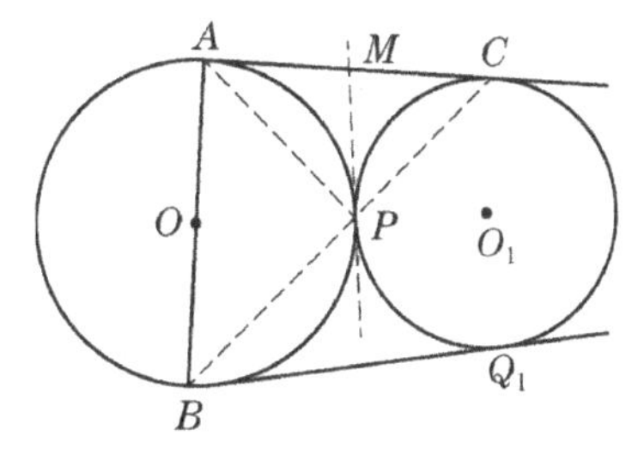

Fig. 19.5

Since AB is the diameter of $\odot O$, we know that $AP \perp PB$, i.e. $\angle APB = 90°$. Then the three points B, P and C are collinear, so $BQ_1^2 = BP \cdot BC$. Because $AB \perp AC$, we have $\angle BAC = 90°$. Since $\angle BAC = \angle APB = 90°$, it's known that $\triangle ABP \backsim \triangle CBA$.

The lengths of the corresponding sides are proportional, so $\frac{AB}{BP} = \frac{BC}{AB}$, i.e. $AB^2 = BP \cdot BC$. Hence, $AB^2 = BQ_1^2$, that is, $AB = BQ_1$.

Remark It is a common basic method to prove the relationship between proportional line segments (including the equal line segments) in a circle by using the segments of secants and tangents theorem and the similarity of triangles.

Example 3. As shown in Figure 19.6, $\triangle ABC$ is inscribed in $\odot O$, P is a point outside $\odot O$. Construct $\angle CPD = \angle A$ so that PD intersects $\odot O$ at the points D and E, and intersects AB and AC at the points M and N respectively. Prove that $DN \cdot NE = MN \cdot NP$.

Analysis Using the intersecting chords theorem to transform the relation of proportional line segments.

Proof. Since

$$\angle CPD = \angle A, \quad \angle MNA = \angle PNC,$$

we know that $\triangle AMN \backsim \triangle PCN$.

The lengths of the corresponding sides are proportional, so $\frac{AN}{PN} = \frac{MN}{NC}$, that is,

$$AN \cdot NC = PN \cdot MN.$$

From the segments of chords theorem, we get $AN \cdot NC = DN \cdot NE$.

Hence,

$$DN \cdot NE = NP \cdot MN.$$

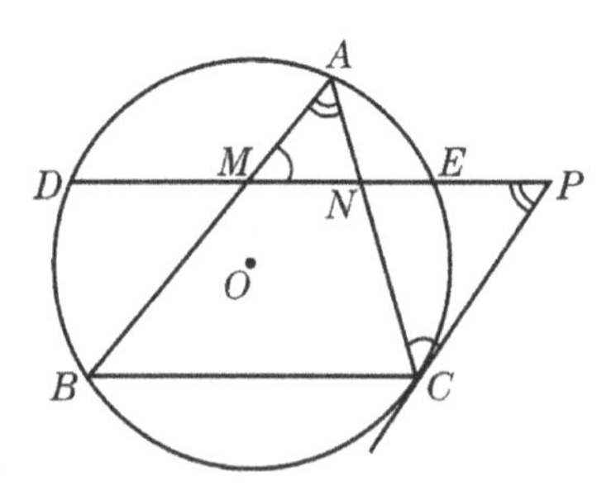

Fig. 19.6

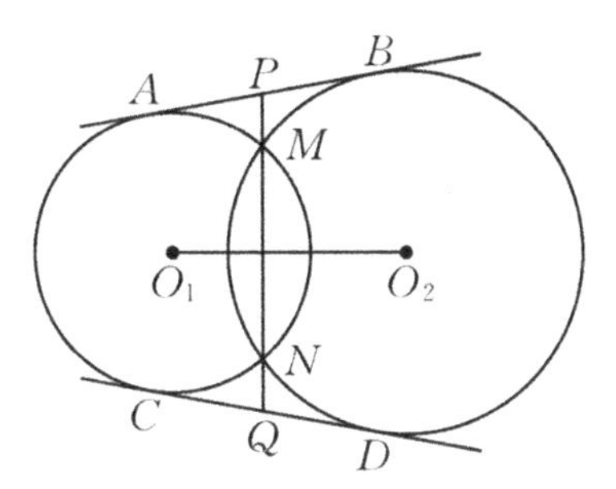

Fig. 19.7

Example 4. As shown in Figure 19.7, $\odot O_1$ and $\odot O_2$ intersect in two points, the common chord is MN, and the external common tangents are AB and CD. Line MN intersects AB at the point P and intersects CD at the point Q. Prove that $PQ^2 = AB^2 + MN^2$.

Proof. Because AB is an external common tangent of $\odot O_1$ and $\odot O_2$, from the secants and tangents theorem, we get

$$PA^2 = PM \cdot PN = PB^2.$$

Then $PA = PB$, so $AB^2 = 4PM \cdot PN$. Similarly, $QC^2 = QN \cdot QM$, $QC = QD$. And since $CD = AB = \sqrt{O_1O_2^2 - (O_2B - O_1A)^2}$, then we have $PA = QC$.

Thus we know that

$$PM \cdot (PM + MN) = QN \cdot (QN + MN),$$

i.e. $PM = QN$.

Hence,

$$\begin{aligned} AB^2 + MN^2 &= 4PM \cdot PN + MN^2 \\ &= 4PM^2 + 4PM \cdot MN + MN^2 \\ &= (2PM + MN)^2 \\ &= (PM + MN + NQ)^2 \\ &= PQ^2. \end{aligned}$$

□

Example 5. As shown in Figure 19.8, $ABCD$ is a square with the side a. The arc with the center D and the radius of DA intersects the semicircle with the diameter of BC at another point P. Extend AP to the point N where AP and BC intersect. Then $\frac{BN}{NC} =$ ____.

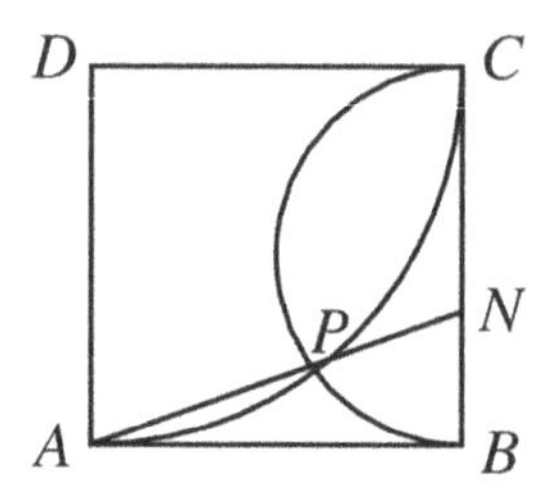

Fig. 19.8

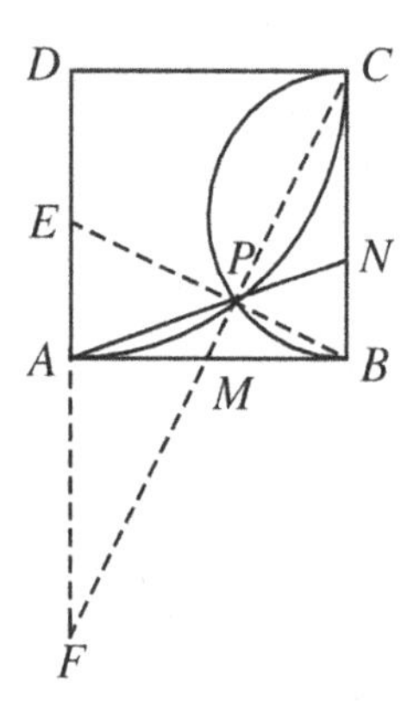

Fig. 19.9

Solution As shown in Figure 19.9, join BP and extend it to intersect AD at the point E. Join CP and extend it to intersect AB at point M and intersect the extended line of DA at the point F.

Since BC is the diameter of the semicircle, we have $\angle BPC = 90°$, that is, $BE \perp CF$. Thus, $\angle ABE = \angle BCM$, then we know that

$$\text{Rt}\triangle EAB \cong \text{Rt}\triangle MBC.$$

The corresponding sides of congruent triangles are equal, so $AE = MB$.

Since

$$MA^2 = MP \cdot MC, \quad MB^2 = MP \cdot MC,$$

we get $MA = MB = \frac{1}{2}a$.

From $\text{Rt}\triangle MAF \backsim \text{Rt}\triangle CDF$, it's obtained that $AF = a$.

Because $\angle EFP = \angle BCP$, we get that

$$\text{Rt}\triangle FPE \backsim \text{Rt}\triangle CPB, \quad \triangle FPA \backsim \triangle CPN.$$

Then $\frac{BC}{EF} = \frac{CP}{PF} = \frac{NC}{AF}$. Thus we obtain that $NC = AF \cdot \frac{BC}{EF} = \frac{2}{3}a$, $BN = a - \frac{2}{3}a = \frac{1}{3}a$. Hence, $\frac{BN}{NC} = \frac{1}{2}$.

Example 6. As shown in Figure 19.10, AB is the diameter of $\odot O$, the chord CD intersects AB at the point E. The tangent of the circle through the point A intersects the extended line of CD at the point F. If $DE = \frac{3}{4}CE$, $AC = 8\sqrt{5}$, D is the midpoint of EF, find the length of AB.

Solution Let $CE = 4x$, $AE = y$, then $DF = DE = 3x$, $EF = 6x$.

As shown in Figure 19.11, join AD and BC. Since AB is the diameter of $\odot O$, AF is a tangent of $\odot O$, we know that $\angle EAF = 90°$, $\angle ACD = \angle DAF$. Since D is the midpoint of the hypotenuse EF of Rt$\triangle AEF$, it can be

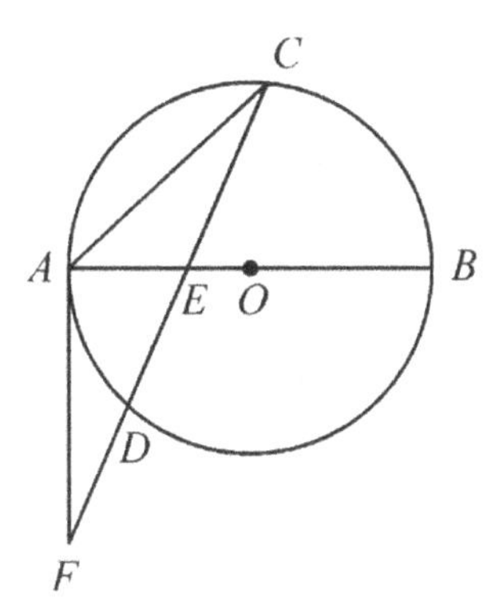

Fig. 19.10

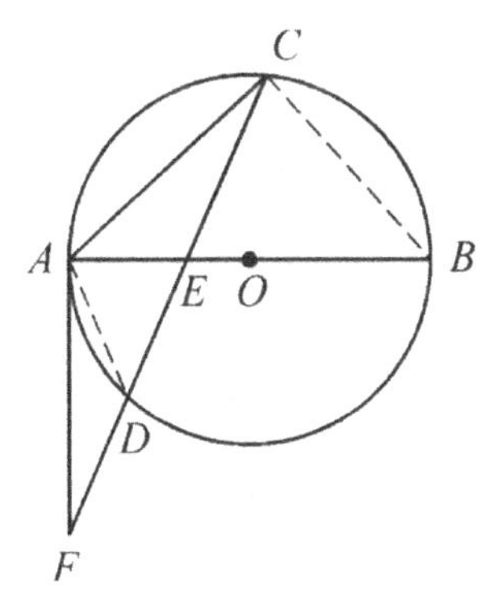

Fig. 19.11

seen that $DA = DE = DF$, which gives $\angle DAF = \angle AFD$. Thus we have $\angle ACD = \angle AFD$, then $AF = AC = 8\sqrt{5}$.

In Rt$\triangle AEF$, from Pythagorean theorem, $EF^2 = AE^2 + AF^2$, i.e. $36x^2 = y^2 + 320$. Let $BE = z$, from the segments of chords theorem, we obtain that $CE \cdot DE = AE \cdot BE$, i.e. $yz = 4x \cdot 3x = 12x^2$.

Thus we get

$$y^2 + 320 = 3yz. \tag{1}$$

Since $AD = DE$, $\angle DAE = \angle AED$ is obtained. And because $\angle DAE = \angle BCE$, $\angle AED = \angle BEC$, we get that $\angle BCE = \angle BEC$, then $BC = BE = z$.

In Rt$\triangle ACB$, from Pythagorean theorem, $AB^2 = AC^2 + BC^2$, so

$$(y + z)^2 = 320 + z^2,$$

i.e. $y^2 + 2yz = 320$. (2)

By combining (1) and (2), we get $y = 8$, $z = 16$.

Hence, $AB = AE + BE = 24$.

Example 7. Prove that the product of the lengths of three angle bisectors of a triangle is smaller than the product of the lengths of its three sides.

Proof. Try to find out the relationship between an angle bisector and two sides of the angle.

As shown in Figure 19.12, let three angle bisectors of $\triangle ABC$ be AD, BE and CF. Construct the circumscribed circle of $\triangle ABC$, extend AD to intersect the circle at the point G, and join CG. Then from

$$\angle ABD = \angle AGC,$$

$$\angle BAD = \angle GAC,$$

we get $\triangle ABD \backsim \triangle AGC$.

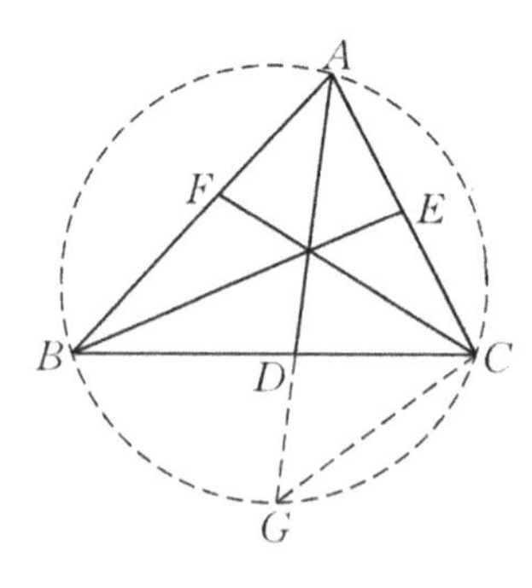

Fig. 19.12

So $\frac{AB}{AG} = \frac{AD}{AC}$, that is,

$$AG \cdot AD = AB \cdot AC.$$

Let $AB = c$, $BC = a$, $AC = b$, then

$$bc = AG \cdot AD > AD^2.$$

Similarly, $BE^2 < ac$, $CF^2 < ab$.

Hence, $AD^2 \cdot BE^2 \cdot CF^2 < a^2b^2c^2$, that is,

$$AD \cdot BE \cdot CF < abc. \qquad \square$$

Exercises

1. As shown in Figure 19.13, AB is the diameter of $\odot O$, C is a moving point on AB (Point C does not coincide with A and B), $CD \perp AB$, AD and CD intersect $\odot O$ at points E and F respectively. Which of the following expressions must be equal to $AB \cdot AC$? ( )

   (A) $AE \cdot AD$ (B) $AE \cdot ED$ (C) $CF \cdot CD$ (D) $CF \cdot FD$

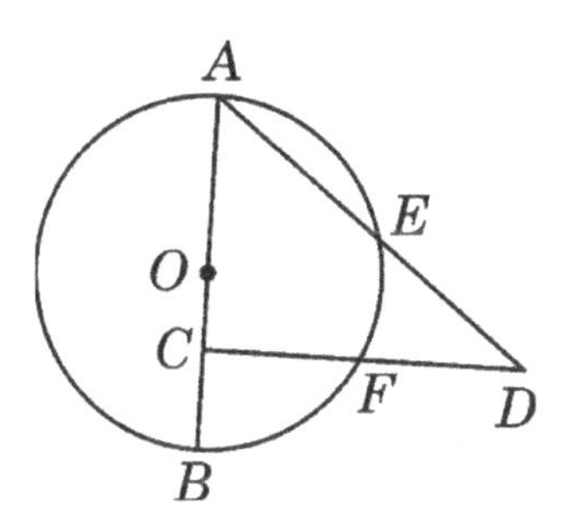

Fig. 19.13

2. As shown in Figure 19.14, construct two tangents PA and PB of $\odot O$ through the point P outside $\odot O$, and the points of tangency are A and B respectively. Join OP which intersects $\odot O$ at the point C, and draw the perpendicular line of AP through the point C with the foot point of E. If $PA = 10\,\text{cm}$, $PC = 5\,\text{cm}$, find the length of CE.

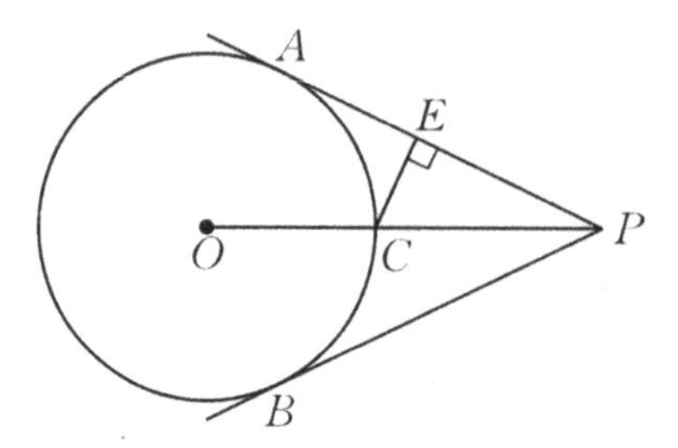

Fig. 19.14

3. As shown in Figure 19.15, in $\triangle ABC$, CM is the angle bisector of $\angle C$, the circumscribed circle of $\triangle AMC$ intersects BC at N. If $AC = \frac{1}{2}AB$, prove that $BN = 2AM$.

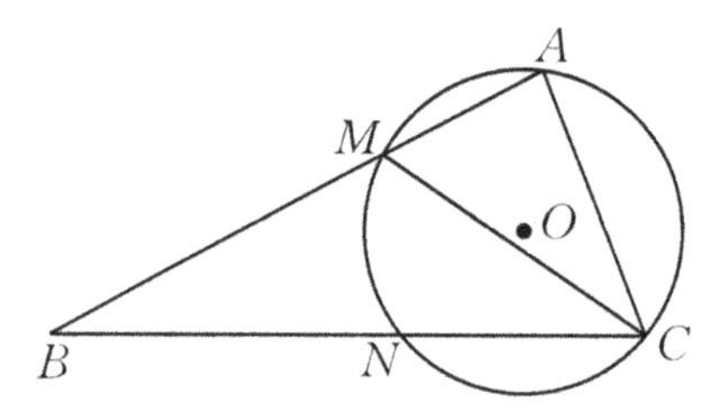

Fig. 19.15

4. As shown in Figure 19.16, the sides AB and AC of $\triangle ABC$ are tangent to $\odot O$ at the points M and N respectively. $\odot O$ and BC intersect at the points E and F. If $BE = EF = FC$, prove that $\angle B = \angle C$.

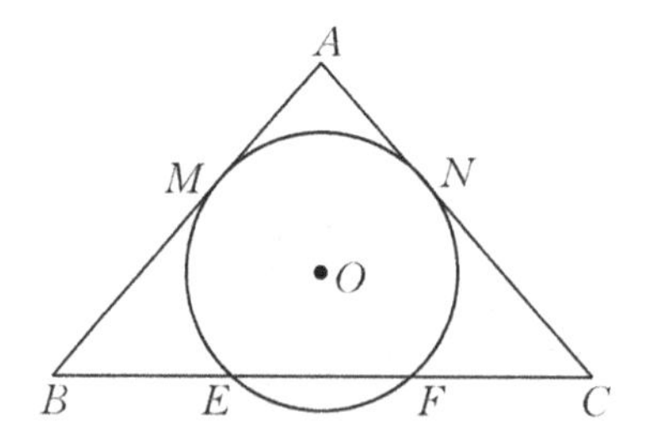

Fig. 19.16

5. As shown in Figure 19.17, the length of the diameter AB of the circle O is a two-digit integer. Exchange the two digits of this two-digit integer, and the new two-digit integer is the length of the chord CD which is perpendicular to the diameter AB. The distance between the intersection H and the center O is a positive rational number. Find the length of AB.

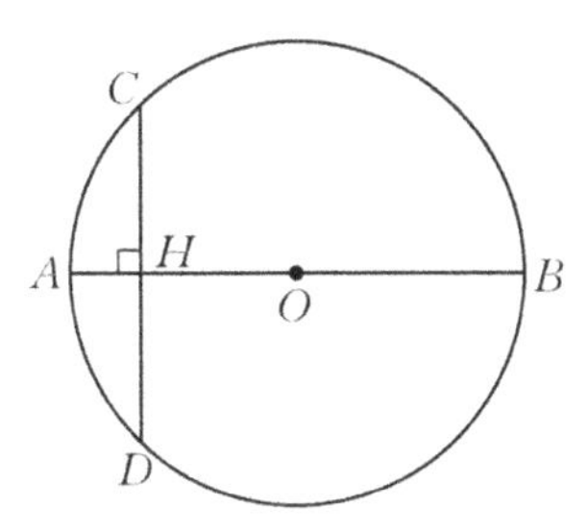

Fig. 19.17

6. Given that the sides of the cyclic quadrilateral $ABCD$ satisfy $AB : BC : CD : DA = 1 : 9 : 9 : 8$, AC and BD intersect at the point P, find $S_{\triangle PAB} : S_{\triangle PBC} : S_{\triangle PCD} : S_{\triangle PDA}$.
7. As shown in Figure 19.18, AB is tangent to $\odot O$ at the point B, M is the midpoint of AB. Draw the secant of the circle through the point M which intersects $\odot O$ at the points C and D. Join AC and extend it to the point E where AC and $\odot O$ intersect. Join AD which intersects $\odot O$ at the point F. Prove that $EF \parallel AB$.

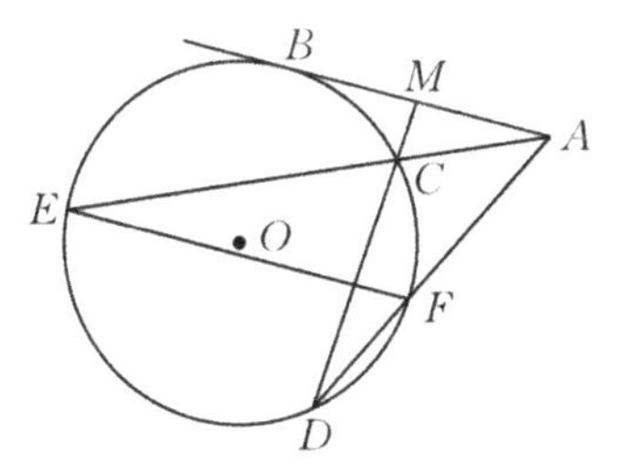

Fig. 19.18

8. As shown in Figure 19.19, A is a point outside the circle O. AB is tangent to the circle O at the point B, AEF is a secant of the circle O. Points E and F are on the circle, $BC \perp AO$ and the foot point is C, CF

intersects OE at the point G. If $CG = 1$, $GF = 4$, $OG = \frac{7}{5}$, find the radius of the circle O.

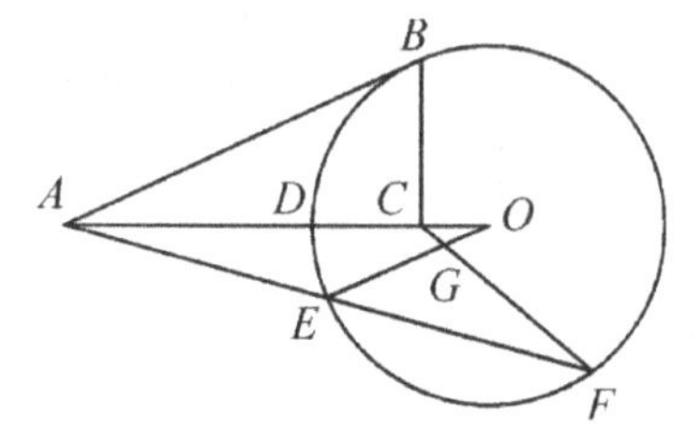

Fig. 19.19

9. As shown in Figure 19.20, $\odot O$ intersects three sides of the regular triangle at 6 points, $AG = 2$, $GF = 13$, $FC = 1$, $HJ = 7$. Find the length of DE.

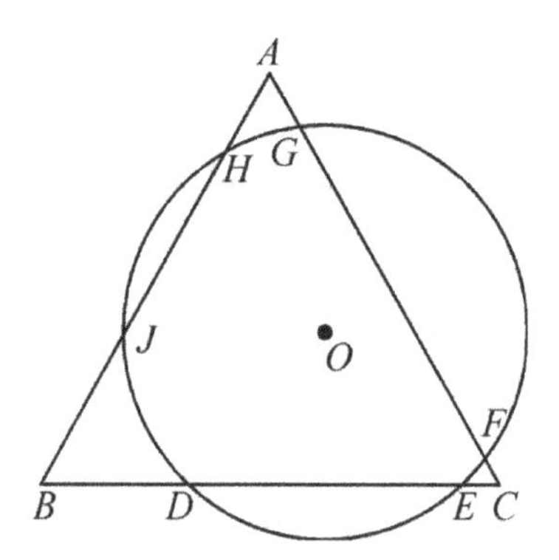

Fig. 19.20

10. As shown in the Figure 19.21, P is a point on the extended line of the side AB of parallelogram $ABCD$. DP intersects AC and BC at the points E and F respectively. EG is a tangent of the circle passing through the three points B, F and P, and the point of tangency is G. Prove that $EG = DE$.

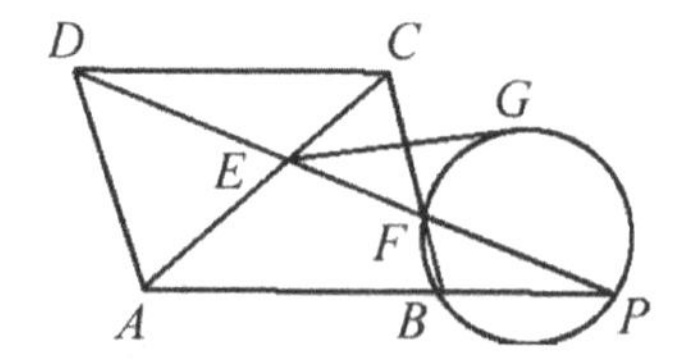

Fig. 19.21

Chapter 20

Four Concyclic Points

As shown in Figure 20.1, if the four points A, B, C and D are concyclic (lie on the same circumference), then the inner opposite angles of the quadrilateral $ABCD$ composed of these four points are supplementary, that is,

$$\angle ABC + \angle ADC = 180^\circ, \quad \angle BAD + \angle BCD = 180^\circ, \quad \text{and}$$
$$\angle ADB = \angle ACB.$$

If in Figure 20.1, extend AB to the point P where AB and DC intersect, join AC which intersects BD at the point Q, then

$$PB \cdot PA = PC \cdot PD, \quad AQ \cdot QC = QB \cdot QD.$$

What about the reverse? We have the following conclusions.

1. If there is a group of inner opposite angles in the quadrilateral $ABCD$ are supplementary, then the quadrilateral $ABCD$ is the inscribed quadrilateral of a circle, that is, the four points A, B, C and D are concyclic.

This is because: as shown in Figure 20.2, $\angle ABC + \angle ADC = 180^\circ$, construct a circle passing through the points A, B and C, connect (or extend) AD to intersect this circle at the point D', then $\angle ABC + \angle AD'C = 180^\circ$. And since $\angle ABC + \angle ADC = 180^\circ$, we get $\angle AD'C = \angle ADC$. It is easy to know that the point D cannot be in (outside) the circle, that is, D and D' coincide, so the conclusion holds.

2. If $\angle ACB = \angle ADB$ in the quadrilateral $ABCD$, then the four points A, B, C and D are concyclic.

As shown in Figure 20.3, similar to the reason why Conclusion 1 holds, Conclusion 2 holds.

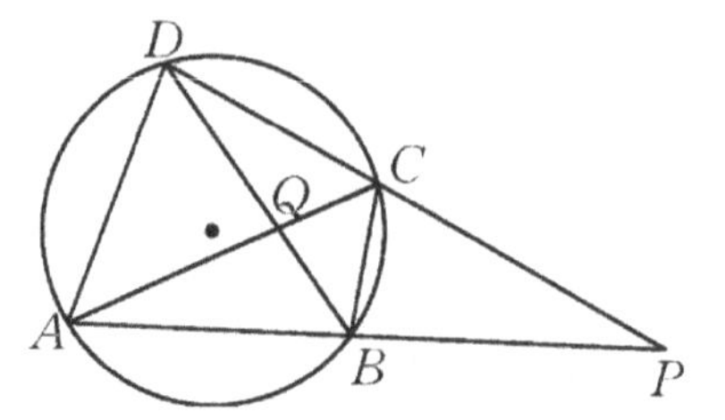

Fig. 20.1

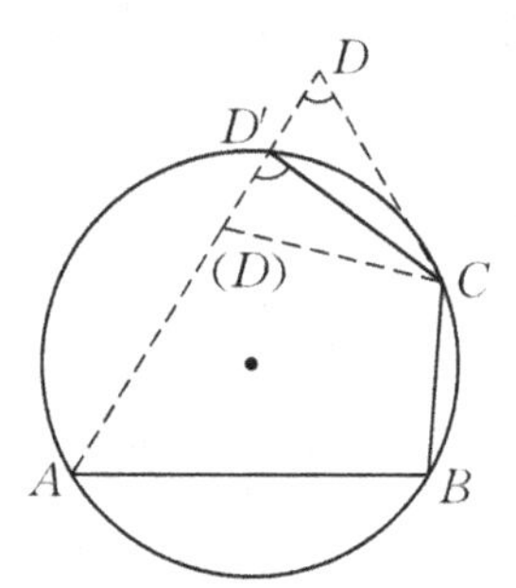

Fig. 20.2

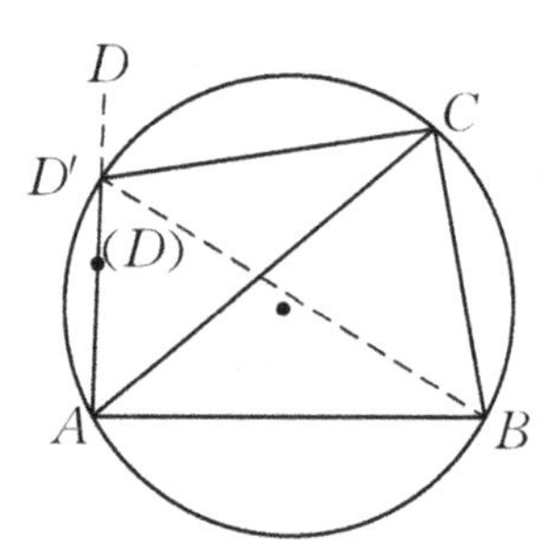

Fig. 20.3

3. If the extended lines of a pair of opposite sides AB and DC of a quadrilateral $ABCD$ intersect at the point P, and $PB \cdot PA = PC \cdot PD$, then the four points A, B, C and D are concyclic.

This is because: from the give condition, we know that $\triangle PBC \backsim \triangle PDA$, then $\angle PBC = \angle PDA$. Since $\angle ABC + \angle PBC = 180°$, we get $\angle ADC + \angle ABC = 180°$. From the conclusion 1, we know that the four points A, B, C and D are concyclic.

In the same way, we can know that:

4. If the diagonals of a quadrilateral $ABCD$ intersect at the point Q, and $QA \cdot QC = QB \cdot QD$, then the four points A, B, C and D are concyclic.

When we know that the four vertices of the quadrilateral $ABCD$ are concyclic, we can get the quantity relations of several angles, especially the equality relations. By using this kind of relations, we can simply get the quantity relations of angles and line segments in the geometric problems, thus solve the related geometric problems.

In fact, when a quadrilateral $ABCD$ is a cyclic quadrilateral, it has not only the relationship of several angles, but also the relationship between line segments, especially the length relationship between the sides and diagonals of quadrilateral, that is, Ptolemy's theorem.

5. (Ptolemy's Theorem) As shown in Figure 20.4, the four vertices of quadrilateral $ABCD$ lie on a common circle, then

$$AB \cdot CD + AD \cdot BC = AC \cdot BD.$$

Proof Take a point M on AC so that $\angle ABM = \angle DBC$, then $\triangle ABM \backsim \triangle DBC$.

Thus we have $\frac{BM}{BC} = \frac{AB}{BD} = \frac{AM}{CD}$, that is, $AB \cdot CD = BD \cdot AM$.

Since $\angle ABD = \angle ABM + \angle MBD = \angle MBD + \angle DBC = \angle MBC$, we know that $\triangle BCM \backsim \triangle BDA$, then $\frac{CM}{AD} = \frac{BC}{BD}$, that is, $AD \cdot BC = BD \cdot CM$.

Hence,

$$AB \cdot CD + AD \cdot BC = BD \cdot (AM + CM) = BD \cdot AC.$$

By using Ptolemy's theorem, we can easily deal with the geometric problems involving proportional line segments (or the equal product of line segments).

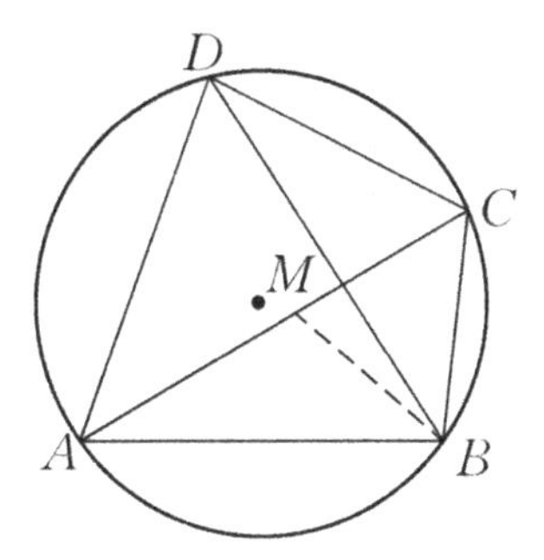

Fig. 20.4

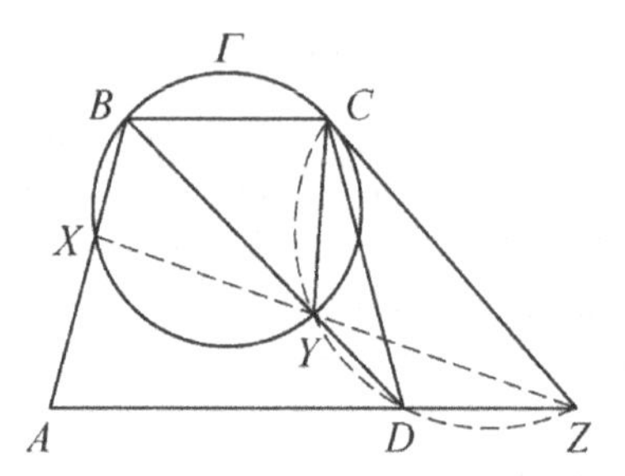

Fig. 20.5

Example 1. In an isosceles trapezoid $ABCD$, $BC \parallel AD$, and AB is not parallel to CD. The circle Γ passing through the points B and C intersects the line segments AB and BD at the points X and Y respectively. The tangent of circle Γ passing through the point C intersects the ray AD at the point Z. Prove the three points X, Y and Z are collinear.

Proof As shown in Figure 20.5, join XY and YZ. Since $BC \parallel AD$, and ZC is a tangent of the circle Γ, we know that $\angle ADB = \angle YBC = \angle YCZ$.

Then we get $\angle YDZ + \angle YCZ = 180°$. Thus the four points C, Y, D and Z are concyclic. Therefore, $\angle CYZ = \angle CDZ$.

From the given condition,

$$\angle XBC = \angle BCD = \angle CDZ = \angle CYZ.$$

Because B, C, Y and X are concyclic, we know that

$$\angle XBC + \angle CYX = 180°.$$

Then we obtain $\angle CYZ + \angle CYX = 180°$.

Hence, the three points X, Y and Z are collinear.

Example 2. Prove that the square of a diagonal of an isosceles trapezoid is equal to the sum of the square of a leg and the product of two bases.

As shown in Figure 20.6, in the trapezoid $ABCD$, $AD = BC$, $AB \parallel CD$, prove that

$$BD^2 = BC^2 + AB \cdot CD.$$

Proof Since a quadrilateral $ABCD$ is an isosceles trapezoid, we know that $\angle BCD = \angle CDA$, $\angle ABC = \angle DAB$. Then we get $\angle DAB + \angle BCD = 180°$, so the four points A, B, C and D are concyclic.

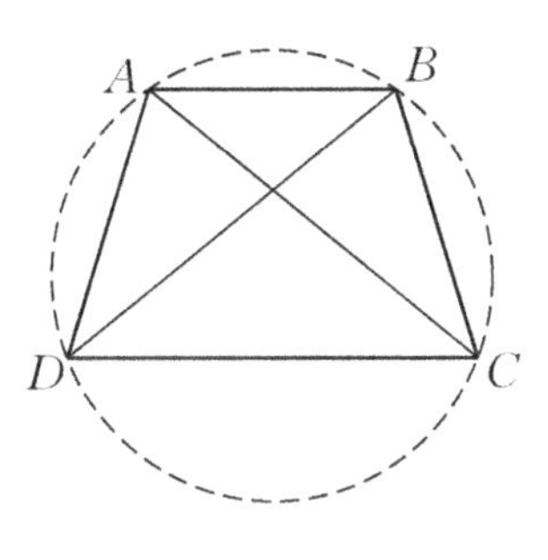

Fig. 20.6

From Ptolemy's theorem,

$$AC \cdot BD = AB \cdot CD + AD \cdot BC.$$

Because $\triangle ADC \cong \triangle BCD$, we know that $AC = BD$. And since $AD = BC$, we obtain

$$BD^2 = AB \cdot CD + BC^2.$$

Think about it Is there any other way to prove it?

Example 3. In $\triangle ABC, \angle A : \angle B : \angle C = 1 : 2 : 4$. Prove that

$$\frac{1}{AB} + \frac{1}{AC} = \frac{1}{BC}.$$

Analysis It can be transformed to prove the equation $AB \cdot BC + AC \cdot BC = AB \cdot AC$. By using Ptolemy's theorem, we think of constructing a cyclic quadrilateral.

Proof As shown in Figure 20.7, construct the circumcircle of $\triangle ABC$, and draw a chord BD which satisfies $BD = BC$, then $\angle BAD = \angle CAB$.

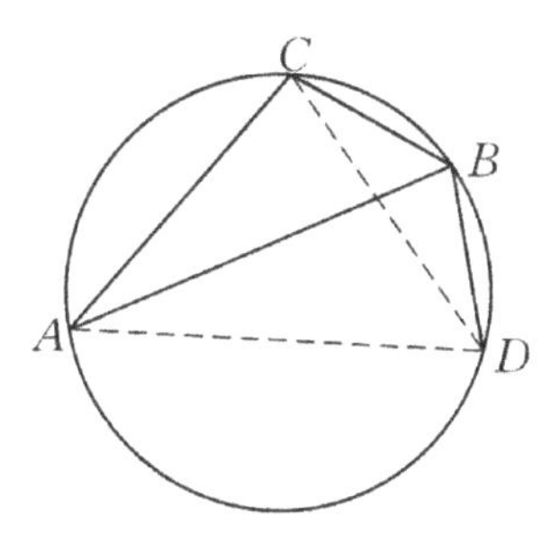

Fig. 20.7

And because $\angle CAB = \angle CDB$, $\angle BCD = \angle CDB$, we get that

$$\begin{aligned}\angle CAD &= 2\angle CAB = \angle CBA = \angle ADC,\\ \angle ABD &= \angle ACD = \angle ACB - \angle DCB\\ &= \angle ACB - \angle CDB = \angle ACB - \angle CAB\\ &= 3\angle CAB = 2\angle CAB + \angle CDB\\ &= \angle CBA + \angle CDB = \angle ADC + \angle CDB\\ &= \angle ADB.\end{aligned}$$

Therefore, $AC = CD$, $AB = AD$.

Since the points A, B, C and D are concyclic, from Ptolemy's theorem, we obtain that

$$AD \cdot BC + AC \cdot BD = AB \cdot CD,$$

i.e. $AB \cdot BC + AC \cdot BC = AB \cdot AC$.

Hence, $\frac{1}{AC} + \frac{1}{AB} = \frac{1}{BC}$.

Example 4. As shown in Figure 20.8, in $\triangle ABC$, $\angle ACB = 90°$, the altitude CH perpendicular to the side AB and the two inner angular bisectors AM and BN of $\triangle ABC$ intersect at P and Q respectively. The midpoints of PM and QN are E and F respectively. Prove that $EF \parallel AB$.

Solution Since BN is the angular bisector of $\angle ABC$, we get $\angle ABN = \angle CBN$.

And because $CH \perp AB$, it can be deduced that

$$\angle CQN = \angle BQH = 90° - \angle ABN = 90° - \angle CBN = \angle CNB.$$

Thus, $CQ = NC$.

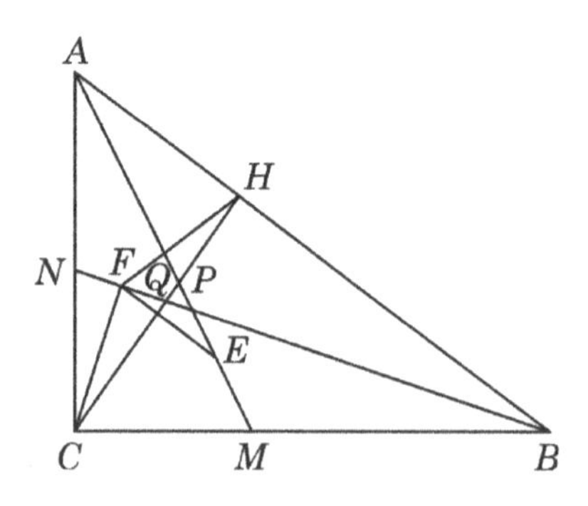

Fig. 20.8

And since F is the midpoint of QN, we know that $CF \perp QN$.

Then we get $\angle CFB = 90° = \angle CHB$, so the points C, F, H and B are concyclic.

And from $\angle FBH = \angle FBC$, it can be seen that $FC = FH$, so the point F is on the perpendicular bisector of CH.

Similarly, it can be proved that the point E is on the perpendicular bisector of CH.

Hence, $EF \perp CH$. And because $AB \perp CH$, we obtain $EF \parallel AB$.

Example 5. As shown in Figure 20.9, $\odot O$ is the circumcircle of $\triangle ABC$, $\angle BAC = 60°$, the altitude BD perpendicular to the side AC and the altitude CE perpendicular to the side AB intersect at the point H. Take a point M on BD so that $BM = CH$.

(1) Prove that $\angle BOC = \angle BHC$.
(2) Prove that $\triangle BOM \cong \triangle COH$.
(3) Find the value of $\frac{MH}{OH}$.

Solution (1) From $\angle BAC = 60°$, it can be known that $\angle BOC = 2\angle BAC = 120°$.

Since $\angle BHC = \angle DHE = 360° - (90° + 90° + \angle BAC) = 120°$, we get $\angle BOC = \angle BHC$.

(2) We know that the points B, C, H and O are concyclic since $\angle BOC = \angle BHC$, then

$$\angle OBM = \angle OCH.$$

And because $BM = CH$, $OB = OC$, we obtain $\triangle BOM \cong \triangle COH$.

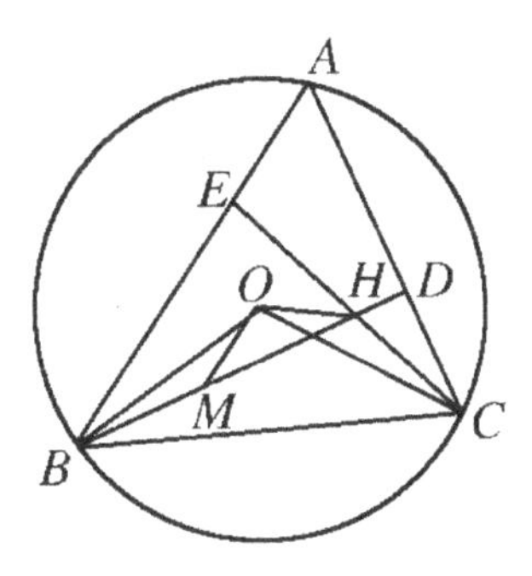

Fig. 20.9

(3) From (2), we know that $OH = OM$, and $\angle COH = \angle BOM$, then

$$\angle OHM = \angle OMH, \angle MOH = \angle BOC = 120°,$$

$$\angle OHM = \frac{1}{2}(180° - 120°) = 30°.$$

In $\triangle OMH$, draw a line segment OP so that $OP \perp MH$ with the foot point of P, then $OP = \frac{1}{2}OH$. From Pythagorean theorem,

$$\left(\frac{1}{2}MH\right)^2 = OH^2 - OP^2 = OH^2 - \left(\frac{1}{2}OH\right)^2.$$

Hence, $\frac{MH}{OH} = \sqrt{3}$.

Remark This question is adapted from the 2002 national high school mathematics league.

Example 6. As shown in Figure 20.10, the circle O_1 and the circle O_2 intersect at the points A and B, P is a point on the extended line of BA. The secant PCD intersects the circle O_1 at the points C and D, and the secant PEF intersects the circle O_2 at the points E and F. Prove that the four points C, D, F and E are concyclic.

Proof As shown in Figure 20.11, join CE and DF.

From the given condition, the four points A, B, D and C are concyclic, then we have

$$PA \cdot PB = PC \cdot PD.$$

Since the four points A, B, F and E are concyclic, then we have

$$PA \cdot PB = PE \cdot PF.$$

Therefore, $PC \cdot PD = PE \cdot PF$, that is, $\frac{PC}{PF} = \frac{PE}{PD}$.

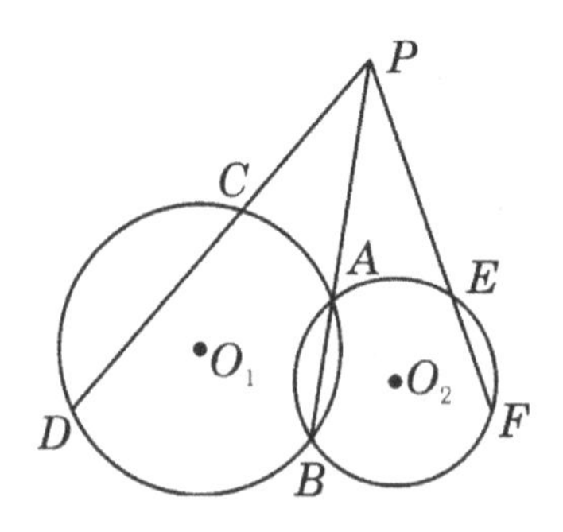

Fig. 20.10

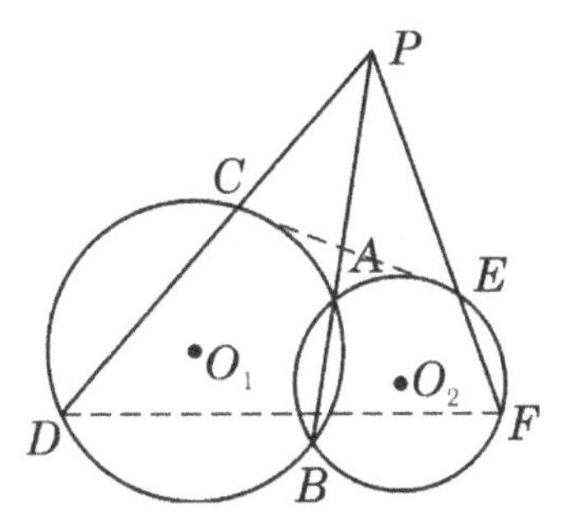

Fig. 20.11

Since $\angle DPF = \angle EPC$, we obtain that $\triangle DPF \backsim \triangle EPC$, then $\angle ECP = \angle PFD$.

And since $\angle ECP + \angle DCE = 180°$, we get $\angle DCE + \angle PFD = 180°$.

Hence, the points C, D, F and E are concyclic.

Remark In other words, if the lines CD, AB and EF intersect at the point P, the four points A, B, D and C are concyclic, and the points A, B, F and E are concyclic, then the points C, D, F and E are concyclic.

Think about it Is the converse of the above proposition true?

Example 7. Given that the points A, B, D and C are concyclic, the points A, B, F and E are concyclic, and the points C, D, F and E are concyclic, then the lines AB, CD and EF are parallel to each other or intersect at one point.

Proof As shown in Figure 20.12, suppose that $AB \parallel CD$, and EF is not parallel to AB, then let the line FE intersect AB at the point P. Join PC which intersects the circle ABC at the point D'. Then we have

$$PC \cdot PD' = PA \cdot PB = PE \cdot PF.$$

From Example 6, the points C, D', F and E are concyclic, that is, the point D' is on the circle passing through the four points C, D, F and E. Therefore, the circle CEF and the circle ABC intersect at the points C, D and D', and these two circles do not coincide, C and D do not coincide, C and D' do not coincide. So the points D and D' coincide, then the lines DC and AB also intersect at the point P.

Hence, when $AB \parallel CD$, there must be $AB \parallel EF$. When AB intersects EF at the point P, the point P must be on the line CD. So the conclusion is true.

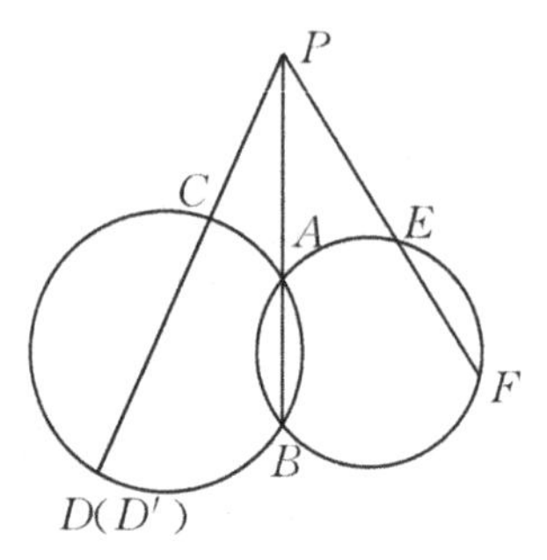

Fig. 20.12

Think about it

The role of "ladder"

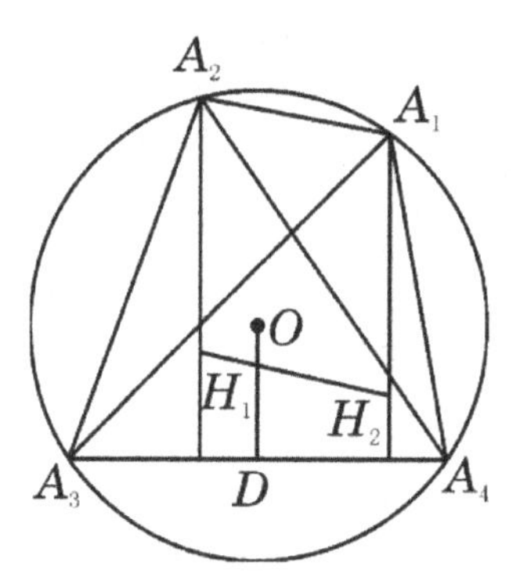

Fig. 20.13

Example Given an cyclic quadrilateral $A_1A_2A_3A_4$, the points H_1, H_2, H_3, and H_4 are the orthocenter of $\triangle A_2A_3A_4$, $\triangle A_3A_4A_1$, $\triangle A_4A_1A_2$ and $A_1A_2A_3$ respectively.

(1) Let D be the midpoint of A_3A_4. Then $A_2H_1 = 2OD$.
(2) Points A_2 and H_2, A_1 and H_1 are all central symmetrical about a fixed point P.
(3) Four points H_1, H_2, H_3 and H_4 are concyclic.

Students, you can try to do the third question directly without considering the first and second questions, and you will realize that the question is more difficult.

Therefore, we usually deal with the problems we encounter, either we have added a ladder, or we try to use the basic conclusions we have mastered and gradually approach them.

Try to solve the above problem.

Exercises

1. If P is any point on the minor arc $\overset{\frown}{BC}$ of circumscribed circle of the regular triangle ABC, prove that $PA = PB + PC$.
2. As shown in Figure 20.14, construct a line OQ through the center of the circumscribed circle of $\triangle ABC$ so that $OQ \perp AB$ and OQ intersects AB, BC and AC at the points F, P and Q respectively. Prove that $OA^2 = OP \cdot OQ$.

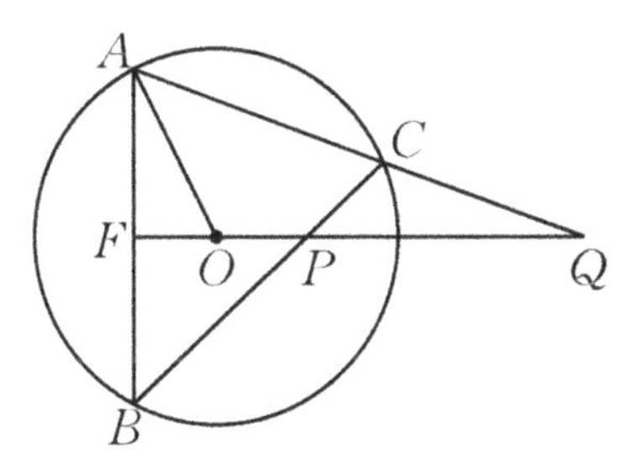

Fig. 20.14

3. In an equilateral triangle ABC, D and E are the points on the sides BC and AC respectively, and $BD = CE = \frac{1}{2}CD$. Join BE and AD that intersect at the point P. Join CP, prove that $CP \perp AD$.
4. If a, b, x and y are all positive real numbers, and $a^2+b^2 = 1$, $x^2+y^2 = 1$, prove that

$$ax + by \leq 1.$$

5. As shown in Figure 20.15, the quadrilateral $ABCD$ is a square, $\angle EAF = 45°$, $EG \perp AC$, $FH \perp AC$. Prove that the center of the circle passing through the three points H, G and B is on BC.

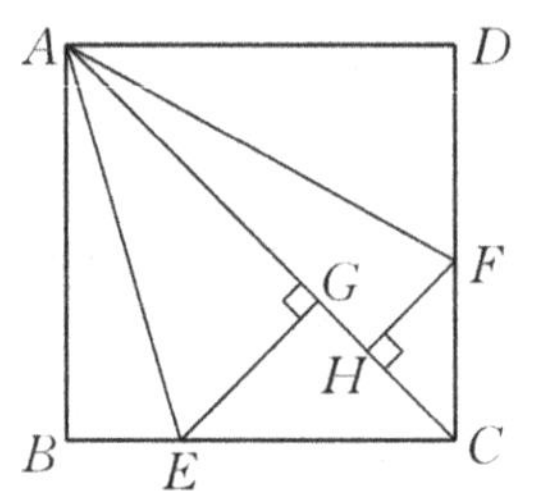

Fig. 20.15

6. As shown in Figure 20.16, in a parallelogram $ABCD$, E is a point on the diagonal BD, and $\angle ECD = \angle ACB$. The extended line of AC and the circumscribed circle of ABD intersect at the point F.Prove that $\angle DFE = \angle AFB$.

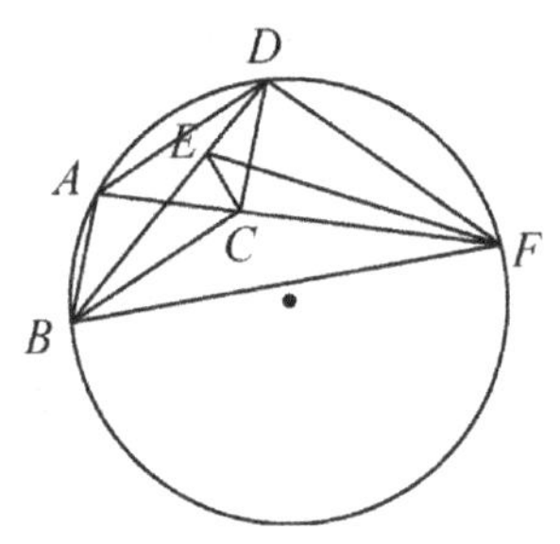

Fig. 20.16

7. As shown in Figure 20.17, P is a point on the minor arc $\widehat{AB}$ of circumscribed circle of the regular triangle, join PC which intersects AB at

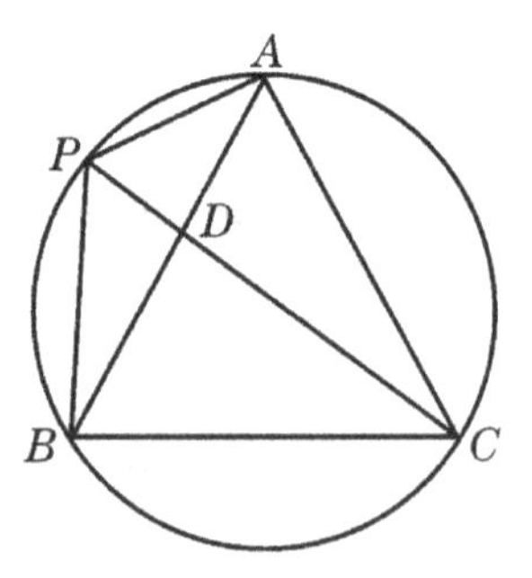

Fig. 20.17

the point D. Prove that

$$\frac{1}{PA}+\frac{1}{PB}=\frac{1}{PD}.$$

8. Let a, b and c be the lengths of three sides of the triangle ABC, which satisfy $a^2 = b(b+c)$. Prove that $\angle A = 2\angle B$.
9. Given that in a cyclic quadrilateral $ABCD, CB = CD$, prove that $CA^2 - CB^2 = AB \cdot AD$.
10. As shown in Figure 20.18, AD and AH are the angular bisector and the height of $\triangle ABC$ (where $AB > AC$) respectively. M is the midpoint of AD, the circumscribed circle of $\triangle MDH$ intersects CM at the point E. Prove that $\angle AEB = 90°$.

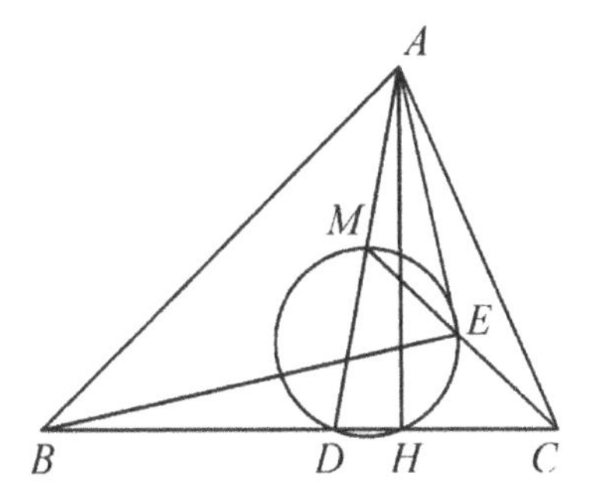

Fig. 20.18

Chapter 21

Problems of Geometric Fixed Value

In plane geometry, there is a certain genre of problems that includes but not limited to: the invariance of the sum, difference, product and quotient between line segments, the quantity invariance between angles, the area invariance of geometric figures, and so on. These problems are called problems of geometric fixed value.

Example 1. As shown in Figure 21.1, let C be the midpoint of the fixed arc $\overset{\frown}{AB}$ on a fixed circle, P be any point on $\overset{\frown}{AB}$, where the points C and P are not on the same side of the line AB. Prove that $\frac{PA+PB}{PC}$ is a fixed value.

Analysis First find out what the fixed value is.

By the arbitrariness of a point P, consider that the point P coincides with the point A, then

$$\frac{PA+PB}{PC}=\frac{AB}{AC}.$$

Since A, B and C are fixed points, then AB and AC are fixed values, so the fixed value in the question may be $\frac{AB}{AC}$.

Try to prove that $\frac{PA+PB}{PC}=\frac{AB}{AC}$, we can transform $PA+PB$, and find similar triangles.

Proof As shown in Figure 21.2, take a point E on the extended line of AP so that $PE=PB$. Join BE and BC, then we have $\angle EBP=\angle BEP$.

Since C is the midpoint of $\overset{\frown}{AB}$, $\angle APB=2\angle CPB$. From $\angle APB=\angle PBE+\angle PEB$, we know $\angle CPB=\angle PEB$.

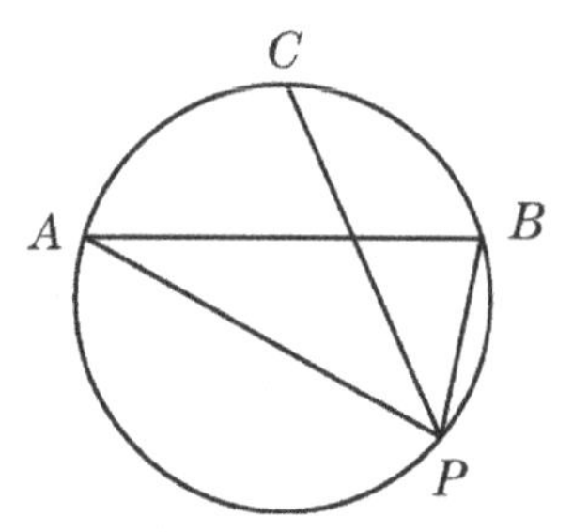

Fig. 21.1

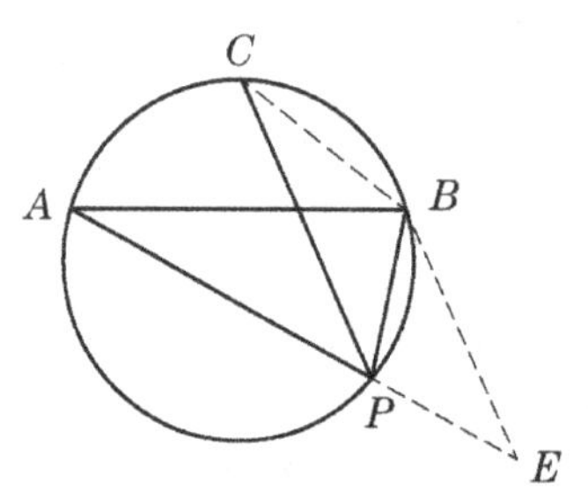

Fig. 21.2

And since $\angle PAB = \angle PCB$, we get $\triangle ABE \backsim \triangle CBP$, then $\frac{AE}{PC} = \frac{AB}{BC}$, that is,

$$\frac{PA + PB}{PC} = \frac{AB}{AC} = a \text{ fixed value.}$$

Remark (1) The above proof is a common way. It is beneficial to use the similar triangles properly for the fixed value problems of the proportion expression (product expression) of the line segments.
(2) In fact, we can also use Ptolemy's theorem to solve this problem. Since C is the midpoint of $\widehat{AB}$, we know that $BC = AC$. For the cyclic quadrilateral $APBC$, it can be seen that $AP \times BC + PB \times AC = PC \times AB$. So $\frac{PA+PB}{PC} = \frac{AB}{AC}$.

As seen from Example 1, in order to solve the problem of geometric fixed value, because it is necessary to know its fixed value, we often consider the invariance of the quantity in special figures to get the fixed value (quantitative, positioning); or we can use the viewpoint of motion change to investigate the fixed value when the motion changes to the limit position to get the fixed value of the general problem.

Once the specific fixed value is known, the problem of fixed value will be transformed into a proof problem, which can be easily solved by using the basic idea of the proof problem.

Example 2. Let PQ be a certain chord in a fixed circle, where the moving chords PA and PB are respectively on both sides of PQ and there always be $\angle APQ = \angle BPQ$. Prove that the line AB has fixed orientation.

Analysis The key lies in the specific orientation.

Using the viewpoint of motion change, the limit state is considered.

When $\angle APQ$ and $\angle BPQ$ are getting smaller and tend to zero, the point A and the point B approach the point Q infinitely along a fixed circle. At this time, the limit position of the line AB is the tangent QT of the circle at the fixed point Q.

Try to prove that $AB \parallel QT$.

Proof As shown in Figure 21.3, construct the tangent QT of the circle through the point Q, and join BQ, then

$$\angle BQT = \angle QPB.$$

Since $\angle APQ = \angle QPB$, $\angle APQ = \angle ABQ$, we obtain $\angle QPB = \angle ABQ$. Then $\angle BQT = \angle ABQ$, that is, the line AB is parallel to QT.

Since QT is a fixed line, the line AB has fixed orientation.

Remark The key to solve the problem in Example 1 is to use special points (figures) to get the possible fixed value; the key to solve the problem in Example 2 is to use the limit state of the change of points or angles or line segments to find the fixed value. The above methods are two basic ideas to solve the problems of geometric fixed value.

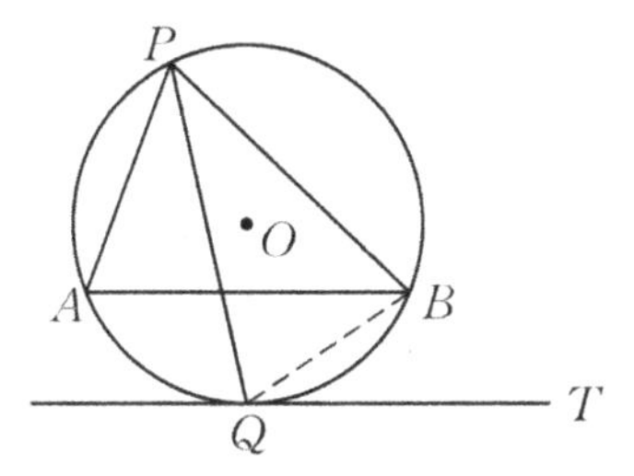

Fig. 21.3

Example 3. Let n be a positive integer. Given n points in $\triangle ABC$, connect these points to the points A, B and C with some line segments that have no common points except the endpoints, such that $\triangle ABC$ is divided into t small triangle. Prove that t is a fixed value, independent of connection of the line segments.

Solution From the given condition, the sum of the interior angles of these t triangles is $t\pi$, the sum of the interior angles of $\triangle ABC$ is π, and the sum of the round angles formed by each point of the n given points is $n \cdot 2\pi$.

Since the sum of the interior angles of these t triangles is equal to the sum of the interior angles of $\triangle ABC$ plus the sum of the round angles formed by each point of the n given points, we know that $t\pi = \pi + n \cdot 2\pi$. Then we get $t = 2n + 1$.

So the conclusion holds.

Example 4. Given that the equilateral triangle ABC is inscribed on the unit circle O (i.e. the radius is 1), and P is any point on the circle O, prove that $PA^2 + PB^2 + PC^2$ is a fixed value.

Analysis As shown in Figure 21.4, when the point P is C, the value of $PA^2 + PB^2 + PC^2$ is equal to 2 BC^2 (a fixed value).

Try to prove that $PA^2 + PB^2 + PC^2 = 2BC^2$.

Proof Since the points A, B, P and C are concyclic, we know that $\angle BPC = 180° - \angle BAC$. Because $\triangle ABC$ is an equilateral triangle, we get $\angle BPC = 120°$.

Construct a line CD through the point C which is perpendicular to BP and intersects BP at the point D. From $\angle BPC = 120°$, we can get $\angle CPD = 180° - \angle BPC = 60°$, so $PD = PC \cdot \cos \angle CPD = \frac{1}{2}PC$.

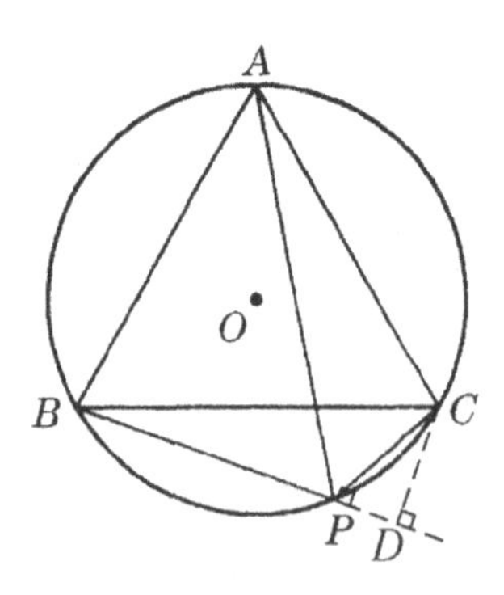

Fig. 21.4

In the right triangle BDC and the right triangle PDC, $BC^2 = CD^2 + BD^2$, $CD^2 + PD^2 = PC^2$.

And since $BD = BP + PD$, $PD = \frac{1}{2}PC$, $CD = \frac{\sqrt{3}}{2}PC$, we get $BC^2 = BP^2 + PC^2 + BP \cdot PC$.

By Ptolemy's theorem, it can be known that $BC \cdot PA = PB \cdot AC + PC \cdot AB$.

Since $AB = BC = AC$, we obtain $PA = PB + PC$, then

$$\begin{aligned} & PA^2 + PB^2 + PC^2 \\ & = (PB + PC)^2 + PB^2 + PC^2 \\ & = 2(PB^2 + PC^2 + PB \cdot PC) \\ & = 2BC^2. \end{aligned}$$

Because the radius of the circle O is 1, $BC = 2 \times 1 \cdot \sin 60° = \sqrt{3}$.

Hence, $PA^2 + PB^2 + PC^2 = 2BC^2 = 6$.

Remark It's easy to prove a basic conclusion that $PA = PB + PC$.

Example 5. Let O be any point inside triangle ABC. The extended lines of AO, BO and CO intersect their opposite sides at the points A_1, B_1 and C_1 respectively. Prove that $\frac{AO}{AA_1} + \frac{BO}{BB_1} + \frac{CO}{CC_1}$ is a fixed value.

Analysis From the proportion formula of the line segment, we can think of the area or the similarity of triangles or the properties of a circle.

Try to transform the area.

Proof As shown in Figure 21.5, the height of the triangle ABO with the base of AO is equal to the height of the triangle ABA_1 with the base of AA_1,

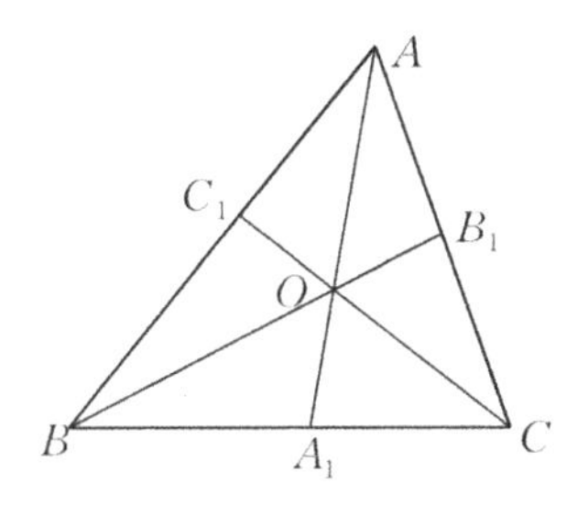

Fig. 21.5

then

$$\frac{S_{\triangle AOB}}{S_{\triangle AA_1B}} = \frac{AO}{AA_1}.$$

Similarly, $\dfrac{S_{\triangle AOC}}{S_{\triangle AA_1C}} = \dfrac{AO}{AA_1}$. Then we get

$$\frac{AO}{AA_1} = \frac{S_{\triangle AOB} + S_{\triangle AOC}}{S_{\triangle AA_1B} + S_{\triangle AA_1C}} = \frac{S_{\triangle AOB} + S_{\triangle AOC}}{S_{\triangle ABC}}.$$

In the same way,

$$\frac{BO}{BB_1} = \frac{S_{\triangle AOB} + S_{\triangle BOC}}{S_{\triangle ABC}},$$

$$\frac{CO}{CC_1} = \frac{S_{\triangle AOC} + S_{\triangle BOC}}{S_{\triangle ABC}}.$$

Hence,

$$\frac{AO}{AA_1} + \frac{BO}{BB_1} + \frac{CO}{CC_1} = \frac{2(S_{\triangle AOB} + S_{\triangle BOC} + S_{\triangle COA})}{S_{\triangle ABC}} = 2.$$

Remark

(1) We can find $\dfrac{A_1O}{AA_1}$. As shown in Figure 21.6, since the bases of $\triangle OBC$ and $\triangle ABC$ are equal and the heights are unequal, but the ratio of the heights is equal to $\dfrac{A_1O}{AA_1}$, we get $\dfrac{S_{\triangle OBC}}{S_{\triangle ABC}} = \dfrac{A_1O}{AA_1}$.

Similarly, $\dfrac{B_1O}{BB_1} = \dfrac{S_{\triangle OAC}}{S_{\triangle ABC}}$, $\dfrac{C_1O}{CC_1} = \dfrac{S_{\triangle AOB}}{S_{\triangle ABC}}$.

Thus we obtain $\dfrac{A_1O}{AA_1} + \dfrac{B_1O}{BB_1} + \dfrac{C_1O}{CC_1} = 1$. Since $\dfrac{AO}{AA_1} = 1 - \dfrac{A_1O}{AA_1}$, $\dfrac{BO}{BB_1} = 1 - \dfrac{B_1O}{BB_1}$, $\dfrac{CO}{CC_1} = 1 - \dfrac{C_1O}{CC_1}$, the conclusion in question holds.

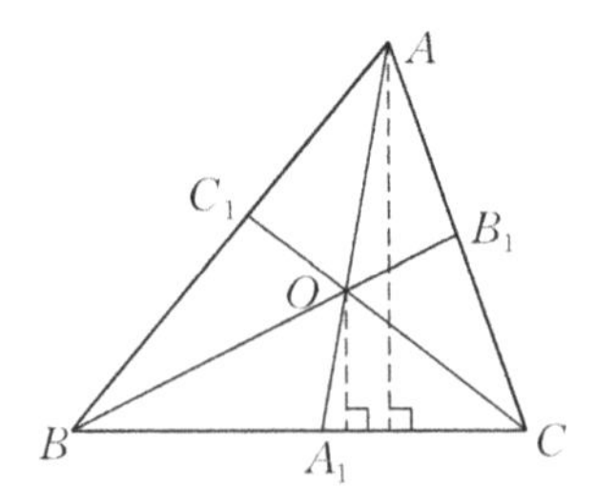

Fig. 21.6

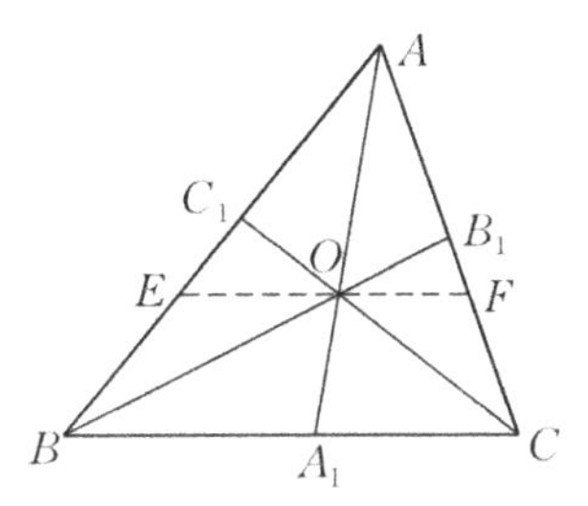

Fig. 21.7

(2) The similarity of triangles can also be used.
As shown in Figure 21.7, construct the parallel line of BC through the point O, which intersects AB at the point E and intersects AC at the point F, then $\triangle AEO \backsim \triangle ABA_1$, $\triangle AOF \backsim \triangle AA_1C$. So from $\frac{AO}{AA_1} = \frac{EO}{BA_1} = \frac{OF}{A_1C}$, we get

$$\frac{AO}{AA_1} = \frac{EO + OF}{BC}.$$

Since $\triangle C_1EO \backsim \triangle C_1BC$, we know that $\frac{C_1O}{C_1C} = \frac{EO}{BC}$, then

$$\frac{CO}{CC_1} = \frac{BC - EO}{BC} = 1 - \frac{EO}{BC}.$$

Similarly, $\frac{BO}{BB_1} = 1 - \frac{OF}{BC}$.
Hence the conclusion holds.

(3) Example 6 can be easily solved by using the results of Example 5.

Example 6. Let O be any point inside triangle ABC. The extended lines of AO, BO and CO intersect their opposite sides at the points A_1, B_1 and C_1 respectively. Given that $AO = 6$, $BO = 9$, $OA_1 = 6$, $OB_1 = 3$, $CC_1 = 20$, find the area of triangle ABC.

Solution As shown in Figure 21.8, let $S_{\triangle AOC_1} = S_2$, $S_{\triangle AOB_1} = S_1$, $S_{\triangle OCB_1} = S_3$.

From the conclusion of Example 5,

$$\frac{AO}{AA_1} + \frac{BO}{BB_1} + \frac{CO}{CC_1} = 2.$$

And since $BO = 9$, $OB_1 = 3$, $AO = 6$, $OA_1 = 6$, $CC_1 = 20$, we get that $OC = 15$, $O_1C = 5$, that is, $OC = 3O_1C$, and $OA_1 = OA$, $OB = 3OB_1$.

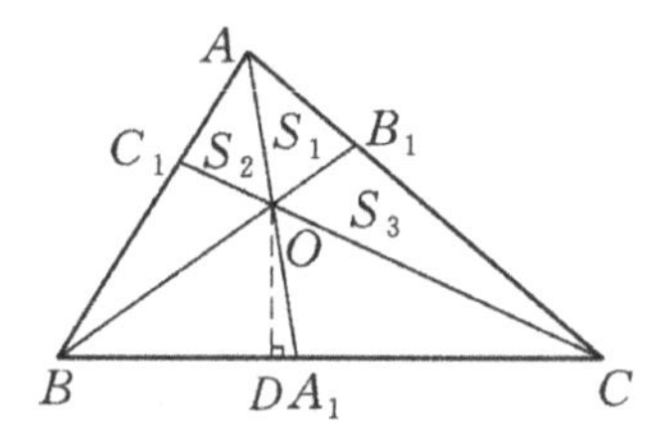

Fig. 21.8

Thus, we know that

$$\begin{aligned}S_{\triangle OBC_1} &= \frac{1}{2}OC_1 \cdot OB \cdot \sin\angle BOC_1 \\ &= \frac{1}{2}OB_1 \cdot OC \cdot \sin\angle B_1OC \\ &= S_{\triangle OB_1C},\end{aligned}$$

i.e. $S_{\triangle OBC_1} = S_3$.

Similarly, $S_{\triangle BOA_1} = 3S_1$, $S_{\triangle OA_1C} = 3S_2$.

From $AO = OA_1$, we obtain that $S_{\triangle AOC} = S_{\triangle COA_1}$, $S_{\triangle AOB} = S_{\triangle A_1OB}$, that is, $S_1 + S_3 = 3S_2$, $S_2 + S_3 = 3S_1$. Then $S_1 = S_2$, $S_3 = 2S_1 = 2S_2$.

Because $S_{\triangle COA_1} = 3S_2$, $S_{\triangle BOA_1} = 3S_1$, we get $S_{\triangle COA_1} = S_{\triangle BOA_1}$, then $BA_1 = CA_1$.

Therefore, in $\triangle OCB$, we have

$$BC^2 = 2(CO^2 + BO^2) - 4OA_1^2 = 2(15^2 + 9^2) - 4 \times 6^2 = 6^2 \times 13.$$

Then $BC = 6\sqrt{13}$.

As shown in Figure 21.8, draw a line segment OD through the point O so that $OD \perp BC$ with the foot point of D.

Let $DA_1 = x$. Then

$$BD = BA_1 - x = \frac{1}{2}BC - x,$$

$$CD = \frac{1}{2}BC + x.$$

Therefore,

$$OD^2 = OB^2 - BD^2 = OC^2 - CD^2,$$

then $x = \frac{(OC-OB)\cdot(OC+OB)}{2BC} = \frac{12}{\sqrt{13}}$.

Thus we get

$$OD^2 = OA_1^2 - A_1D^2 = 6^2 - \frac{12^2}{13} = \frac{9 \times 6^2}{13},$$

i.e. $OD = \frac{18}{\sqrt{13}}$.
Then it can be obtained that $S_{\triangle COB} = \frac{1}{2}BC \cdot OD = 54$.
Hence, $S_{\triangle ABC} = 2S_{\triangle COB} = 108$.

Exercises

1. Let A be a fixed point outside the circle, and P be a fixed point on the circle. Construct any secant ABC from the point A to the circle. If the midpoints of PB and PC are the points M and N respectively, prove that the straight line MN always passes a fixed point Q.
2. Given that the tangents PA and PB of a circle O are drawn from a point P outside the circle O, and the points of tangency are the points A and B respectively. Take any point C on the minor arc $\overset{\frown}{AB}$, and construct the tangent of the circle O through the point C which intersects PA and PB at the points D and E respectively. Prove that

   (1) the perimeter of $\triangle PDE$ is a fixed value;
   (2) the size of $\angle DOE$ is a fixed value.

3. Let P be a fixed point on the angular bisector of a fixed angle AOB. Construct a circle with the chord of OP which intersects OA and OB at the points C and D respectively. Prove that $OC + OD$ is a fixed value.
4. As shown in Figure 21.9, AD is the median to the base BC of a given triangle ABC, P is any point on the line segment BC, and construct the parallel line of AD from the point P which intersects the lines AB

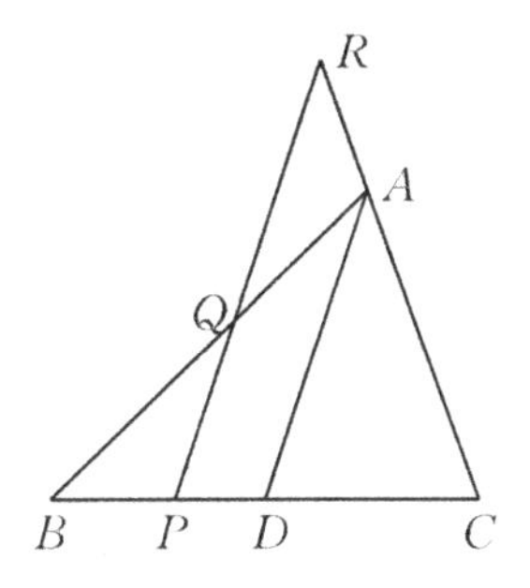

Fig. 21.9

and CA at the points Q and R respectively. Prove that $PQ + PR$ is a fixed value.

5. Given that AB and CD are two fixed diameters of a circle O, P is any point on the circumference, prove that the distance between the projections of the point P on AB and CD is a fixed value.
6. As shown in Figure 21.10, O is the midpoint of the base BC (with fixed length) of the isosceles triangle ABC. Construct a semicircle with the center of point O which is tangent to the legs of triangle ABC at the points D and E respectively. Draw the tangent of the semicircle through any point F on the semicircle, which intersects AB and AC at the points M and N respectively. Prove that $BM \cdot CN$ is a fixed value.

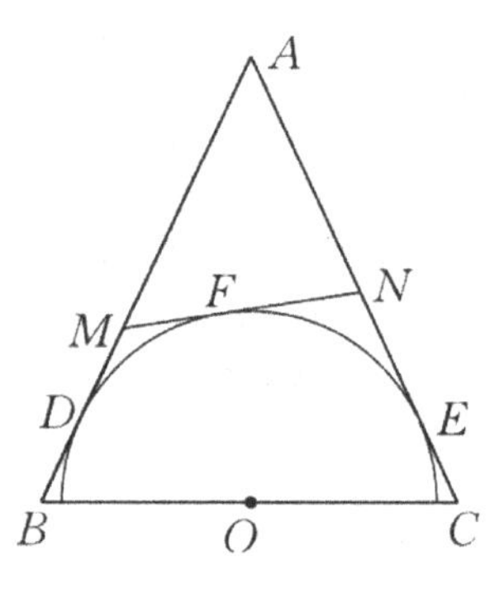

Fig. 21.10

7. As shown in Figure 21.11, A is a fixed point on the fixed circle O. Take any point B on the tangent passing through the point A, and make any secant CDE through the midpoint C of the line segment AB, which intersects the circle O at the points D and E. The lines BD and BE intersect the circle O at the points P and Q respectively. Prove that the chord PQ has fixed orientation.

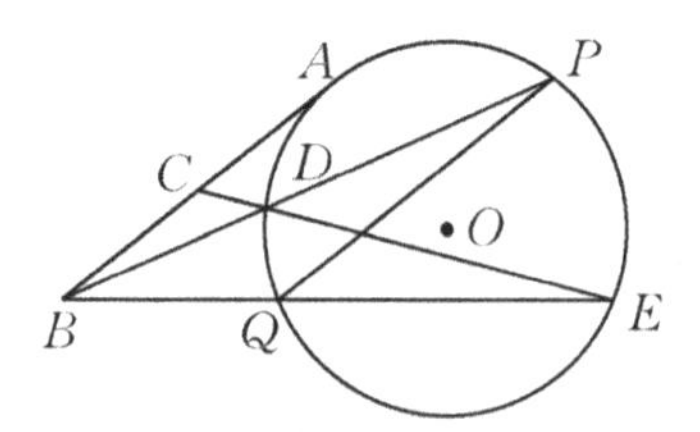

Fig. 21.11

8. If the points M and N are on the top base and the bottom base of the trapezoid $ABCD$ respectively, and MN divide the area of the trapezoid equally, prove that MN must pass through a fixed point.
9. Construct a straight line arbitrarily through the fixed point A on the angular bisector of $\angle XOY$, which intersects OX and OY at the points P and Q respectively. Prove that $\frac{1}{OP} + \frac{1}{OQ}$ is a fixed value.
10. In the quadrilateral $ABCD$, if the diagonals satisfy $AC = BD$, prove that the perimeter of the quadrilateral which is inscribed in the quadrilateral and each side is parallel to the two diagonal lines is a fixed value.
11. As shown in Figure 21.12, let PMN be a secant passing through the center of the circle O, PAB be another secant, and M, N, A, B be the intersection points of these two secants and the circle O. Prove that $\frac{AM \cdot BM}{AN \cdot BN}$ is a fixed value.

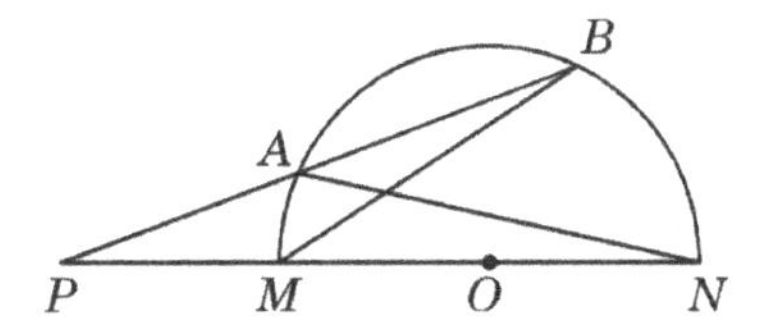

Fig. 21.12

Chapter 22

Five Centers of Triangle

There are five important points in a triangle: barycenter, circumcenter, orthocenter, incenter and escenter.

The barycenter is the intersection of the three medians of a triangle. Each median is divided by this point into two parts with a ratio of 2:1.

The circumcenter is the intersection of perpendicular bisectors of the three sides of a triangle. The distances from this point to the three vertices of the triangle are equal. A circle with this point as its center and the distance from this point to one vertex as its radius passes through the three vertices of the triangle, that is, this point is the center of the circumscribed circle of the triangle.

The orthocenter is the intersection of the three altitudes made from each vertex of a triangle to its opposite side. Six four-point circles can be obtained from the seven points including three vertices, three perpendicular feet and the orthocenter of the triangle.

The incenter is the intersection of three inner angular bisectors of a triangle. The distances from this point to the three sides of the triangle are equal. The circle with this point as the center and the distance from this point to one side of the triangle as the radius is tangent to the three sides of the triangle. The area of the triangle is equal to the product of the radius of the inscribed circle and half of the circumference of the triangle.

The escenter is the intersection of outer angular bisectors of any two angles of a triangle and the inner angular bisector of the third angle. The distances from this point to the straight line of the three sides of the triangle are equal. The circle with this point as the center and the distance from it to the straight line of one side of the triangle as the radius is tangent to one side of the triangle and the extended lines of the other two sides respectively. The triangle has three escenters.

There are two basic contents about the problem of "five centers" of a triangle, the one is about the common points of lines, and the other is about some applications of the "five centers". By dealing with the problem of "five centers" of the triangle, we can get the basic method of dealing with the problem of lines sharing one point, and master the basic idea of using "five centers" to deal with relationships between lines and angles.

Example 1. Prove that the three medians, three inner angular bisectors and three altitudes of a triangle intersect at the same point respectively.

Proof (1) Prove that the three medians of a triangle intersect at the same point.

Let D, E and F be the midpoints of three sides BC, CA and AB of $\triangle ABC$ respectively, then we have

$$\frac{BD}{DC}\cdot\frac{CE}{EA}\cdot\frac{AF}{FB}=\frac{BD}{BD}\cdot\frac{CE}{CE}\cdot\frac{AF}{AF}=1.$$

According to the converse theorem of Ceva's theorem, three lines AD, BE and CF intersect at the same point. (You can try to prove it with area method.)

(2) Prove that the three inner angular bisectors intersect at the same point I.

As shown in Figure 22.1, let D, E and F be the intersection of three inner angular bisectors AD, BE and CF of $\triangle ABC$ and the sides BC, CA and AB respectively. Then

$$\frac{BD}{DC}=\frac{AB}{AC},\quad \frac{CE}{EA}=\frac{BC}{AB},\quad \frac{AF}{FB}=\frac{AC}{BC}.$$

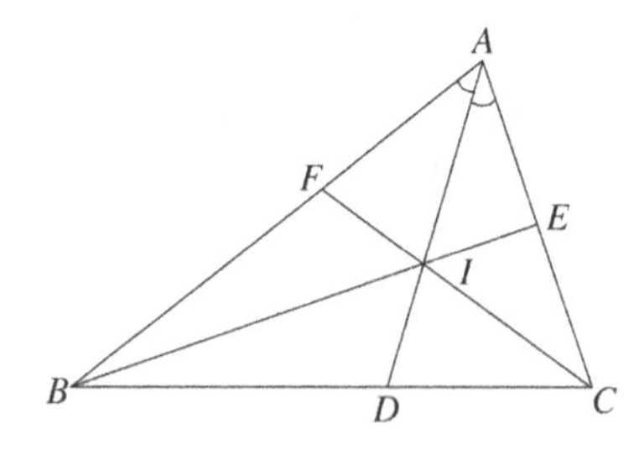

Fig. 22.1

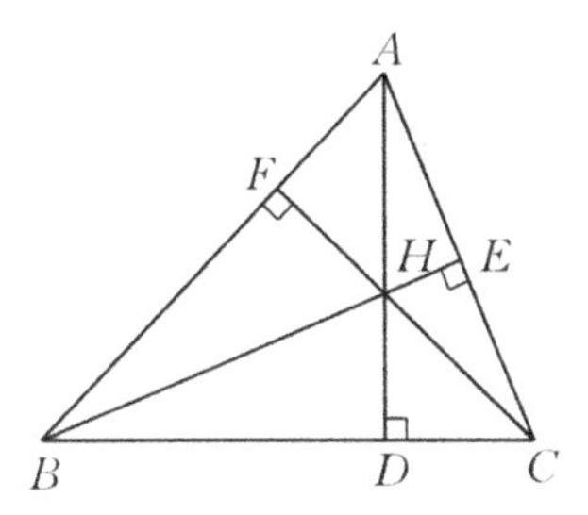

Fig. 22.2

Thus we get

$$\frac{BD}{DC} \cdot \frac{CE}{EA} \cdot \frac{AF}{FB} = 1.$$

So the lines AD, BE and CF intersect at the same point I.

(3) Prove that three altitudes of a triangle intersect at the same point H.

When $\triangle ABC$ is an acute triangle, as shown in Figure 22.2, then

$$BD = AB \cdot \cos B, CD = AC \cdot \cos C,$$

$$CE = BC \cdot \cos C, AE = AB \cdot \cos A,$$

$$AF = AC \cdot \cos A, FB = BC \cdot \cos B.$$

Thus we get $\frac{BD}{DC} \cdot \frac{CE}{EA} \cdot \frac{AF}{FB} = 1.$

So the lines AD, BE and CF intersect at the same point.

When $\triangle ABC$ is an obtuse triangle, might as well let $\angle A$ be an obtuse angle, as shown in Figure 22.3, then

$$BD = AB \cdot \cos \angle ABC, CD = AC \cdot \cos \angle ACB,$$

$$CE = BC \cdot \cos \angle ACB, EA = AB \cdot (-\cos \angle BAC),$$

$$AF = AC \cdot (-\cos \angle BAC), FB = BC \cdot \cos \angle ABC.$$

Thus we get

$$\frac{BD}{DC} \cdot \frac{CE}{EA} \cdot \frac{AF}{FB} = 1.$$

So the lines AD, BE and CF intersect at the same point.

When $\triangle ABC$ is a right triangle, the altitudes AD, BE and CF pass through the right vertex, that is, they share a same point.

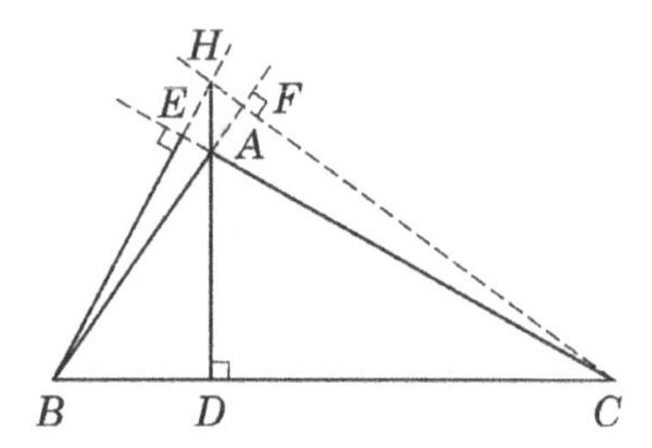

Fig. 22.3

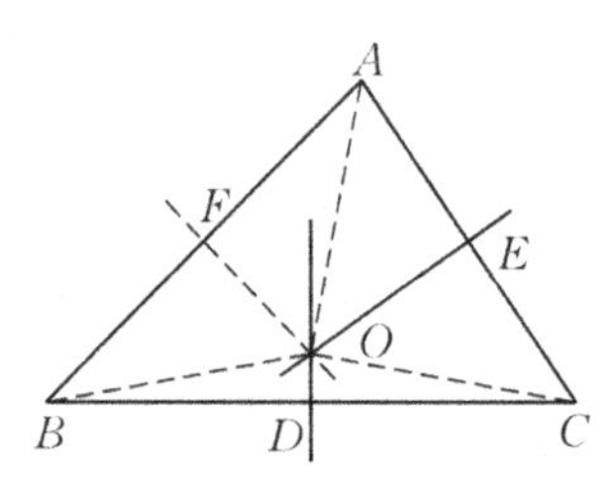

Fig. 22.4

Remark

(1) In order to prove that three altitudes of a triangle intersect at the same point, the knowledge of four concyclic points and the same method can also be used.

(2) For the three perpendicular bisectors of the triangle ABC, as shown in Figure 22.4, suppose that the perpendicular bisectors OE and OD of the sides AC and BC intersect at the point O, then we have that $OB = OC$, $OA = OC$, so

$$OB = OA.$$

Let the midpoint of AB be F. Then in $\triangle AOB$, $OB = OA$, so $OF \perp AB$, that is, OF is the perpendicular bisector of AB.
Hence, the three perpendicular bisectors of a triangle intersect at the same point.

(3) Similar to the method of proving that three inner angular bisectors of a triangle intersect at one point, we can get that two outer angular bisectors of a triangle and the inner angular bisector of the third angle intersect at a same point, that is, the escenter.

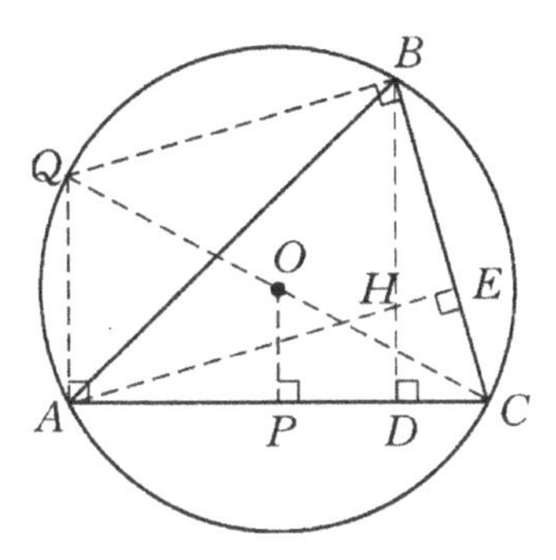

Fig. 22.5

Example 2. Prove that in any triangle ABC, the distance from the orthocenter to the vertex B is twice the distance from the center of the circumscribed circle of $\triangle ABC$ to the side AC.

Proof As shown in Figure 22.5, draw BD through the point B so that $BD \perp AC$ at the point D, and draw AE through the point A so that $AE \perp BC$ at the point E, BD and AE intersect at the point H. Construct OP through the point O so that $OP \perp AC$ at the point P. Connect CO and extend it to intersect the circumscribed circle O of $\triangle ABC$ at the point Q, and join BQ and AQ.

Because CQ is the diameter, we know that $\angle QAC = 90°$, then $AQ \parallel OP$. And since the point O is the midpoint of QC, it can be deduced that $AQ = 2OP$.

Because $BD \perp AC$, $QA \perp AC$, we get $AQ \parallel BD$. Because $\angle QBC = 90°$, $AE \perp BC$, we obtain $BQ \parallel AE$. Thus, the quadrilateral $AQBH$ is a parallelogram, then

$$BH = AQ = 2OP.$$

Remark As shown in Figure 22.6, if the intersection of OH and BP is G, from $BD \parallel OP$, we get

$$\frac{BH}{OP} = \frac{BG}{GP} = \frac{GH}{OG} = 2.$$

That is, G is the point where BP is divided into 2:1 and P is the midpoint of AC. Hence, the point G is the barycenter of $\triangle ABC$.

From the above, we can get that the orthocenter, the barycenter and the circumcenter of a triangle are collinear, and the barycenter divides the

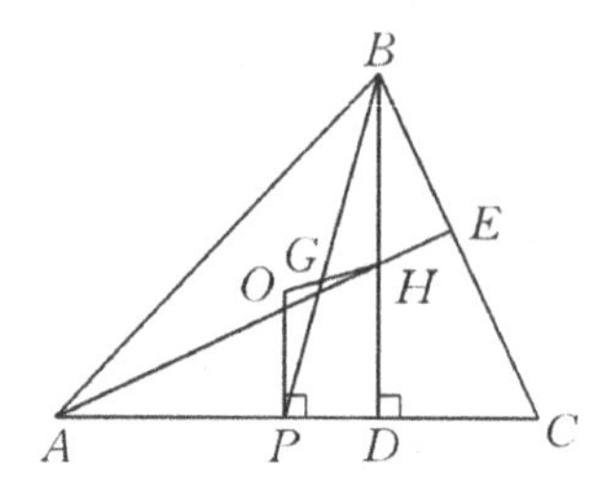

Fig. 22.6

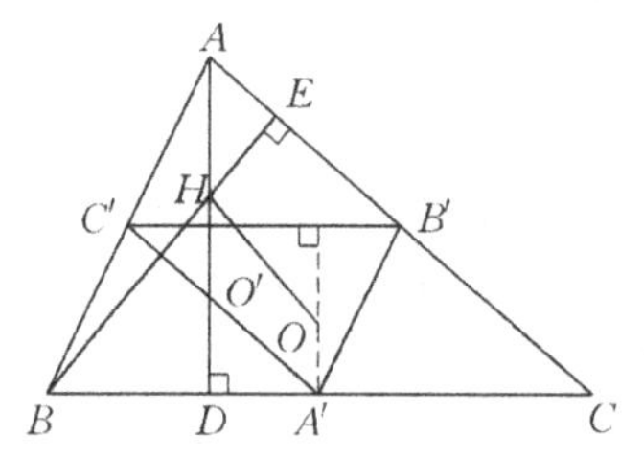

Fig. 22.7

line segment between the orthocenter and the circumcenter into 2:1. This line is called Euler Line.

With regard to Euler Line, we can also get the following conclusion.

As shown in Figure 22.7, the circumcenter O' of the median triangle (the triangle formed by connecting the midpoints of the three sides of the original triangle) is the midpoint of the segment HO on Euler line of the original triangle, and the radius of the circumscribed circle of the median triangle is equal to half of the radius of the circumscribed circle of the original triangle.

Example 3. Given that $\triangle ABC$ is an acute triangle, $\odot O$ passes through the points B and C, and intersects the sides AB and AC at the points D and E respectively. If the radius of $\odot O$ is equal to the radius of the circumscribed circle of $\triangle ADE$, then which center of ΔABC must be passed through by $\odot O$? ( )

(A) incenter (B) circumcenter (C) barycenter (D) orthocenter

Solution As shown in Figure 22.8, join BE.

$\angle BAC$ and $\angle ABE$ are acute angles because $\triangle ABC$ is an acute triangle. Since the radius of $\odot O$ is equal to the radius of the circumscribed circle of

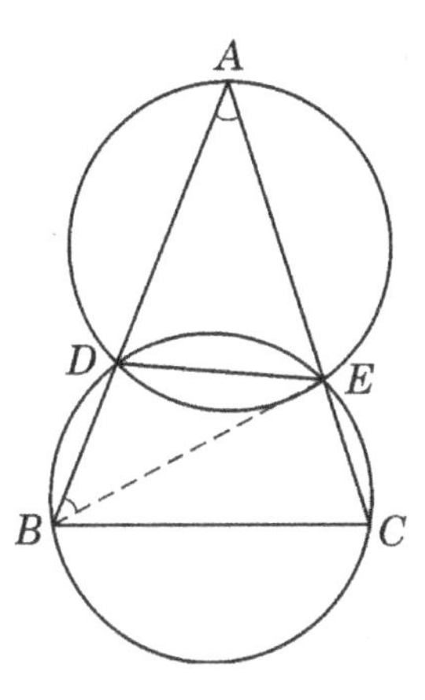

Fig. 22.8

$\triangle ADE$, and DE is the common chord of the two circles, we get

$$\angle BAC = \angle ABE.$$

Thus, $\angle BEC = \angle BAC + \angle ABE = 2\angle BAC$.

If the circumcenter of $\triangle ABC$ is O_1, then $\angle BO_1C = 2\angle BAC$, $\angle BO_1C = \angle BEC$. Thus we can know that B, C, E and O_1 are concyclic.

Because B, C and E are on the circle O, $\odot O$ must pass through the circumcenter of $\triangle ABC$.

So the answer is B.

Example 4. As shown in Figure 22.9, in $\triangle ABC$, $\angle BAC = 90°$, I is the incenter of the triangle. Prove that:

(1) $\angle AIC = 90° + \frac{1}{2}\angle ABC$;
(2) $IB \cdot IC = IA \cdot BC$.

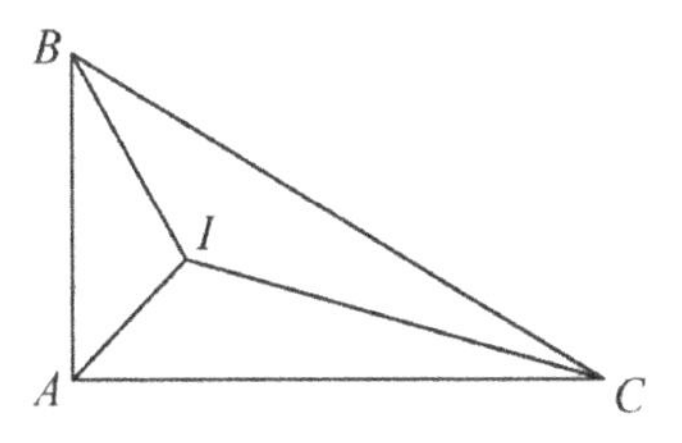

Fig. 22.9

Proof (1) Because I is the incenter of the triangle, that is, I is the intersection of the bisectors of $\angle BAC$, $\angle ABC$ and $\angle BCA$, we have:

$$\begin{aligned}\angle AIC &= 180^\circ - \frac{1}{2}\angle BAC - \frac{1}{2}\angle ACB \\ &= 180^\circ - \frac{1}{2}(180^\circ - \angle ABC) \\ &= 90^\circ + \frac{1}{2}\angle ABC,\end{aligned}$$

i.e. $\angle AIC = 90^\circ + \frac{1}{2}\angle ABC$.

(2) **Proof 1** As shown in Figure 22.10, construct a line IE through the point I so that $IE \perp IA$, and IE intersects AC at the point E, then

$$\angle EIC = \angle AIC - 90^\circ.$$

Because $\angle AIC = 90^\circ + \dfrac{1}{2}\angle ABC$, we get

$$\angle EIC = \frac{1}{2}\angle ABC = \angle IBC.$$

In $\triangle BIC$ and $\triangle IEC, \angle IBC = \angle EIC$, $\angle BCI = \angle ICE$, so $\triangle BIC \backsim \triangle IEC$. Then we get that $\dfrac{BI}{IE} = \dfrac{BC}{IC}$, i.e. $IB \cdot IC = IE \cdot BC$.

Since $\angle BAC = 90^\circ$, and IA is the bisector of $\angle BAC$, it can be seen that $\angle IAC = 45^\circ$. Then we have that $\angle IEA = 45^\circ$, i.e. $IA = IE$.

Hence, $IB \cdot IC = IA \cdot BC$.

Proof 2 As shown in Figure 22.11, extend BI to the point E at which BI and AC intersect.

Because I is on the bisector of $\angle BCE$, we get

$$\frac{BC}{EC} = \frac{BI}{IE}. \tag{1}$$

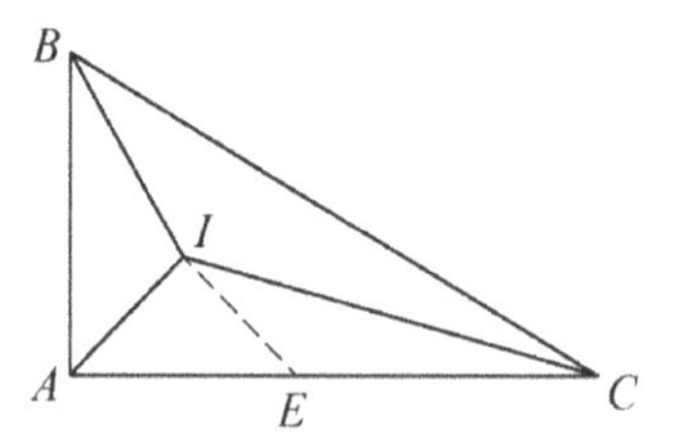

Fig. 22.10

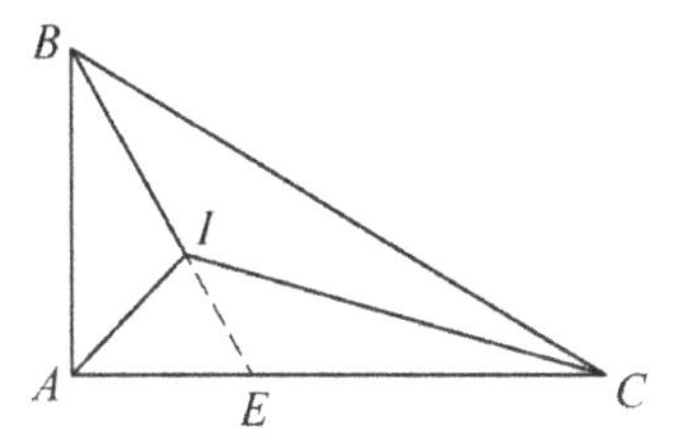

Fig. 22.11

Because I is the intersection of the bisectors of $\angle BAC$, $\angle ABC$ and $\angle BCA$, and $\angle BAC = 90°$, we obtain that

$$\angle EIC = \angle IBC + \angle ICB = \frac{1}{2}(\angle ABC + \angle ACB) = 45°,$$

i.e. $\angle EIC = \angle IAC$.

In $\triangle EIC$ and $\triangle IAC, \angle ICE = \angle ACI$, so $\triangle EIC \backsim \triangle IAC$, then we have

$$\frac{EC}{IC} = \frac{EI}{IA}. \tag{2}$$

From (1) $\times$ (2), it can be deduced that $\dfrac{BC}{IC} = \dfrac{BI}{IA}$, that is, $IB \cdot IC = IA \cdot BC$.

Example 5. In $\triangle ABC$, $AB = 7$, $BC = 8$, $CA = 9$, construct a line DE parallel to BC passing through the center I of inscribed circle of $\triangle ABC$, and DE intersects AB and AC at the points D and E respectively. Find the length of DE.

Solution As shown in Figure 22.12, let a, b and c be the lengths of three sides BC, CA and AB of $\triangle ABC$ respectively, r be the radius of the inscribed circle I, and h_a be the altitude to side BC. Then

$$\frac{1}{2}ah_a = S_{\triangle ABC} = \frac{1}{2}(a+b+c)r.$$

So we get $\dfrac{r}{h_a} = \dfrac{a}{a+b+c}$.

Since $\triangle ADE \backsim \triangle ABC$, the lengths of the corresponding sides are proportional, we obtain

$$\frac{h_a - r}{h_a} = \frac{DE}{BC}.$$

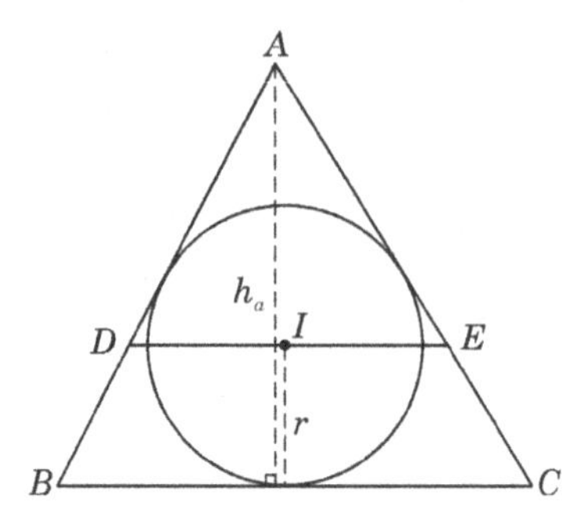

Fig. 22.12

Then we know that

$$DE = \frac{h_a - r}{h_a} \cdot a = \left(1 - \frac{r}{h_a}\right) a = \left(1 - \frac{a}{a+b+c}\right) a = \frac{a(b+c)}{a+b+c}.$$

Hence, $DE = \frac{8\times(7+9)}{8+7+9} = \frac{16}{3}$.

Remark It can also be obtained by Heron's formula $S_\triangle = \sqrt{p(p-a)(p-b)(p-c)}$ (where $p = \frac{1}{2}(a+b+c)$) and the transformation of equal areas.

Example 6. Prove that the orthocenter of the acute triangle coincides with the center of the inscribed circle of the triangle with three perpendicular feet as its vertices.

Proof As shown in Figure 22.13, $BD \perp AC$, $AF \perp BC$, $CE \perp AB$, and the perpendicular feet are the points D, F and E respectively, then the points E, H, D and A are concyclic, the points A, B, F and D are concyclic, and the points B, F, H and E are concyclic. So we get $\angle HED = \angle HAD = \angle HBF = \angle FEH$. It can be seen that EH is the inner angular bisector of $\angle FED$.

Similarly, HF is the inner angular bisector of $\angle DFE$, HD is the inner angular bisector of $\angle EDF$.

Thus, H is the incenter of $\triangle DEF$.

Since H is the orthocenter of $\triangle ABC$, the conclusion holds.

Remark

(1) In Example 6, we can also get that $\frac{S_{\triangle ADE}}{S_{\triangle ABC}} = \cos^2 A$, $R = 2R'$, where R is the radius of the circumcircle of $\triangle ABC$, and R' is the radius of the circumcircle of $\triangle DEF$.

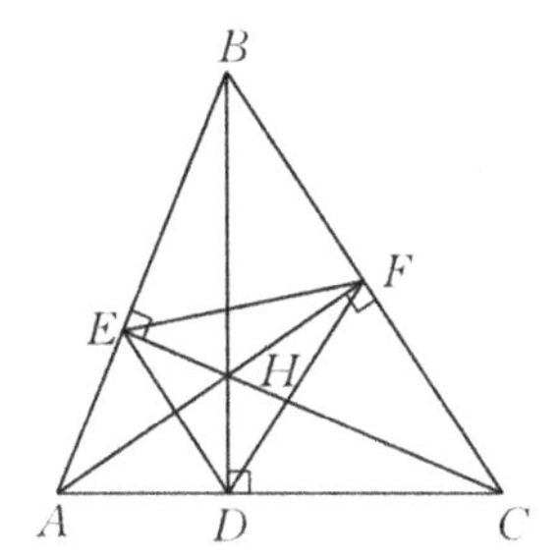

Fig. 22.13

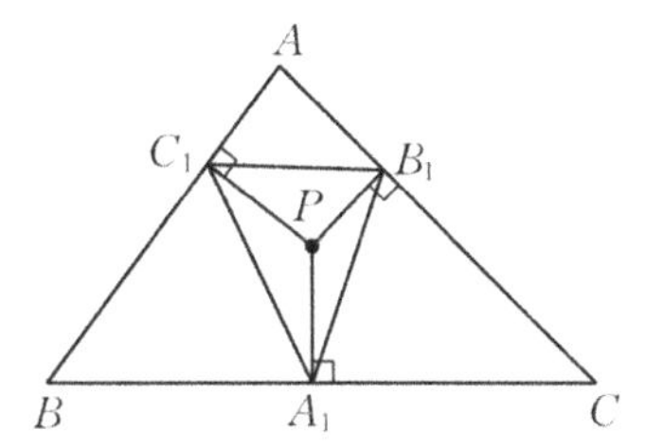

Fig. 22.14

(2) The triangle DEF whose vertices are located at the feet of the altitudes of $\triangle ABC$ is called the pedal triangle of $\triangle ABC$.

Generally, considering any point P inside $\triangle ABC$, draw the perpendicular lines PA_1, PB_1 and PC_1 to three sides BC, CA and AB, and the triangle $A_1B_1C_1$ with their perpendicular feet as the vertices is called the pedal triangle of $\triangle ABC$ generated by "Point P" (as shown in Figure 22.14).

When P is the orthocenter or circumcenter, we get the pedal triangle or median triangle.

(3) The pedal triangle $\triangle A_1B_1C_1$ (the first one) of $\triangle ABC$ is generated from the point P, and the pedal triangle $A_2B_2C_2$ of $\triangle\triangle A_1B_1C_1$ (which is called the second pedal triangle of $\triangle ABC$) is generated from the point P. In this way, the third pedal triangle $A_3B_3C_3$ is obtained, and $\triangle A_3B_3C_3$ is similar to $\triangle ABC$.

This is because: $PB_1 \perp AB_1$, $PC_1 \perp AC_1$, then the point P is on the circumscribed circle of $\triangle AB_1C_1$. For the same reason, the point P is also on the circumscribed circle of $\triangle A_2B_1C_2$, $\triangle A_3B_3C_2$, $\triangle A_2B_2C_1$ and

$\triangle A_3B_2C_3$. Thus we know that $\angle C_1AP = \angle C_1B_1P = \angle A_2B_1P = \angle A_2C_2P = \angle B_3C_2P = \angle B_3A_3P$.

And since $\angle PAB_1 = \angle PC_1B_1 = \angle PC_1A_2 = \angle PB_2A_2 = \angle PB_2C_3 = \angle PA_3C_3$, we get $\angle C_1AP + \angle PAB_1 = \angle B_3A_3P + \angle PA_3C_3$. Therefore, $\angle BAC = \angle B_3A_3C_3$. Similarly, we obtain that $\angle ACB = \angle A_3C_3B_3$, $\angle CBA = \angle C_3B_3A_3$.

Hence, $\triangle ABC \backsim \triangle A_3B_3C_3$. (as shown in Figure 22.15)

Example 7. The circumcenter of $\triangle ABC$ is the point O, the incenter is the point I, R and r are the radius of its circumscribed circle and the radius of its inscribed circle respectively, then $OI^2 = R^2 - 2Rr$.

Proof As shown in Figure 22.16, join AI which intersects the circle O at the point D, join DO which intersects the circle O at the point E, join BD and BE, extend OI so that it intersects the circle O at the points G and H, and draw IF through the point I so that $IF \perp AB$ at the point F, then $IF = r$, $DE = 2R$.

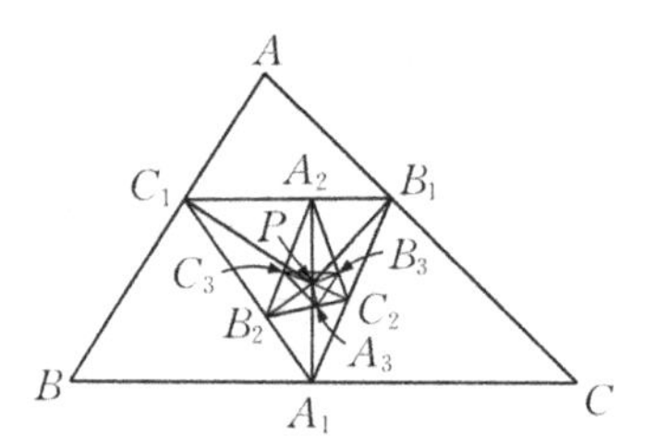

Fig. 22.15

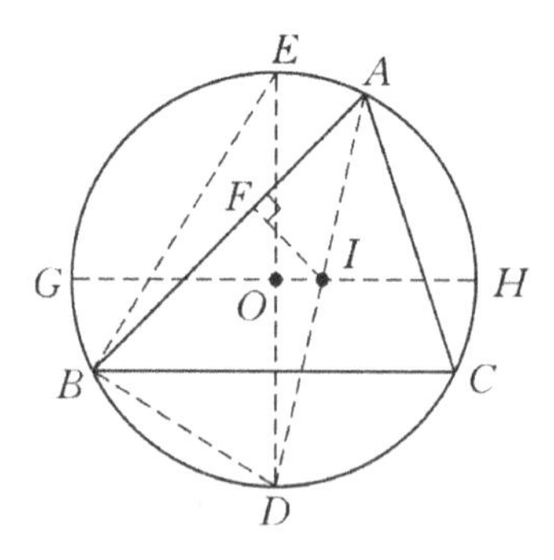

Fig. 22.16

Because $GI \cdot IH = AI \cdot ID$, and $GI = R + OI$, $IH = R - OI$, we get that $GI \cdot IH = R^2 - OI^2$, then

$$R^2 - OI^2 = AI \cdot ID.$$

From $\angle BAD = \angle BED$, $\angle AFI = \angle EBD = 90°$, it can be deduced that $\triangle AFI \backsim EBD$.

Then we have $\dfrac{AI}{DE} = \dfrac{IF}{BD}$, that is,

$$AI \cdot BD = DE \cdot IF = 2Rr.$$

Since $\angle IBD = \frac{1}{2}\angle ABC + \angle DBC = \frac{1}{2}\angle ABC + \frac{1}{2}\angle BAC = \angle BID$, we get $ID = BD$.

Hence, $R^2 - OI^2 = AI \cdot BD = 2Rr$, i.e. $OI^2 = R^2 - 2Rr$.

Remark (1) From $OI^2 \geq 0$, $R \geq 2r$ is obtained. Thus we can get that the radius of the circumscribed circle of a triangle is not less than the diameter of the inscribed circle.
(2) The conclusion of Example 7 is called Euler Theorem. Inequality $R \geq 2r$ is also called Euler Inequality.

Example 8. As shown in Figure 22.17, given that the escenter contained by the inner angle A of the triangle ABC is I_a, the escenter contained by the inner angle B is I_b, the escenter contained by the inner angle C is I_c, then the pedal triangle of $\triangle I_a I_b I_c$ is $\triangle ABC$.

Proof Because CI_c is the inner angular bisector of $\angle ACB$, $I_a I_b$ is the outer angular bisector of $\angle ACB$, we obtain

$$I_c C \perp I_a I_b.$$

Similarly, we have that $I_a A \perp I_b I_c$, $I_b B \perp I_c I_a$.

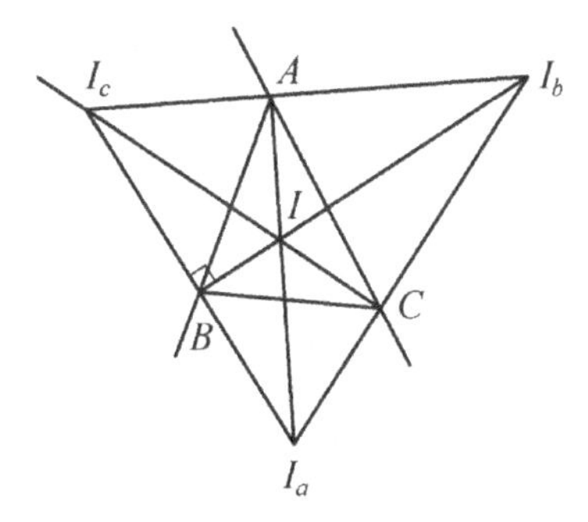

Fig. 22.17

Hence, A, B and C are three perpendicular feet of $\triangle I_aI_bI_c$, then we know that $\triangle ABC$ is the pedal triangle of $\triangle I_aI_bI_c$.

Example 9. Let CD be the altitude to the hypotenuse AB of the right triangle ABC. Suppose that I_1 and I_2 are the incenters of $\triangle ADC$ and $\triangle BDC$ respectively, $AC = 3$, $BC = 4$, find the length of I_1I_2.

Solution As shown in Figure 22.18, I_1E is drawn to make $I_1E \perp AB$ at the point E, and I_2F is drawn to make $I_2F \perp AB$ at the point F.

In right triangle ABC, $AC = 3$, $BC = 4$,

$$AB = \sqrt{AC^2 + BC^2} = 5.$$

Since $CD \perp AB$, according to the Projection Theorem, $AD = \dfrac{AC^2}{AB} = \dfrac{9}{5}$, we get that

$$\begin{aligned} BD &= AB - AD = \frac{16}{5}, \\ CD &= \sqrt{AC^2 - AD^2} \\ &= \frac{12}{5}. \end{aligned}$$

Because I_1E is the radius of the inscribed circle of the right triangle ACD, we get

$$I_1E = \frac{1}{2}(AD + CD - AC) = \frac{3}{5}.$$

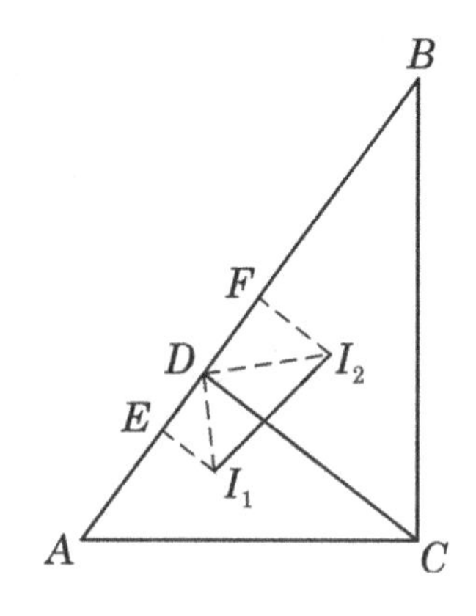

Fig. 22.18

Join DI_1 and DI_2, then DI_1 and DI_2 are the angle bisectors of $\angle ADC$ and $\angle BDC$, so we know that $\angle I_1DC = \angle I_1DA = \angle I_2DC = \angle I_2DB = 45°$.

Thus, $\angle I_1DI_2 = 90°$, i.e. $I_1D \perp I_2D$.

Then we obtain $DI_1 = \dfrac{I_1E}{\sin \angle ADI_1} = \dfrac{\frac{3}{5}}{\sin 45°} = \dfrac{3\sqrt{2}}{5}$.

Similarly, we get that $I_2F = \dfrac{4}{5}$, $DI_2 = \dfrac{4\sqrt{2}}{5}$.

Hence, $I_1I_2 = \sqrt{DI_1^2 + DI_2^2} = \sqrt{2}$.

Remark I_1D and I_2D can also be obtained by using the area. Let r_1 be the radius of the inscribed circle of Rt$\triangle ACD$. Then from the area transformation we can get that

$$\frac{1}{2}(AC + CD + AD) \cdot r_1 = \frac{1}{2} \times AD \times CD,$$

i.e. $\frac{1}{2} \times \left(3 + \frac{12}{5} + \frac{9}{5}\right) r_1 = \frac{1}{2} \times \frac{9}{5} \times \frac{12}{5}$.

Solving it gives $r_1 = \frac{3}{5}$.

And since Rt$\triangle I_1ED$ is an isosceles triangle, I_1D can be obtained. Similarly, I_2D is obtained, then I_1I_2 is obtained. (This area method is given by Mr. Dai Chiwen of Guangzhou)

Reading

Integrated Block and Combined Boxing

A student asked me a question about plane geometry as follow:

Let the quadrilateral $ABCD$ be inscribed on the circle. BA and CD intersect at the point P, AD and BC intersect at the point Q, AC and BD intersect at the point M. Point O is the center of the circle. Prove that M is the orthocenter of $\triangle OPQ$.

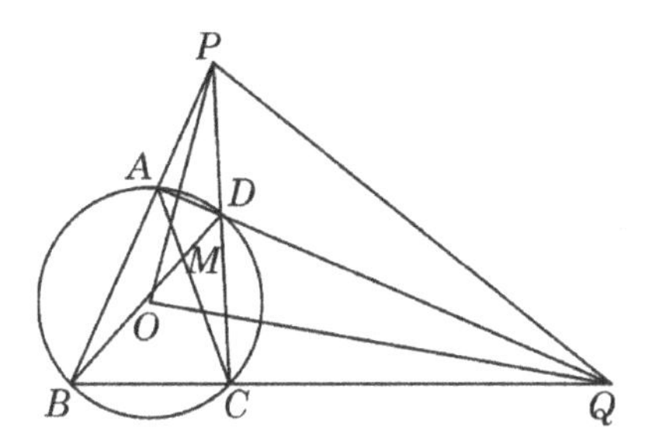

Fig. 22.19

Points P, Q and M are symmetrical (equal status), so it is better to replace the conclusion of this question with the following (equivalent) statement:

O is the orthocenter of $\triangle MPQ$. (1)

I haven't done such a geometric problem for a long time, and I don't know whether the proposition is right or not for a moment. So I called my friend, Mr. Ye Zhonghao from Shanghai Education Press.

Mr. Ye is a real expert. When he heard the question, he immediately answered me that the proposition is correct, and he gave two solutions.

First of all, the second solution depends on a simpler problem (lemma):

If the extended lines of the opposite sides of the cyclic quadrilateral $ABCD$ interest at the points P and Q, then PQ^2 = (power of P to circle) + (power of Q to circle). (2)

Here, the power of a point outside the circle refers to the square of the length of the tangent from this point to the circle.

In order to prove (2), PQ^2 is divided into two parts, which are equal to the power of P to the circle and the power of Q to the circle respectively.

For this reason, a circle is drawn through the points P, A and D, and it intersects PQ at the point N. Let the given circle be $\odot O$. It's easy to know that

$$QN \times PQ = QD \times QA = \text{power of } Q \text{ to } \odot O.$$

Since $\angle PND = \angle BAD = \angle DCQ$, the four points N, D, C and Q are concyclic. Then we get that

$$PN \times PQ = PD \times PC = \text{power of } P \text{ to } \odot O.$$

Adding the above two equalities gives (2).

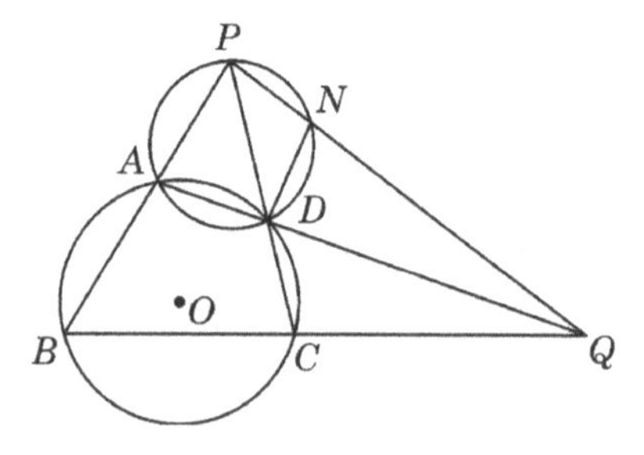

Fig. 22.20

Power of P to $\odot O$ (the square of the length of tangent) is $OP^2 - R^2$, where R is the radius of $\odot O$.

For the point M inside $\odot O$, we can also define its power to $\odot O$, that is $OM^2 - R^2$. This power is negative (it is equal to the opposite number of the product of the two parts AM and MC into which the chord AC is divided by the point M).

In Figure 22.19, for the same reason, it can be proved that(draw auxiliary circles through the points P, A, C and Q, A, C respectively)

$$PM^2 = (\text{power of } P \text{ to } \odot O) + (\text{power of } M \text{ to } \odot O), \tag{3}$$

$$QM^2 = (\text{power of } Q \text{ to } \odot O) + (\text{power of } M \text{ to } \odot O). \tag{4}$$

Now go back to the proof of (1).

Just need to prove that $OM \perp PQ$ (for the same reason, $OP \perp MQ, OQ \perp MP$).

This is equivalent to

$$OP^2 - OQ^2 = MP^2 - MQ^2. \tag{5}$$

From (3) and (4) above, we can get that

$$\begin{aligned} MP^2 - MQ^2 &= (\text{power of } P \text{ to } \odot O) - (\text{power of } Q \text{ to } \odot O) \\ &= (OP^2 - R^2) - (OQ^2 - R^2) = OP^2 - OQ^2. \end{aligned}$$

So proposition (1) holds.

It can be seen from the above proof that to solve a complex problem, it is often necessary to do a simple problem first. Conversely, simple problems are done very well, and complex problems are not difficult. For an experienced problem solver, many simple problems have been integrated, such as the integrated block in the TV or speaker, the set in go, and the combined boxing of a boxer, which can launch a continuous attack in an instant without much thought. Of course, we need to pay more attention to and accumulate more at ordinary times, and turn various basic routines into our own skills.

(Excerpted from: Shan Zun. *Problem solving research*. Shanghai: Shanghai Education Press, 2016, 146~147.)

Exercises

1. There are two points P and Q on the plane. The number of triangles with P as the circumcenter and Q as the incenter is ( ).
(A) only 1 (B) 2 (C) at most 3 (D) countless
2. Given that three sides of an acute triangle ABC are a, b and c, the distances from its circumcenter to the three sides are m, n and p, then $m : n : p =$ ( ).
(A) $\frac{1}{a} : \frac{1}{b} : \frac{1}{c}$ (B) $a : b : c$
(C) $\cos A : \cos B : \cos C$ (D) $\sin A : \sin B : \sin C$
3. As shown in Figure 22.21, the point O is inside the acute triangle ABC, construct $EF \parallel BC$, $PQ \parallel CA$ and $HG \parallel AB$ through the point O. When $\frac{EF}{BC} = \frac{PQ}{CA} = \frac{HG}{AB}$, the point O is the ( ) of $\triangle ABC$.
(A) orthocenter (B) barycenter
(C) circumcenter (D) incenter

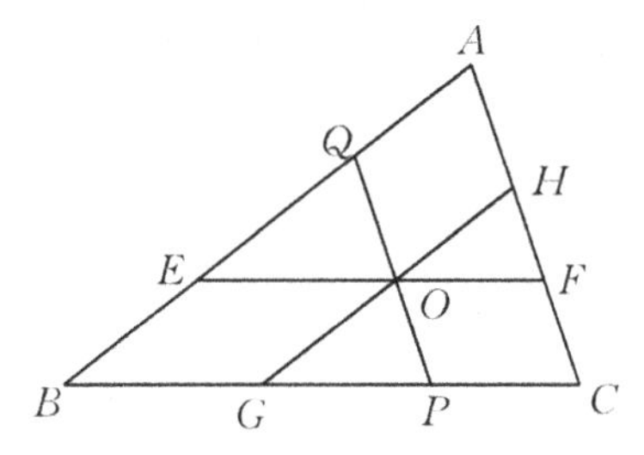

Fig. 22.21

4. Let O be a point on the plane where the equilateral triangle ABC lies. And make $\triangle ABO$, $\triangle OBC$ and $\triangle OCA$ all be isosceles triangles. How many points O satisfying this condition?
5. In $\triangle ABC$, the medians to the sides AC and AB are BD and CE respectively. If BD and CE intersect at the point G, prove that $S_{\triangle BCG} = S_{ADGE}$.
6. Given that the distance from the vertex A of the acute triangle ABC to the orthocenter H is equal to the radius of its circumscribed circle, find the size of $\angle A$.
7. As shown in the Figure 22.22, in the isosceles triangle ABC, $AB = AC$, $BC = 4$, the radius of the inscribed circle is 1. Find the length of each leg of the isosceles triangle.

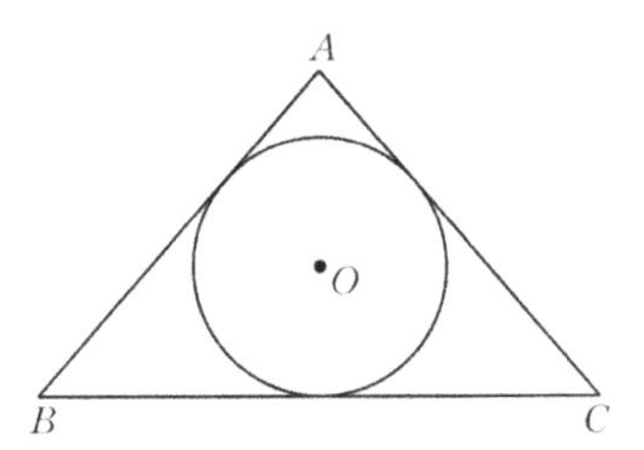

Fig. 22.22

8. Given that two altitudes AD and BE of $\triangle ABC$ intersect at the point H, the extended line of AD intersects the circumcircle at the point K, prove that $HD = DK$.
9. Given that O is the circumcenter of the acute triangle ABC, AO intersects BC at the point D, the points E and F are the circumcenters of $\triangle ABD$ and $\triangle ACD$. If $AB > AC$, $EF = BC$, then $\angle C - \angle B =$ __.
10. Given that O is the incenter of equilateral triangle ABC, and line m passes through the point O. Construct the perpendicular lines to the line m through the points A, B and C, and the perpendicular feet are D, E and F respectively.

    When the straight line m is parallel to BC (as shown in Figure 22.23(1)), it's easy to prove that $BE + CF = AD$.

    When the straight line m rotates around the point O to be not parallel to BC, in the two cases of Figures 22.23(2) and 22.23(3), is the above conclusion still true? If it is true, please prove it; if not, what is the quantitative relationship between the line segments AD, BE and CF? Please write down your conjecture without proof.

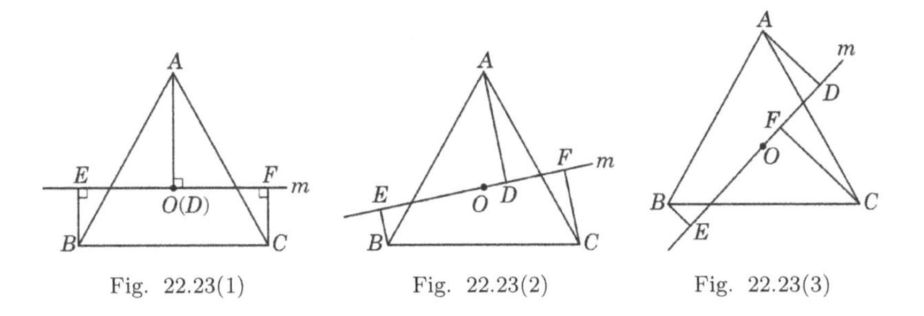

Fig. 22.23(1) Fig. 22.23(2) Fig. 22.23(3)

11. Let K be any point inside $\triangle ABC$. The barycenters of $\triangle KAB$, $\triangle KBC$ and $\triangle KCA$ are D, E and F respectively, then the value of $S_{\triangle DEF}:S_{\triangle ABC}$ is ( ).

    (A) $\frac{1}{9}$ (B) $\frac{2}{9}$ (C) $\frac{4}{9}$ (D) $\frac{2}{3}$

12. Given that the three sides of an acute triangle ABC satisfy the inequality $AB < AC < BC$. If I is the center of the inscribed circle of the triangle ABC, and O is the center of the circumcircle, prove that the line IO intersects the line segments AB and BC.

13. As shown in Figure 22.24, M, N and P are the barycenters of the regular triangles $\triangle ABC$, $\triangle BFE$ and $\triangle EDC$ respectively. Prove that $\triangle MNP$ is a regular triangle.

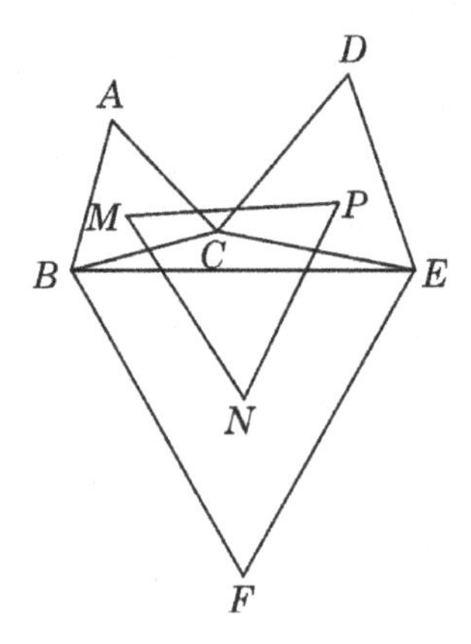

Fig. 22.24

Chapter 23

Geometric Inequalities

The inequality relation of some quantities in geometry is called geometric inequality.

The line segment connecting two points is the shortest, which is a basic inequality relation. From this basic inequality, we can know that for any three points in the plane, the sum of the line segments from one point to the other two points is greater than or equal to the line segment connecting these two points. The equal sign holds when the point is on the line segment connecting these two points. Thus we have the following conclusion.

In a triangle, the sum of any two sides is greater than the third side, that is, the difference between any two sides is less than the third side.

Therefore, we know that:

Among the line segments connecting a fixed point and each point on a fixed line in a plane, the vertical line segment is the shortest.

In a triangle, the exterior angle is larger than the interior angles which are not adjacent to it.

In a triangle, the angle opposite the longer side is greater, and the side opposite the greater angle is longer.

In two triangles with two groups of equal corresponding sides, the third side is longer if the included angle is greater, and the included angle is greater if the third side is longer.

The geometric inequalities in junior high school mainly include line segment inequalities, angle inequalities and area inequalities.

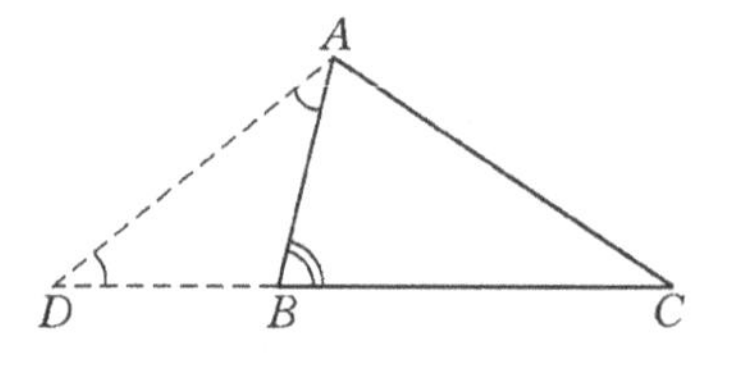

Fig. 23.1

Example 1. In $\triangle ABC$, $\angle B = 2\angle C$, prove that $AC < 2AB$.

Proof As shown in Figure 23.1, extend CB to the point D so that $BD = AB$, and join AD, then

$$\angle BDA = \angle BAD,$$

$$\angle ABC = 2\angle ADB.$$

Because $\angle ABC = 2\angle ACB$, we know that

$$\angle ADB = \angle ACB.$$

Therefore, $AD = AC$.

In $\triangle ABD, BD + AB > AD$, so $2AB > AD$, i.e. $AC < 2AB$.

Remark

(1) It is one of the common ideas to concentrate the inequality to a triangle (or two closely related triangles) as much as possible and solve it by using the inequality relation of line segments (or the inequality relation of angles) in triangles.
(2) Can we change "$\angle B = 2\angle C$" to "$\angle B = n\angle C$" in this question? Then we have: in $\triangle ABC, \angle B = n\angle C$($n$ is a positive integer greater than 1), prove that $AC \leq nAB$.

This is because as shown in Figure 23.2, on the circumscribed circle of the triangle ABC, we divide the arc $\overset{\frown}{AC}$ opposite to $\angle B$ into n equal parts, and join the adjacent points to obtain n equal chords $AA_1, A_1A_2, \ldots, A_{n-2}A_{n-1}$, $A_{n-1}C$, all of which are equal to AB.

Since the length of the polyline $AA_1A_2A_3 \cdots A_{n-1}C$ is greater than AC, it can be known that $nAB > AC$.

This is to concentrate inequality relations into polygons (or circles).

Example 2. As shown in Figure 23.3, in the isosceles triangle $ABC, AB = AC$, D is the midpoint of BC, E is any point inside $\triangle ABD$, join AE, BE and CE. Prove that $\angle AEB > \angle AEC$.

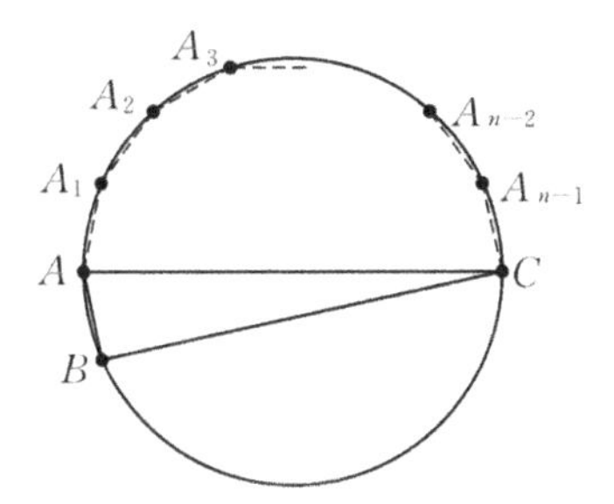

Fig. 23.2

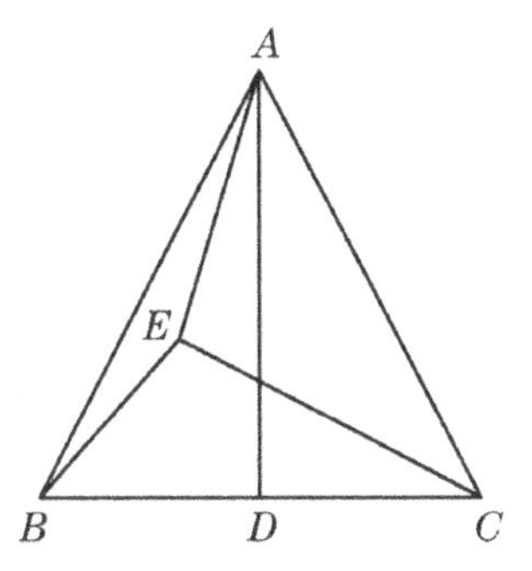

Fig. 23.3

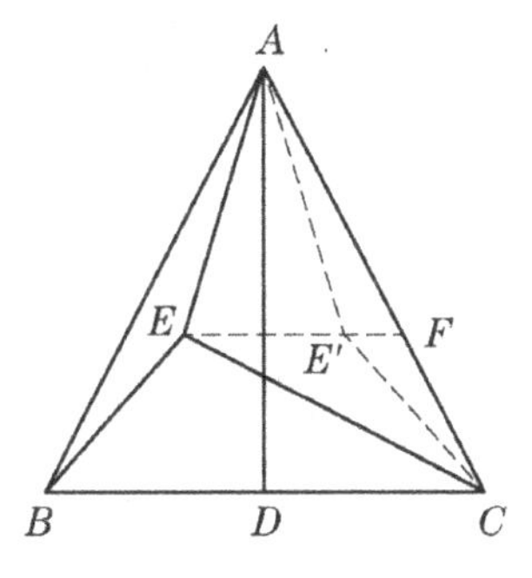

Fig. 23.4

Proof As shown in Figure 23.4, we draw a symmetry point E' of the point E about AD, join AE', $E'C$ and EE', extend EE' to the point F at which EE' and AC intersect.

Because E is inside $\triangle ABD$, $AB = AC$, D is the midpoint of BC, we can obtain

$$\triangle AEB \cong \triangle AE'C.$$

Thus, $\angle AEB = \angle AE'C$.

Since

$$\angle AE'F = \angle AEF + \angle EAE' > \angle AEF,$$

$$\angle CE'F = \angle CEF + \angle ECE' > \angle CEF,$$

we know that $\angle AE'F + \angle CE'F > \angle AEF + \angle CEF$, that is, $\angle AE'C > \angle AEC$.

Hence, $\angle AEB > \angle AEC$.

Remark It is one of the ways of dealing with geometric inequality, which is to consider symmetry, and to concentrate angles or line segments into a figure for easy discussion.

Example 3. Prove that in a triangle, the sum of the long side and the height on this side is greater than or equal to the sum of the short side and the height on this side.

As shown in Figure 23.5, in $\triangle ABC, AB < BC$, $AD \perp BC$ at the point $D, CE \perp AB$ at the point E. Prove that

$$BC + AD > AB + CE.$$

Proof Since $CE \perp AB$, $AD \perp BC$, we know that

$$\triangle CBE \backsim \triangle ABD.$$

Thus we have $\dfrac{AB}{BC} = \dfrac{AD}{CE}$, i.e. $\dfrac{AB}{AD} = \dfrac{BC}{CE} = \dfrac{BC - AB}{CE - AD}$.

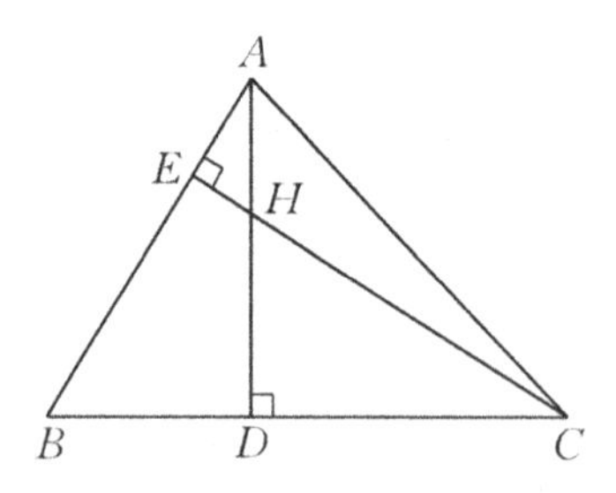

Fig. 23.5

From $AB > AD$, it can be seen that

$$\frac{AB}{AD} = \frac{BC - AB}{CE - AD} > 1.$$

Since $BC > AB$, that is, $BC - AB > 0$, then we get $CE - AD > 0$.

Therefore, $BC - AB > CE - AD$,

i.e. $BC + AD > AB + CE$.

Remark By using the similarity of triangles, we obtain that the ratio between the proportion line segments is greater than 1(less than 1), then we get that the inequality holds. This is an idea to solve the problem of geometric inequality.

In Example 3, you can also make a point A' on side BC so that $BA' = BA$. Draw a line segment $A'F$ through the point A' so that $A'F \perp CE$ and $A'F$ intersect CE at the point F. And draw a line segment $A'G$ through the point A' so that $A'G \parallel CE$ and $A'G$ intersect BA at the point G (as shown in Figure 23.6).

It's easy to see that $\triangle ABD \cong A'BG$. Then

$$AD = A'G = FE.$$

Since $A'C > CF$, we get

$$BC - AB > CE - AD.$$

This is to concentrate the unequal quantities into a right triangle by making the difference, and use the theorem that the hypotenuse is larger than the leg of the right triangle.

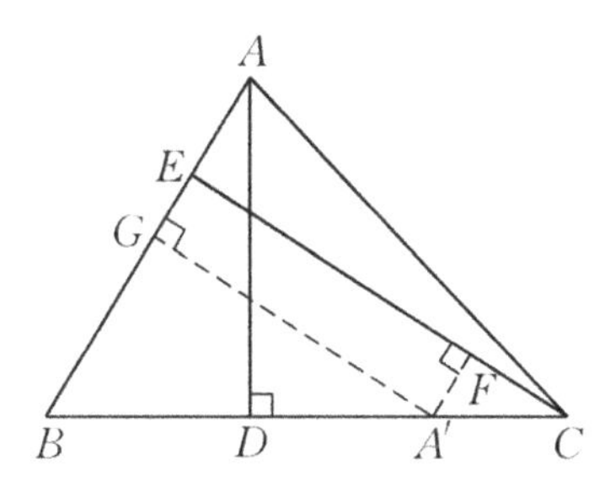

Fig. 23.6

Example 4. Prove that in the convex quadrilateral $ABCD$, we have

$$AB^2 + BC^2 + CD^2 + DA^2 \geq AC^2 + BD^2.$$

Proof As shown in Figure 23.7, let E be the midpoint of AC, F be the midpoint of BD. Join DE, BE and EF, and draw a line segment DH so that DH is perpendicular to AC (produced at H).

Then for Rt $\triangle AHD$, Rt $\triangle EHD$ and Rt $\triangle CHD$, from the Pythagorean theorem, we can get

$$AD^2 = AH^2 + DH^2, \tag{1}$$

$$CD^2 = CH^2 + DH^2, \tag{2}$$

$$DE^2 = DH^2 + EH^2.$$

Since $AH = AE + EH$, $CH = CE - EH$, $AE = EC = \frac{1}{2}AC$, from (1) + (2), we get

$$DA^2 + CD^2 = \frac{1}{2}AC^2 + 2DE^2. \tag{3}$$

Similarly, we can prove that

$$AB^2 + BC^2 = \frac{1}{2}AC^2 + 2BE^2, \tag{4}$$

$$2(BE^2 + DE^2) = BD^2 + 4EF^2.$$

Then from (3) + (4), we obtain

$$AB^2 + BC^2 + CD^2 + DA^2 = AC^2 + BD^2 + 4EF^2.$$

Hence,

$$AB^2 + BC^2 + CD^2 + DA^2 \geq AC^2 + BD^2.$$

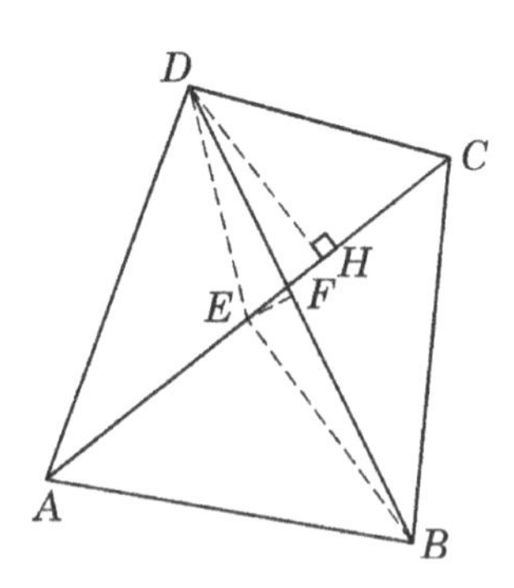

Fig. 23.7

Remark

(1) When $EF = 0$, $AB^2 + BC^2 + CD^2 + DA^2 = AC^2 + BD^2$, the points E and F coincide, that is, the diagonal lines AC and BD bisect each other, and the quadrilateral $ABCD$ is a parallelogram. So we have the following conclusion:

in a parallelogram, the sum of the squares of the lengths of four sides is equal to the sum of the squares of the lengths of the diagonals.

Conversely, a quadrilateral is a parallelogram if the sum of the squares of the lengths of four sides is equal to the sum of the squares of the lengths of the diagonals.

(2) In the proof of Example 4, the relationship between the length of the median drawn to one side of the triangle and the three sides is obtained.

That is, in $\triangle ABC$, let $BC = a$, $AC = b$, $AB = c$, the length of the median drawn to side AB be m_c. Then

$$a^2 + b^2 = \frac{c^2}{2} + 2m_c^2.$$

This conclusion is also called Pappus Law (or Apollonius Theorem).

Example 5. Prove that for any triangle, there must exist two sides, whose lengths are u and v satisfying $1 \leq \frac{u}{v} < \frac{1+\sqrt{5}}{2}$.

Proof Let the three sides of any $\triangle ABC$ be a, b and c, where $a \geq b \geq c$.

If the conclusion is not true, there must be

$$\frac{a}{b} \geq \frac{1+\sqrt{5}}{2}, \tag{1}$$

$$\frac{b}{c} \geq \frac{1+\sqrt{5}}{2}. \tag{2}$$

Let $b = c + s$, $a = b + t = c + s + t$. It's obvious that $s, t > 0$. Substituting these into (1) gives

$$\frac{c+s+t}{c+s} \geq \frac{1+\sqrt{5}}{2},$$

i.e. $\dfrac{1+\frac{s}{c}+\frac{t}{c}}{1+\frac{s}{c}} \geq \dfrac{1+\sqrt{5}}{2}$.

Let $x = \frac{s}{c}$, $y = \frac{t}{c}$, then

$$\frac{1+x+y}{1+x} \geq \frac{1+\sqrt{5}}{2}. \tag{3}$$

From $a < b+c$, we get $c+s+t < c+s+c$, that is, $t < c$, then $y = \frac{t}{c} < 1$. From inequality (2), we have

$$\frac{b}{c} = \frac{c+s}{c} = 1 + x \geq \frac{1+\sqrt{5}}{2}. \tag{4}$$

From (3) and (4), we obtain that

$$y \geq \left(\frac{1+\sqrt{5}}{2} - 1\right)(1+x) \geq \frac{\sqrt{5}-1}{2} \cdot \frac{1+\sqrt{5}}{2} = 1,$$

which contradicts $y < 1$.

So the original proposition is proved.

Example 6. As shown in Figure 23.8, the radius of the circumscribed circle O of $\triangle ABC$ is 1, and P is a point inside $\triangle ABC$, where $OP = a$.

(1) Prove that one of the following three inequalities must be true:

$$\angle ABP + \angle ACP \geq \angle BAC,$$
$$\angle BAP + \angle BCP \geq \angle ABC,$$
$$\angle CBP + \angle CAP \geq \angle ACB.$$

(2) Prove that $PB \cdot PC \leq 1 - a^2$.

(3) Prove that $PA \cdot PB \cdot PC \leq (1-a)(1+a)^2$.

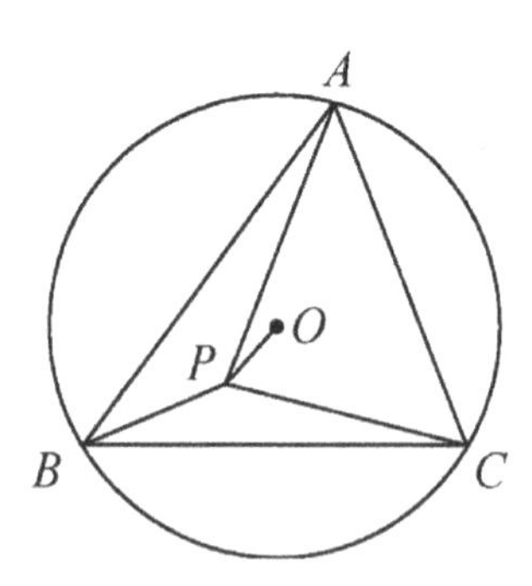

Fig. 23.8

Proof (1) Suppose that none of the three inequalities are true, then

$$\angle ABP + \angle ACP < \angle BAC,$$

$$\angle BAP + \angle BCP < \angle ABC,$$

$$\angle CBP + \angle CAP < \angle ACB.$$

Adding three inequalities gives $\angle BAC + \angle ABC + \angle ACB < \angle BAC + \angle ABC + \angle ACB$, which is a contradiction.

Hence, one of the three inequalities in the question must be true.

(2) It can be seen from (1) that we might as well let $\angle ABP + \angle ACP \geq \angle BAC$. As shown in Figure 23.9, we extend BP to the point Q where BP and $\odot O$ intersect, and join the line segments OA, OB, OQ and QC.

In an isosceles triangle OBQ, by using Pythagorean Theorem or Intersecting Chords Theorem, it is easy to get $PB \cdot PQ = OA^2 - a^2 = 1 - a^2$. (This is a well-known conclusion that readers can prove by yourselves.)

And from $\angle ABP = \angle ACQ$, we obtain that

$$\angle ACQ + \angle ACP = \angle PCQ \geq \angle BAC = \angle PQC.$$

Therefore, in $\triangle PCQ$, $PC \leq PQ$, then

$$PB \cdot PC \leq PB \cdot PQ = 1 - a^2.$$

(3) For three points A, O and P, there is always $PA \leq AO + OP = 1 + a$. When the point P coincides with the point O, $a = 0$ and the equal sign holds.

Hence, $PA \cdot PB \cdot PC \leq (1 + a)(1 - a^2) = (1 - a)(1 + a)^2$, the equal sign holds when $a = 0$.

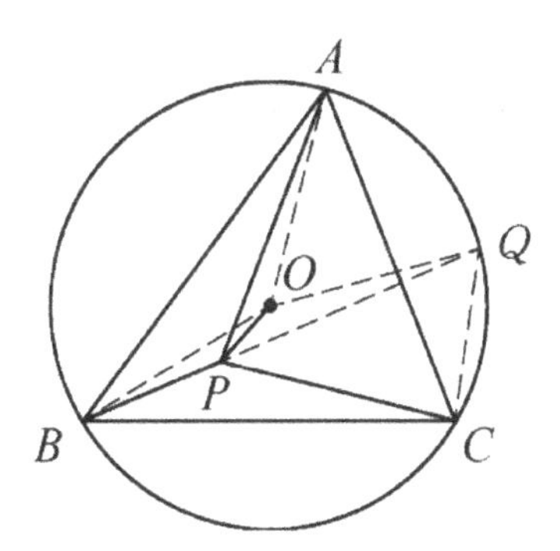

Fig. 23.9

Remark

(1) In Example 6, if the questions (1) and (2) are removed, it is a geometric question raised by the famous mathematician Erdös.
(2) The method used in the question (1) is called reductio ad absurdum. For details, please refer to Chapter 25.
(3) After learning the knowledge of senior high school, it is easy to know that the maximum value of $(1-a)(1+a)^2 (0 \leq a < 1)$ is $\frac{32}{27}$.

Exercises

1. Let D be a point on the side BC of $\triangle ABC$ (excluding the points B and C). Prove that AD is less than the longest side of $\triangle ABC$.
2. Let P and Q be any two points inside $\triangle ABC$ or on the side of $\triangle ABC$. Prove that the length of the line segment PQ is less than or equal to the longest side of ABC.
3. Prove that the median drawn to one side of a triangle is less than half of the sum of the other two sides.
4. Prove that the median drawn to the long side of a triangle is shorter than the median drawn to the short side.
5. As shown in the Figure 23.10, BC is the longest side of $\triangle ABC$, $AB = AC(<BC)$, $CD = BF$, $BD = CE$, then the size range of $\angle DEF$ is ( ).

   (A) $0° < \angle DEF < 45°$
   (B) $45° < \angle DEF < 60°$
   (C) $60° < \angle DEF < 90°$
   (D) none of the above

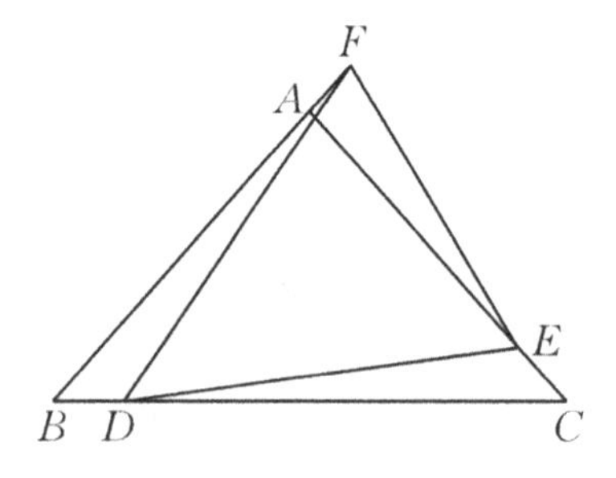

Fig. 23.10

6. Let a, b and c be the three sides of an acute triangle ABC, and h_a, h_b and h_c be the three heights perpendicular to the corresponding sides. Prove that
$$\frac{1}{2} < \frac{h_a + h_b + h_c}{a + b + c} < 1.$$
7. Let a, b and c be the three sides of a triangle, and m_a, m_b and m_c be the three medians drawn to the corresponding sides. Prove that
$$bc - \frac{a^2}{4} \le m_a^2 \le bc + \frac{a^2}{4},$$
$$ca - \frac{b^2}{4} \le m_b^2 \le ca + \frac{b^2}{4},$$
$$ab - \frac{c^2}{4} \le m_c^2 \le ab + \frac{c^2}{4}.$$
8. As shown in Figure 23.11, hang an electric light bulb B at the fixed point A. If the bulb B is pulled up from a certain point C on the lamp rope AB to make B be closest to a fixed point D on the desktop, then where should the point C be pulled (AD is greater than the light rope AB)?

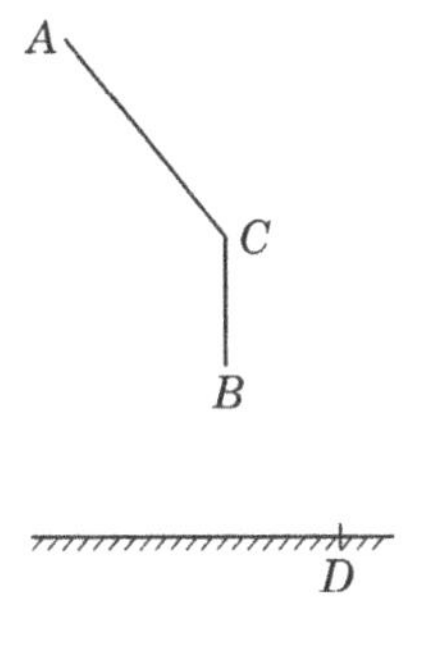

Fig. 23.11

9. Prove that the circumference of an isosceles triangle is the shortest among the triangles with the same base and the equal area.
10. In $\triangle ABC, \angle ABC < \angle ACB$, BD and CE are two angular bisectors. Prove that $CE < BD$.

11. Someone use a triangular piece of paper ABC with the area of S to cut out a parallelogram $DEFG$. The area of $\square DEFG$ is denoted as T.

    (1) As shown in Figure 23(1), if the vertices of $\square DEFG$ are all on the sides of $\triangle ABC$, while D and G are the midpoints of AB and AC respectively, find T (denoted by S).
    (2) As shown in Figure 23(2), if the vertices of $\square DEFG$ are all on the sides of $\triangle ABC$, prove that $T \leq \frac{1}{2}S$.
    (3) For any $\square DEFG$ obtained by paper-cut, does the inequality $T \leq \frac{1}{2}S$ still hold? Please explain the reason.

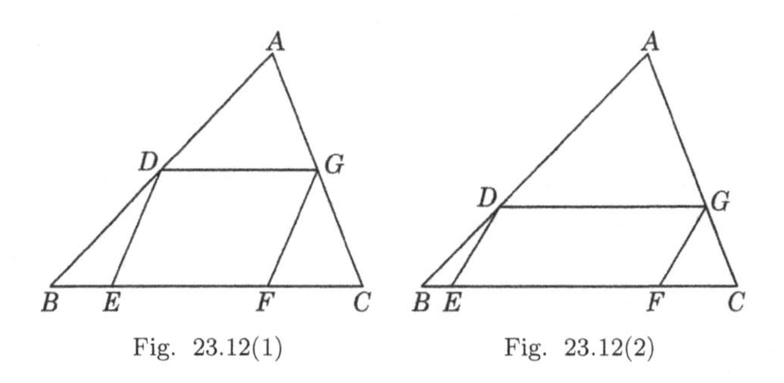

Fig. 23.12(1) Fig. 23.12(2)

Chapter 24

Indefinite Equation

An equation with more unknowns than equations is called an indefinite equation. Generally, only its integer solutions or positive integer solutions are discussed.

The indefinite equation is one of the oldest branches of the integer theory. As early as the beginning of the third century AD, Diophantine, an ancient Greek mathematician, studied this kind of the equation. People also call it Diophantine equation.

For the linear equation $ax + by = c$ with two unknowns, where a, b and c are integers and both a and b are not zero, we have the following conclusions:

If $(a, b) = d > 1$, but d can't divide c, that is, $d \nmid c$, then the equation has no integer solutions.

If an integer solution of equation $ax + by = c$ is $x = x_0$, $y = y_0$, where $(a, b) = 1$, then all the solutions of this equation are $x = x_0 + bt$, $y = y_0 - at$, where $t = 0, \pm 1, \pm 2, \ldots$.

Example 1. Find the positive integer solutions to the equation $xy - 3x^2 = 12$.

Analysis By observing the equation, we know that the index of y is 1. Consider x in the equation as a letter coefficient, and get the linear equation about y.

Solution From $xy - 3x^2 = 12$, it's seen that $x \neq 0$, then

$$y = \frac{3x^2 + 12}{x} = 3x + \frac{12}{x}.$$

Because x and y are positive integers, x must be a positive factor of 12, then $x = 1, 2, 3, 4, 6, 12$. So we get y as 15, 12, 13, 15, 20, 37.

Hence, the positive integer solutions to the original equation are

$$\begin{cases} x_1 = 1, \\ y_1 = 15, \end{cases} \quad \begin{cases} x_2 = 2, \\ y_2 = 12, \end{cases} \quad \begin{cases} x_3 = 3, \\ y_3 = 13, \end{cases}$$

$$\begin{cases} x_4 = 4, \\ y_4 = 15, \end{cases} \quad \begin{cases} x_5 = 6, \\ y_5 = 20, \end{cases} \quad \begin{cases} x_6 = 12, \\ y_6 = 37. \end{cases}$$

Remark

(1) The original equation can also be regarded as a quadratic equation with one unknown x, where x and y are positive integers, then $\Delta = y^2 - 12^2$ is a perfect square number (see Chapters 6 and 7), that is, $y^2 - 12^2 = p^2$. So we have $(y + p)(y - p) = 12^2$, and then we can get y and x.
(2) In addition, we can directly use the basic properties of integer division. The original equation is converted into $x(y - 3x) = 12$. Since $x \mid 12$ and x is a positive integer, we can get that $x = 1, 2, 3, 4, 6, 12$. Then we can get the corresponding y.

Example 2. Find all positive integers x and y that satisfy $\frac{1}{x} - \frac{1}{y} = \frac{1}{4}$.

Solution The original equation $\frac{1}{x} - \frac{1}{y} = \frac{1}{4}$ can be converted into $4(y - x) = xy$.

Consider factorization, then we get $(4 - x)(y + 4) = 16$.

Because x and y are both positive integers, we know that $1 \leq 4 - x \leq 3$. Then $4 - x = 1$ or 2, that is, $x = 3$ or 2. Thus we obtain that when $x = 3$, $y = 12$; when $x = 2$, $y = 4$.

Therefore, x and y satisfying the given conditions are $x = 3$, $y = 12$; $x = 2$, $y = 4$.

Remark

(1) In the question, if "positive integers x and y" is changed to "integers x and y", what is the result?
(2) It can be seen from Example 2 that in general, the integers x and y satisfying $\frac{1}{x} \pm \frac{1}{y} = \frac{1}{a}$ (where a is a positive integer) can be obtained by using the method of solving Example 2.

Example 3. Xiao Wang and Xiao Li play the game of "stone, scissors and cloth". The rules are as follows: 3 points for a win, 1 point for a draw and

0 points for a loss. There are 7 games in total. At the end of the game, the one with the highest score is a winner.

(1) If Xiao Wang scores 10 points at the end of the game, then how many wins, draws and losses does he get in 7 games?
(2) If the result of Xiao Wang's first three games is one win, one draw and one loss, then what kind of results should he achieve in the last four games to ensure his victory over Xiao Li?

Solution

(1) Let Xiao Wang win x times and draw y times, where x and y are natural numbers. Then

$$\begin{cases} 3x + y = 10, \\ x + y \leq 7, \\ 0 < x < 4. \end{cases}$$

So we get $x = 3$, $y = 1$ or $x = 2$, $y = 4$.
That is, Xiao Wang's result is 3 wins, 1 draw and 3 losses, or 2 wins, 4 draws and 1 loss.
(2) Xiao Wang scored 4 points in the first three games with one win, one draw and one loss, while Xiao Li scored 4 points in the first three games with one win, one draw and one loss too.

Suppose that in the last four games, Xiao Wang wins x times and draws y times, then Xiao Wang's score is $3x+y$ and Xiao Li's score is $3(4-x-y)+y$.
If Xiao Wang wins the game, then

$$\begin{cases} 3x + y > 3(4 - x - y) + y, \\ x + y \leq 4, \\ 0 < x \leq 4. \end{cases}$$

Solving the inequalities gives $\begin{cases} x = 4, \\ y = 0, \end{cases}$ $\begin{cases} x = 3, \\ y = 1, \end{cases}$ $\begin{cases} x = 3, \\ y = 0, \end{cases}$ $\begin{cases} x = 2, \\ y = 2, \end{cases}$ $\begin{cases} x = 2, \\ y = 1 \end{cases}$ or $\begin{cases} x = 1, \\ y = 3. \end{cases}$

That is, to ensure Xiao Wang's victory over Xiao Li, Xiao Wang should achieve 4 wins, 3 wins and 1 draw, 3 wins and 1 loss, 2 wins and 2 draws, 2 wins 1 draw and 1 loss, or 1 win and 3 draws in the last four games.

Example 4. Let p be a prime number, and m be an integer, which satisfy

$$p^3 + m(p-2) = m^2 - p + 1.$$

Find all such number pairs(p, m).

Solution From $p^3 + m(p-2) = m^2 - p + 1$, we get

$$p(p^2 + m + 1) = (m+1)^2. \tag{1}$$

So $(m+1)^2$ is divisible by p.

Because p is a prime number, we get that $|m+1|$ is divisible by p.

Let $m + 1 = kp$ (where k is an integer). Substituting this into the equation (1) gives $p = k(k-1)$.

Since $k(k-1)$ is an even number, it is seen that p is an even number. At the same time, p is a prime number, so $p = 2$.

Thus we have $k(k-1) = 2$. Solving the equation gives $k = -1$ or 2. Then we get $m = -3$ or 3.

Hence, the number pairs $(p, m) = (2, 3)$, $(2, -3)$.

Example 5. Find all three positive integers satisfying that the sum of these three numbers is equal to their product.

Solution From the given condition, might as well let these three positive integers be x, y and z, where $1 \le x \le y \le z$, then we get

$$x + y + z = xyz. \tag{1}$$

How to find the positive integer solutions of this equation? Try to use the inequality relation.

Since $1 \le x \le y \le z$, from the equation (1), we know that

$$xyz = x + y + z \le z + z + z = 3z.$$

So $1 \le xy \le 3$, that is, xy can be 1, 2 or 3.

When $xy = 1$, we have $x = 1$, $y = 1$. Then from $2 + z = z$, we get that there is no solution.

When $xy = 2$, we have $x = 1$, $y = 2$. Then from $3 + z = 2z$, we get $z = 3$.

When $xy = 3$, we have $x = 1$, $y = 3$. Then from $4 + z = 3z$, we get $z = 2$. However, $z < y = 3$, which does not meet the requirements.

Thus, the positive integers x, y, z satisfying the equation (1) are $x = 1$, $y = 2$, $z = 3$.

So the three numbers in the question are 1, 2, 3.

Remark

(1) The method of "estimation" in Example 5 can be used to solve Example 2.
(2) If we change "three positive integers" to "four positive integers" or "five positive integers" in Example 5, what is the result?

If the sum of n positive integers is equal to their product, can you directly answer one set of the solutions?

Example 6. Whether there is a triangle whose three sides are exactly three continuous positive integers, and one of its interior angles is twice as large as the other? Prove your conclusion.

Solution There is a triangle that satisfies the conditions.

As shown in Figure 24.1, when $\angle A = 2\angle B$, we extend BA to the point D so that $AD = AC = b$ and join CD, then $\triangle ACD$ is an isosceles triangle.

Because $\angle BAC$ is an exterior angle of $\triangle ACD$, it is known that $\angle BAC = 2\angle D$.

From the given condition, $\angle BAC = 2\angle B$, so $\angle B = \angle D$. Then we know that $\triangle CBD$ is an isosceles triangle.

And since $\angle D$ is a common angle between $\triangle ACD$ and $\triangle CBD$, we get that $\triangle ACD \backsim \triangle CBD$, then $\frac{AD}{CD} = \frac{CD}{BD}$, i.e. $\frac{b}{a} = \frac{a}{b+c}$. So $a^2 = b(b+c)$.

Starting with the easy numbers, we know that $6^2 = 4 \times (4+5)$. So when the three sides of $\triangle ABC$ are $a = 6$, $b = 4$, $c = 5$, $\angle A = 2\angle B$, that is, this triangle satisfies the given conditions.

Hence, there is a triangle satisfying the conditions.

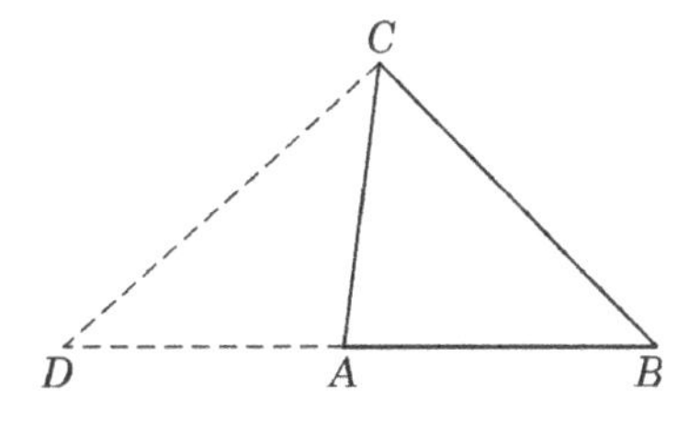

Fig. 24.1

Remark

(1) The triangle satisfying the condition is unique.
If $\angle A = 2\angle B$, we can get $a^2 = b(b+c)$. There are three situations as follows:

(i) When $a > c > b$, let $a = n+1$, $c = n$, $b = n-1$(n is a positive integer greater than 1). Substituting these values into $a^2 = b(b+c)$ gives $(n+1)^2 = (n-1)(2n-1)$. Then we get $n = 5$. So $a = 6$, $b = 4$, $c = 5$.

(ii) When $c > a > b$, let $c = n+1$, $a = n$, $b = n-1$(n is a positive integer greater than 1). Substituting these values into $a^2 = b(b+c)$ gives $n^2 = (n-1)\cdot 2n$. Then we get $n = 2$. So $a = 2$, $b = 1$, $c = 3$, which cannot form a triangle.

(iii) When $a > b > c$, let $a = n+1$, $b = n$, $c = n-1$(n is a positive integer greater than 1). Substituting these values into $a^2 = b(b+c)$ gives $(n+1)^2 = n(2n-1)$, i.e. $n^2-3n-1 = 0$. There are no integer solutions to this equation.

Hence, there is a triangle whose three sides are exactly three consecutive positive integers, and one of its interior angles is twice as large as the other. Meanwhile, only the triangle with the three sides of 4, 5 and 6 meets the condition.

(2) From Exercise 8 in Chapter 20, it is seen that in $\triangle ABC$, if and only if $a^2 = b(b+c)$, $\angle A = 2\angle B$.

Example 7. Let a be a positive integer. Consider a quadratic function $y = x^2 + (a+17)x + 38 - a$ and a inversely proportional function $y = \frac{56}{x}$, if the intersection points of the graphs of these two functions are all integral points (a point whose x-coordinate and y-coordinate are both integers), find the value of a.

Solution Consider the simultaneous equations

$$\begin{cases} y = x^2 + (a+17)x + 38 - a, \\ y = \frac{56}{x}. \end{cases}$$

Eliminating y gives

$$x^2 + (a+17)x + 38 - a = \frac{56}{x},$$

that is,

$$x^3 + (a+17)x^2 + (38-a)x - 56 = 0.$$

Factoring the equation gives

$$(x-1)[x^2+(a+18)x+56]=0. \quad (1)$$

It is obvious that $x_1=1$ is a root of the equation (1), so (1, 56) is an intersection point of the graphs of these two functions.

Because a is a positive integer, the discriminant of the equation

$$x^2+(a+18)x+56=0 \quad (2)$$

is $\Delta=(a+18)^2-224>0$, then the equation must have two distinct real roots.

Since the intersection points of the graphs of the two functions are integral points, the roots of the equation (2) are integers. So its discriminant $\Delta=(a+18)^2-224$ should be a perfect square number.

Let $(a+18)^2-224=k^2$ (where k is a nonnegative integer). Then $(a+18)^2-k^2=224$, i.e.

$$(a+18+k)(a+18-k)=224.$$

Obviously, $a+18+k$ and $a+18-k$ are both odd or even, and $a+18+k\geq 18$.

From $224=112\times 2=56\times 4=28\times 8$, we can get that

$$\begin{cases} a+18+k=112, \\ a+18-k=2, \end{cases} \quad \begin{cases} a+18+k=56, \\ a+18-k=4, \end{cases} \quad \text{or} \quad \begin{cases} a+18+k=28, \\ a+18-k=8. \end{cases}$$

Solving the simultaneous inequalities gives $\begin{cases} a=39, \\ k=55, \end{cases}$ $\begin{cases} a=12, \\ k=26, \end{cases}$ or $\begin{cases} a=0, \\ k=10. \end{cases}$

Since a is a positive integer, the answers can only be $\begin{cases} a=39, \\ k=55 \end{cases}$ or $\begin{cases} a=12, \\ k=26. \end{cases}$

When $a=39$, the equation (2) is $x^2+57x+56=0$, whose roots are -1 and -56. At this time, the graphs of the two functions have two intersection points $(-1, -56)$ and $(-56, -1)$.

When $a=12$, the equation (2) is $x^2+30x+56=0$, whose roots are -2 and -28. At this time, the graphs of the two functions have two intersection points $(-2, -28)$ and $(-28, -2)$.

Hence, $a=12$ or 39.

Example 8. Given that $[x]$ is the maximum integer not more than x, for example $[2.3] = 2$, $[-3.1] = -4$, find the minimum positive integer n that makes

$$[\sqrt{n} + \sqrt{n+1}] = [\sqrt{4n+2018}].$$

Solution Let $k = [\sqrt{n} + \sqrt{n+1}] = [\sqrt{4n+2018}]$. Then k is a positive integer, and

$$\begin{cases} k \le \sqrt{n} + \sqrt{n+1} < k+1, \\ k \le \sqrt{4n+2018} < k+1. \end{cases}$$

Thus we get

$$\begin{cases} k^2 \le 2 + 1 + 2\sqrt{n^2+n} < (k+1)^2, & (1) \\ k^2 \le 4n + 2018 < (k+1)^2. & (2) \end{cases}$$

Since $n < \sqrt{n^2+n} < n+1$, combining inequality (1) gives

$$k^2 < 2n + 1 + 2(n+1) = 4n + 3. \quad (3)$$

From (2) and (3), we know that

$$k^2 < 4n + 3 < 4n + 2018 < (k+1)^2.$$

Since k^2 is the perfect square number, it can be known that the remainder of k^2 divided by 4 is 0 or 1. Thus we obtain

$$k^2 \le 4n + 1 < 4n + 3 < 4n + 2018 < (k+1)^2,$$

then

$$(4n + 2018) - (4n + 1) < (k+1)^2 - k^2.$$

Simplifying the inequality gives $2017 < 2k + 1$. So $k > 1008$, that is, $k \ge 1009$.

Therefore, we get $4n + 1 \ge k^2 \ge 1018081$, then $n \ge 254520$.

When $n = 254520$,

$$[\sqrt{n} + \sqrt{n+1}] = 1009 = [\sqrt{4n+2018}].$$

So the minimum value of n is 254520.

In junior high school, there are usually two basic solutions to the indefinite equation problems. One is to use some basic properties of integers, the other is to use the basic knowledge and methods of junior high school, such as factorization, completing the square, the discriminant of the quadratic equation and the relationship between the roots and coefficients, inequality relation and estimation, etc. In general specific operation, these two basic solutions are often integrated to solve the problems related to indefinite equations.

Reading

Essential Reductio ad Absurdum

Once upon a time, there was a man who sold both spears and shields. He held up the spear and said, "My spear is very sharp. No matter how strong a shield is, my spear can pierce it". After that, he held up the shield and said, "My shield is very strong. No matter how sharp a spear is, it cannot pierce my shield". Now someone came up and asked him, "If you hold up your spear to pierce through your shield, what will happen?" The man could not answer the question because whatever he said was contradictory.

In mathematics, contradiction is often used to prove a conclusion. This method of proof is called "Reductio ad Absurdum".

For example, all the natural numbers $1, 2, 3, \ldots$ are divided into two groups randomly, there must be a group in which the sum of two numbers is a square number. To prove this conclusion, we can use "Reductio ad Absurdum".

Suppose that the above conclusion is not true, then all natural numbers is divided into two groups, the sum of any two numbers in each group is not a square number.

The group containing 1 is called the group A and the other is called the group B.

Since the sum of every two numbers in the group A is not a square number, then 3 is not in the group A because $3 + 1 = 4$. In the same way, $8(= 9 - 1)$, $15(= 16 - 1)$, $24(= 25 - 1), \ldots$ are not in the group A either.

Since 3 is in the group B, $6(= 9 - 3)$ is in the group A.

Since 15 is in the group B, $10(= 25 - 15)$ is in the group A.

However, $6 + 10 = 16$ is a square number, which conflicts with "the sum of any two numbers in the group A is not the square number".

This contradiction shows that the hypothesis does not hold, which proves the conclusion of the original question.

It can be seen from this example that to prove a problem with "Reductio ad Absurdum" is to assume that the conclusion is not tenable first, that is to say, the opposite conclusion is true; then try to derive the contradiction, and get that the hypothesis is wrong while the original conclusion is true.

"Reductio ad Absurdum" is indispensable. It's like attacking a fortress, it's a good way to attack from the back when it's hard for the front attack to work.

"To prohibit mathematicians from using reductio ad absurdum is just like banning boxers from using fists", said German mathematician Albert D. (1862–1943).

(Excerpted from: Shan Zun. *Train of thought.* Nanjing: Jiangsu Education Press, 2006, 36~37.)

Exercises

1. Find the integer solutions to the equation $x^2 = 5 - xy$.
2. Let a be a real number. If $\sqrt{23 - a}$ and $\sqrt{6 - a}$ are both integers, find the value of a.
3. In the plane rectangular coordinate system xOy, a point whose x-coordinate and y-coordinate are both integers is called an integral point. Let k be an integer. If there is at least one integral point in the intersections of the graphs of the inversely proportional function $y = \frac{6}{x}$ and the function $y = kx - 4$, then the number of all possible values of k is ( ).

   (A) 1 (B) 2 (C) 3 (D) 4

4. The number of positive integer solutions (x, y) to the indefinite equation $x^2 - 2y^2 = 5$ is ( ).

   (A) 0 (B) 2 (C) 4 (D) an infinite number

5. The group number of integer solutions (x, y) to the indefinite equation $x^2 + 2xy + 3y^2 = 34$ is ( ).

   (A) 3 (B) 4 (C) 5 (D) 6

6. (1) Find the positive integer solutions to the equation $\dfrac{1}{x} + \dfrac{1}{y} = \dfrac{1}{2}$.

   (2) Find the number of positive integer solutions (x, y) to the equation $\dfrac{1}{x} + \dfrac{1}{y} = \dfrac{3}{7}$.
7. Find the integer solutions to the equation $xy - 10(x + y) = 1$.
8. If the minimum value of x, y and z is not less than 3, find the integer solutions to the equation $\dfrac{1}{x} + \dfrac{1}{y} - \dfrac{1}{z} = \dfrac{1}{2}$.

9. Find the integer solutions to the equation $14x^2 - 24xy + 21y^2 + 4x - 12y - 18 = 0$.
10. Find the integer solutions to the equation $y^3 = x^2 + x$.
11. (1) Consider an identity $x^3 - x^2 - x + 1 = (x - 1)(x^2 + kx - 1)$, find the value of k.
    (2) If x is an integer, prove that $\frac{x^3-x^2-x+1}{x^2-2x+1}$ is an integer.
12. Find the positive integer solutions to the equation $m^2 - 2mn + 14n^2 = 217$.
13. Find the positive integer solutions to the simultaneous equations $\begin{cases} x^3 - y^3 - z^3 = 3xyz, \\ x^2 = 2(y + z). \end{cases}$
14. Find the minimum positive integer n so that the average of quadratic sum of the first n positive integers $1, 2, \ldots, n$ $(n > 1)$ is an integer.
15. Find all the positive integer solutions to the equation $x^2 + y^2 = 208(x - y)$.
16. Given that n different numbers are randomly selected from the continuous natural numbers $1, 2, 3, L, \ldots$, 2008.

    (1) Prove that when $n = 1007$, no matter how to select these n numbers, there always exist 4 numbers of them whose sum is equal to 4017.
    (2) When $n \leq 1006$(n is a positive integer), is the above conclusion true? Please explain the reason.

Chapter 25

Reductio ad Absurdum

Reductio ad absurdum is the basic logical proof in mathematics. We know that one of the two opposite judgments of "s is p" and "s is not p" must be true and the other is false. This is the law of excluded middle in the law of logical thinking, which is the logical basis of the law of reductio ad absurdum.

About the effect of reductio ad absurdum, please refer to "Reading" at the end of Chapter 24.

Example 1. Let a, b and c be odd numbers. Prove that the equation $ax^2 + bx + c = 0$ has no rational roots.

Analysis It is complicated to prove the conclusion directly. Reductio ad absurdum can be used.

Proof Suppose that this equation has a rational root, and let this rational root be $\frac{q}{p}$ (where p and q are two coprime integers).

Then we have

$$a\left(\frac{q}{p}\right)^2 + b\left(\frac{q}{p}\right) + c = 0,$$

i.e.

$$aq^2 + bpq + cp^2 = 0.$$

Since p and q are coprime, p and q are both odd, or one of them is odd and the other is even.

As we know, a, b and c are all odd numbers. If p and q are both odd, then $aq^2 + bpq + p^2c$ is odd, which contradicts the fact that 0 is an even number. If p and q are one odd and one even, then $aq^2 + bpq + cp^2$ is also an odd number, which contradicts the fact that 0 is an even number. The hypothesis does not hold.

Hence, the equation $ax^2 + bx + c = 0$ has no rational roots.

Remark The above problem-solving process uses the reductio ad absurdum. The steps to deal with the problem are as follows:

(1) Counter assumption: to make a hypothesis contrary to the conclusion of verification.
(2) Reduction to absurdity: starting from the counter assumption, the result which contradicts with axioms, definitions, given theorems or conditions is deduced.
(3) Make a conclusion: prove that the counter assumption cannot hold, thus affirming that the original conclusion has to be established.

The characteristic of the reductio ad absurdum is to prove a proposition by deriving the contradiction and reducing it to fallacy. Reductio ad absurdum is also called reduction to absurdity.

Example 2. Given a quadrilateral $ABCD$, E and F are the midpoints of AD and BC respectively, and $EF = \frac{1}{2}(AB + CD)$ (as shown in Figure 25.1). Prove that $AB \parallel CD$.

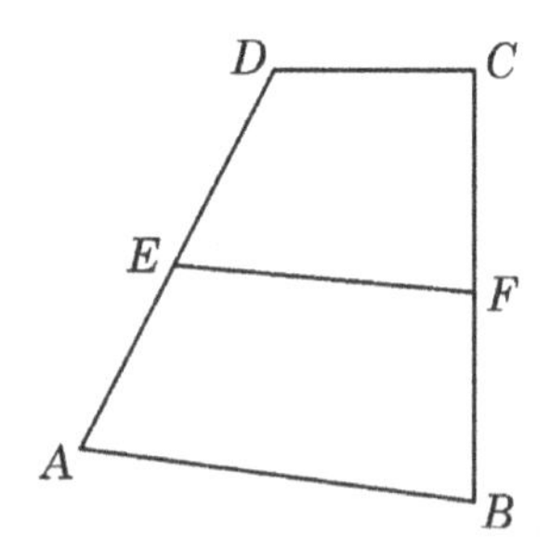

Fig. 25.1

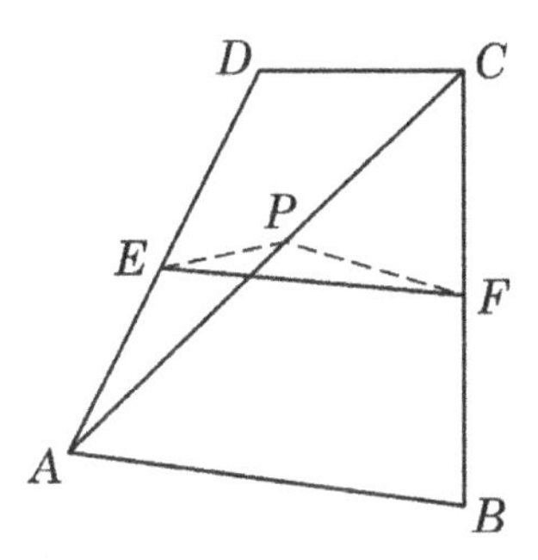

Fig. 25.2

Proof Suppose $AB \nparallel CD$. As shown in Figure 25.2, join AC. Let P be the midpoint of AC. Join EP and FP. Then

$$EP \parallel DC, FP \parallel AB.$$

However, $AB \nparallel CD$, so EPF is not a straight line. Then we can get

$$EP + PF > EF. \tag{1}$$

Since $EP = \frac{1}{2}DC, PF = \frac{1}{2}AB$, substituting these into (1) gives

$$EP + PF = \frac{1}{2}DC + \frac{1}{2}AB = \frac{1}{2}(DC + AB) > EF.$$

This contradicts the given condition $EF = \frac{1}{2}(AB + CD)$. So the hypothesis $AB \nparallel CD$ does not hold. Hence, $AB \parallel CD$.

Example 3. Given that a_1, a_2, ..., a_8 are all positive numbers, and

$$a_1 + a_2 + \cdots + a_8 = 20, \tag{1}$$

$$a_1 a_2 \cdots a_8 < 4. \tag{2}$$

prove that at least one of the numbers a_1, a_2, ..., a_8 is less than 1.

Analysis It is more convenient to use reductio ad absurdum to prove that "at least one number is less than 1".

Proof Suppose that all of $a_1, a_2, \ldots, a_8$ are not less than 1, then let $a_i = 1 + b_i (b_i \geq 0,\ i = 1, 2, \ldots, 8)$. From the equality (1), we have

$b_1 + b_2 + \cdots + b_8 = 12$. Thus, we get

$$\begin{aligned} a_1 a_2 \ldots a_8 &= (1+b_1)(1+b_2)\cdots(1+b_8) \\ &= 1 + (b_1 + b_2 + \cdots + b_8) + \cdots + b_1 b_2 \cdots b_8 \\ &\geq 1 + (b_1 + b_2 + \cdots + b_8) \\ &= 1 + 12 = 13. \end{aligned}$$

This contradicts the condition (2). So it is impossible that all of a_1, a_2, ..., a_8 are not less than 1, that is, at least one of them is less than 1.

Think about it In Example 3, what number can replace the number 4 on the right side of the inequality "$a_1 a_2 \ldots a_8 < 4$", and the proposition conclusion still holds.

Example 4. Given that a_1, a_2, ..., a_n and b_1, b_2, ..., b_n are $2n$ positive numbers, and $a_1^2 + a_2^2 + \cdots + a_n^2 = b_1^2 + b_2^2 + \cdots + b_n^2$. prove that the minimum number of n fractions $\frac{a_1}{b_1}, \frac{a_2}{b_2}, \ldots, \frac{a_n}{b_n}$ is not more than 1, and the maximum number is not less than 1.

Proof Suppose that the minimum number of these n fractions is greater than 1, then all the fractions are greater than 1.

Thus, $\frac{a_1}{b_1} > 1, \frac{a_2}{b_2} > 1, \ldots, \frac{a_n}{b_n} > 1$.

Since b_1, b_2, ..., b_n are all positive numbers, we get that $a_1 > b_1$, $a_2 > b_2$, ..., $a_n > b_n$, i.e. $a_1^2 > b_1^2$, $a_2^2 > b_2^2$, ..., $a_n^2 > b_n^2$.

Add all the inequalities to get $a_1^2 + a_2^2 + \cdots + a_n^2 > b_1^2 + b_2^2 + \cdots + b_n^2$, which contradicts the given condition $a_1^2 + a_2^2 + \cdots + a_n^2 = b_1^2 + b_2^2 + \cdots + b_n^2$.

Therefore, the minimum number of these n fractions must be not greater than 1. Similarly, it can be proved that the maximum number of these n fractions must be not less than 1.

Example 5. Prove that the number of prime numbers is infinite.

Proof Suppose the number of prime numbers is limited, then the finite prime numbers are denoted as p_1, p_2, ..., p_n in order of values.

Make a new number $N = p_1 p_2 \ldots p_n + 1$. It's obvious that $N \neq p_i$ $(i = 1, 2, \ldots, n)$. Now let's prove that N is a prime number.

If not, N must be divisible by one number of $p_1, p_2, \ldots, p_n$. But from the construction of N, we can see that when N is divided by any prime number of $p_1, p_2, \ldots, p_n$, the remainder is 1, that is, N cannot be divisible by $p_1, p_2, \ldots, p_n$. This shows that N is not a composite number, that is,

N is a prime number. In this way, besides the original n prime numbers, a new prime number N is added. Therefore, there are at least $n+1$ prime numbers, which contradicts the hypothesis.

So the number of prime numbers is infinite.

Example 6. Prove that the product of any three consecutive positive integers is not the integer power of a positive integer greater than 1.

Proof Suppose that the conclusion of the original proposition is incorrect, then there must be three consecutive positive integers whose product is exactly the integer power of a positive integer greater than 1. Might as well assume that there are positive integers n, a and k, so that

$$(n-1)n(n+1) = a^k (k > 1). \tag{1}$$

Obviously, both $n-1$ and $n+1$ are prime to n, then $(n-1)(n+1)$ is prime to n. Thus, from (1) we can get

$$(n-1)(n+1) = b^k, n = c^k \text{ (where } b \text{ and } c \text{ are positive integers)}, \tag{2}$$

and $a = bc$.

From (2), we obtain $n^2 = (c^2)^k$ and $(c^2)^k - b^k = n^2 - (n-1)(n+1) = 1$.

However, the difference of $k(k > 1)$ power of two different positive integers is obviously greater than 1. Then we get $1 > 1$, which is obviously unreasonable, contradictory. So the hypothesis does not hold.

Hence, the original proposition holds.

Remark

(1) We can also use

$$a^k - b^k = (a-b)(a^{k-1} + a^{k-2}b + \cdots + ab^{k-2} + b^{k-1})$$

to get $(c^2)^k - b^k > 1$ in this question.

(2) Think about it: is the conclusion still true when "any three consecutive positive integers" is changed to "any two consecutive positive integers" or "any four consecutive positive integers"?

(3) The essence of the problem involves the prime factorization of positive integers, and the conclusion has negativity and infinity, so we can consider using the reductio ad absurdum.

Example 7. There are nine points in ΔABC. Try to prove that three points can be selected from them, and the area of the triangle formed by these three points is not greater than $\frac{1}{4}$ of the area of ΔABC.

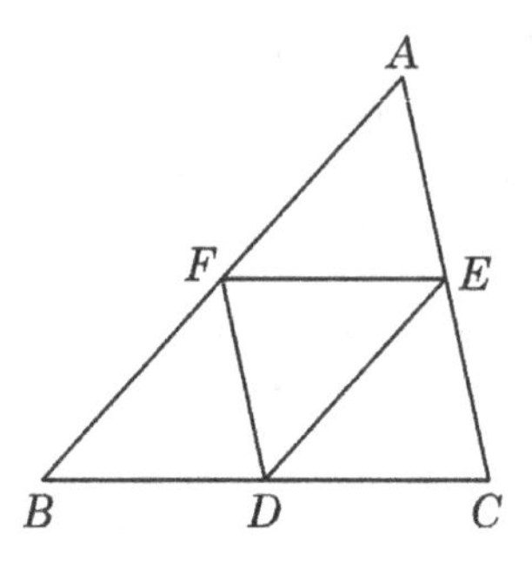

Fig. 25.3

Proof As shown in Figure 25.3, let D, E and F be the midpoints of three sides of ABC. Obviously, the area of the four small triangles in Figure 25.3 is equal to $\frac{1}{4}$ of the area of ΔABC.

Now we prove that there must exist such a small triangle (including its edge) containing at least three of the nine known points.

Suppose that each triangle (including its edge) contains at most two known points, then the whole triangle ABC contains at most 8 known points, which contradicts the given condition. So there must be such a small triangle (including its edge), which contains at least three of the nine known points.

Because these three points are in this small triangle, the area of the triangle formed by these three points is certainly not greater than the area of the small triangle where they are, that is, it is not greater than $\frac{1}{4}$ of the area of ΔABC.

Remark This example is to prove that "not more than...". After analysis, the essence is the problem of "at least...". In the questions of the drawer principle, this kind of characteristic should be noticed.

Example 8. Given that there are six circles in the plane, and the center of each circle is outside the rest circles, prove that no point in the plane will be inside the six circles at the same time.

Analysis It's difficult to solve this problem by direct proof, but it's easier to solve it by reductio ad absurdum.

Proof Suppose that there is a point M in the plane which is inside the six circles at the same time. Let's mark the center of the six circles around

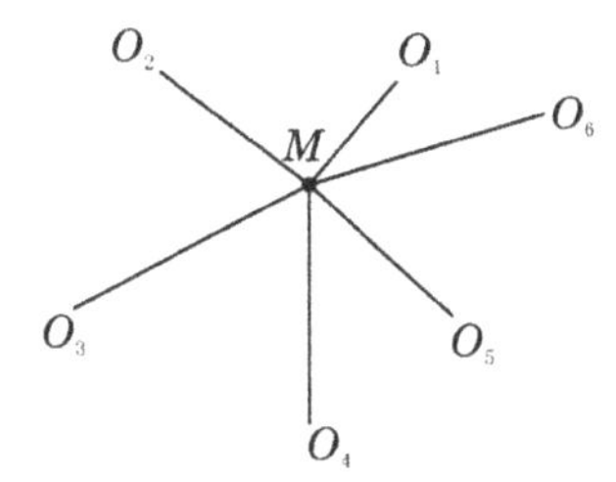

Fig. 25.4

M in a counterclockwise direction as O_1, O_2, O_3, $\ldots$, O_6 in turn, and join M to these six centers (as shown in Figure 25.4).

Firstly, consider $\triangle O_1MO_2$. Since M is in $\odot O_1$ and O_2 is outside $\odot O_1$, we know that $O_1O_2 > O_1M$. Similarly, $O_1O_2 > O_2M$. Thus, in $\triangle O_1MO_2$, O_1O_2 is larger than other two sides. According to the theorem that the angle opposite the longer side is greater in the triangle, we obtain that $\angle O_1MO_2 > \angle O_1O_2M$ and $\angle O_1MO_2 > \angle O_2O_1M$.

Therefore,

$$3\angle O_1MO_2 > \angle O_1MO_2 + \angle O_1O_2M + \angle O_2O_1M = 180°,$$

i.e. $\angle O_1MO_2 > 60°$.

Similarly, it is proved that $\angle O_2MO_3$, $\angle O_3MO_4$, $\angle O_4MO_5$, $\angle O_5MO_6$ and $\angle O_6MO_1$ are also more than $60°$ respectively. So

$$\begin{aligned}&\angle O_1MO_2 + \angle O_2MO_3 + \angle O_3MO_4 + \angle O_4MO_5 + \angle O_5MO_6\\&\quad + \angle O_6MO_1 > 6\times 60° = 360°.\end{aligned}$$

However, it can be seen from Figure 25.4 that the sum of the above six angles is a round angle, namely $360°$. So the result inferred is in contradiction with the definition of a round angle. Then the hypothesis does not hold.

Hence, the original proposition holds.

Example 9. In a rectangle with the area of 5 square units, nine rectangles with the area of 1 square unit are placed. Prove that no matter how these nine rectangles are placed, there are always 2 rectangles whose overlapping area is greater than or equal to $\frac{1}{9}$.

Proof Suppose that the area of overlapping part of any two rectangles is less than $\frac{1}{9}$.

Let the rectangle with 5 square units be A and nine rectangles with 1 square unit be

$$A_1, A_2, A_3, \ldots, A_9.$$

If the overlapping area of A_1 and A_2 is less than $\frac{1}{9}$, then the part of A_2 which is outside A_1 occupies the area of more than $1 - \frac{1}{9} = \frac{8}{9}$ square units in A.

Therefore, A_1 and A_2 cover the area of more than $\left(1 + \frac{8}{9}\right)$ square units in A. If each overlapping area of A_3 and A_1, A_2 is less than $\frac{1}{9}$, then the overlapping area of A_3 and A_1, A_2 is less than $\frac{2}{9}$ square units. Thus the part of A_3 which is outside A_1 and A_2 occupies the area of more than $1 - \frac{2}{9} = \frac{7}{9}$ square units in A.

Therefore, A_1, A_2 and A_3 cover the area of more than $\left(1 + \frac{8}{9} + \frac{7}{9}\right)$ square units in A. Generally, if each overlapping area of A_i and A_1, A_2, $\ldots$, A_{i-1} is less than $\frac{1}{9}$, then the overlapping area of A_i and A_1, A_2, $\ldots$, A_{i-1} is less than $\frac{i-1}{9}$ square units. Thus the part of A_i which is outside A_1, A_2, $\ldots$, A_{i-1} occupies the area of more than $1 - \frac{i-1}{9}$ square units in A.

In this way, 9 rectangles occupy the area of more than $1 + \frac{8}{9} + \frac{7}{9} + \cdots + \frac{1}{9} = 5$ square units in A.

That is to say, the area covered by the nine rectangles with the area of 1 square unit is greater than 5, which contradicts the condition that the nine rectangles are known to be placed in the rectangles with the area of 5 square units.

So the hypothesis does not hold.

Hence, the original proposition holds.

Reading

Don't abuse the reductio ad absurdum

The reductio ad absurdum is a very powerful method. In order to prove "$A \to B$", there was only one condition A. By using the reductio ad absurdum, we suppose that B did not hold, there would be another condition $\overline{B}$ (not B), and the conclusion to be derived from conditions A and $\overline{B}$ is only "contradiction". Therefore, the reductio ad absurdum can often solve some problems which are difficult to prove directly.

However, the reductio ad absurdum should not be abused. Many problems that can be proved directly do not need to be proved by the reductio ad absurdum. Especially in expressions, it is clearer to speak directly than to speak in the opposite direction.

Let's see the following example, in which the original solution can be greatly modified.

Prove that if n is an even number greater than 4, then $2^n - 1$ is the product of at least three positive integers greater than 1.

Proof As we know, $2^2 - 1 = 3$, $2^4 - 1 = 3{\cdot}5$, $2^{2k} - 1 = (2^k - 1){\cdot}(2^k + 1)$. If $n = 2k > 4$, the number $2^{2k} - 1$ is equal to the product of 2 prime numbers, then the numbers $2^k - 1$ and $2^k + 1$ are prime numbers, which is impossible. When k is an odd number, $3|2^k + 1$; when k is an even number, $3|2^k - 1$, while $2^k - 1 > 3$. Therefore, when n is an even number greater than 4, $2^n - 1$ is the product of at least three natural numbers greater than 1.

The above proof adopts the reductio ad absurdum. In fact, the first half of the proof only needs the remain the part of "as we know, $2^{2k} - 1 = (2^k - 1)(2^k + 1)$.

And the following is followed by "when k is an odd number...". All the parts of the reductio ad absurdum should be deleted.

(Excerpted from: Shan Zun. *Problem solving research*. Shanghai: Shanghai Education Press, 2016, 105~106.)

Exercises

1. Given that the real numbers m, n, p and q satisfy the following conditions: (1) $m + n = 1$; (2) $p + q = 1$; (3) $mp + nq > 1$, prove that there must be a number less than 0 among m, n, p and q.
2. Prove that $\sqrt{24}$ is an irrational number.
3. Prove that the number of acute angles cannot be more than 3 among all the interior angles of a convex polygon.
4. In a convex quadrilateral $ABCD$, if $AB + BD \leq AC + CD$, prove that $AB \leq AC$.
5. Given that E and F are any point on the two sides AB and AC of $\triangle ABC$ respectively, prove that BF and CE cannot bisect each other.
6. In ABC, $AB = AC$, P is a point inside $\triangle ABC$, and $\angle APB > \angle APC$, prove that $\angle BAP < \angle CAP$.
7. Given that AB and CD are any two chords without passing through the center in $\odot O$, prove that AB and CD cannot bisect each other.

8. Given that p, q and r are all positive numbers, prove that at least one of the three equations of $x^2 - \sqrt{p}x + \frac{q}{8} = 0$, $x^2 - \sqrt{q}x + \frac{r}{8} = 0$ and $x^2 - \sqrt{r}x + \frac{p}{8} = 0$ has two distinct positive roots.
9. Prove that there are no such four points A, B, C and D in the plane that make all of ΔABC, ΔBCD, ΔCDA and ΔDAB be acute triangles.
10. Let the points E, F and G be on the three sides AB, BC and CA of ΔABC (except the end point). Prove that at least one triangle of ΔAEG, ΔBEF and ΔCFG has the area which is not greater than $\frac{1}{4}$ of the area of ΔABC.
11. Prove that a triangle with two equal angular bisectors is an isosceles triangle.
12. Let p and q be positive prime numbers. Does the equation $x^2 + p^2x + q^3 = 0$ have a rational root? If not, prove the conclusion; if yes, find the rational roots.
13. On a $4 \times n$ chessboard, can a horse start from any point, jump over every check of the chessboard and return to its original position (horse's step follows the rules of Chinese chess)?
14. Prove that there is no lattice point regular triangle.
15. A difficult mathematics competition consists of two parts: the first test and the second test. There are 28 questions in these two tests, and each contestant has just solved 7 questions. For each of these 28 questions, there are exactly two contestants who solved it. Prove that there must be one contestant who cannot solve the questions in the first test, or can solve at least 4 questions of them.

Chapter 26

Extreme Principle

As we all know, in any number of positive integers, there must be a minimum value. Among a finite number of real numbers, there must be a maximum and a minimum.

This basic fact, embodied in mathematical problems, is the extremum in quantity, such as the maximum and the minimum numbers, the maximum and the minimum angles (areas), the shortest and the longest line segments, the limit position of the movement change of the point and the extreme situation of the state change of various things (such as the best, the worst, etc.).

By considering the extreme situation, we can get the enlightenment and methods to solve the general problems, which often give people a sense of being unexpectedly rescued from a desperate situation. Therefore, people call the above basic methods and ideas used in dealing with problems as extreme principles.

Example 1. For a given equilateral triangle, prove that the sum of the vertical distances from any point in the triangle to three sides is a fixed number.

Proof Consider the extreme situation and get the fixed value.

As shown in Figure 26.1, when a point P coincides with the vertex A, the distances from P to the sides AB and AC are both equal to zero, while the distance from P to the side BC is exactly equal to the height of the triangle ABC.

In this way, the problem is converted into the following problem.

For a given equilateral triangle ABC, P is any point in ΔABC. Prove that the sum of the distances from P to the three sides of the triangle is equal to the height of the triangle.

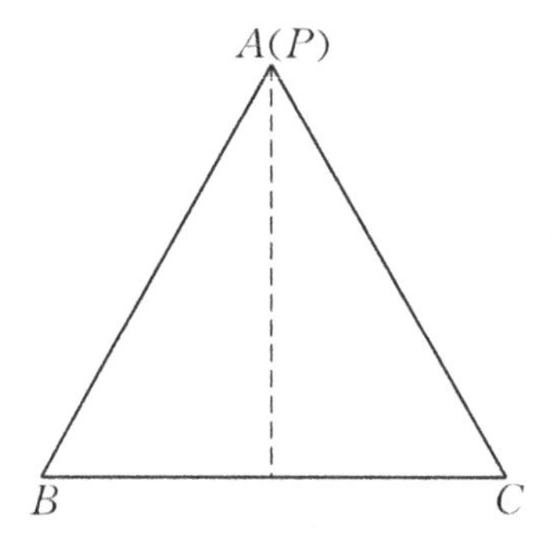

Fig. 26.1

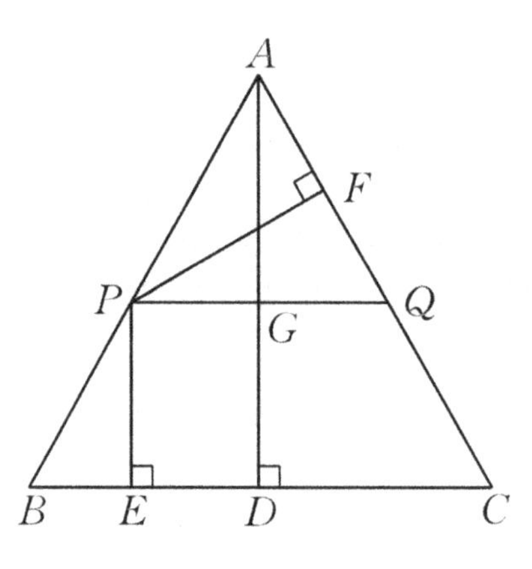

Fig. 26.2

Then consider the special position of the point P to find the general solution.

Consider the point P on the side of a triangle. Let P be on side AB, as shown in Figure 26.2. Draw a line AD perpendicular to BC through the point A (produced at D). Through the point P, draw a line PE perpendicular to BC (produced at E) and a line PF perpendicular to AC (produced at F).

Draw a line PQ through the point P so that $PQ \parallel BC$ and PQ intersects AC at the point Q and intersects AD at the point G. Then ΔAPQ is an equilateral triangle, so $PF = AG$. And since $PE = GD$, we obtain $PF + PE = AD$.

To sum up, we should try our best to transform the general case into the above case, whose breakthrough is to make parallel lines on one side.

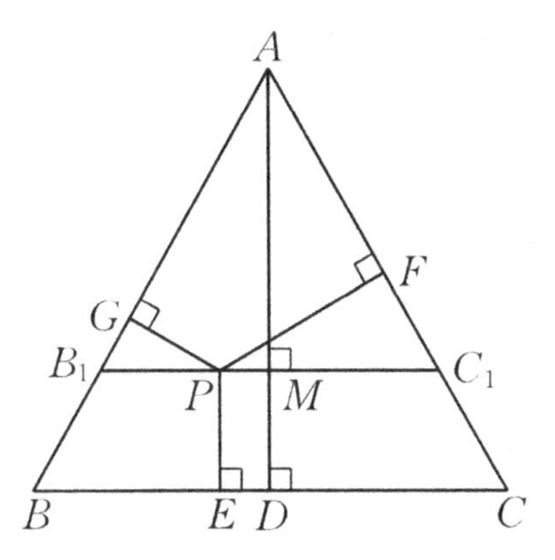

Fig. 26.3

As shown in Figure 26.3, when P is a point in ΔABC, draw a line B_1C_1 through the point P so that $B_1C_1 \parallel BC$, B_1C_1 intersects AD at the point M, intersects AB at the point B_1 and intersects AC at the point C_1.

In the equilateral triangle AB_1C_1, P is a point on the side B_1C_1, then we have $PG + PF = AM$. And because $PE = MD$, we get

$$PE + PG + PF = PE + AM = MD + AM = AD.$$

Remark

(1) Example 1 shows the idea of using the extreme principle to solve problems step by step.

(2) There is also a simpler general proof of this problem, which is to use the relation of the areas of triangles.

Because $S_{\Delta_{ABC}} = S_{\Delta_{APB}} + S_{\Delta_{APC}} + S_{\Delta_{BPC}}$, we can get

$$\frac{1}{2}BC \cdot h_a = \frac{1}{2} \cdot PE \cdot BC + \frac{1}{2}PF \cdot AC + \frac{1}{2}PG \cdot AB.$$

As we know, $AB = BC = CA$, so $PE + PF + PG = h_a$(h_a is the height perpendicular to BC).

Using this solution, we get the general conclusion:

If the lengths of three sides of ΔABC are a, b and c respectively, the distances from any point P in the triangle to the corresponding three sides are d_a, d_b and d_c, the height perpendicular to AB is h_c, then $\frac{a}{c}d_a + \frac{b}{c}d_b + d_c = h_c =$ a fixed value.

In particular, if $a = b = c$, this is the result of Example 1. If $d_a = 0$, $b = c$, then $d_b + d_c = h_c$, that is, the sum of the distances from a point on the base of an isosceles triangle to two waists is equal to the height perpendicular to one waist.

(3) If the point P is outside the equilateral triangle ABC, what is the result? You can use the relation of the areas to get the result by yourself.

Can the conclusion of Example 1 be extended to "n Regular Polygon"?

Example 2. Given 101 points in the plane, the distance between any two points is less than 10, and any three points are the vertices of an obtuse triangle. Prove that there are circles with a diameter of not more than 10, covering these 101 points.

Proof Among these 101 points, let A and B be the two points with the largest distance to each other and $AB < 10$. The circle with a diameter of AB covers all the 101 points.

This is because, as shown in Figure 26.4, if the vertical lines l_1 and l_2 of AB are made respectively through A and B, the given points cannot be outside strip area enclosed by the lines l_1 and l_2. Otherwise, the distance between the points P and B (or A) is greater than AB, which contradicts with the maximum of AB.

Meanwhile, the given points cannot be outside the circle inside the strip area. Otherwise, the points P', A and B cannot form an obtuse triangle, which contradicts with the given conditions.

So the conclusion is true.

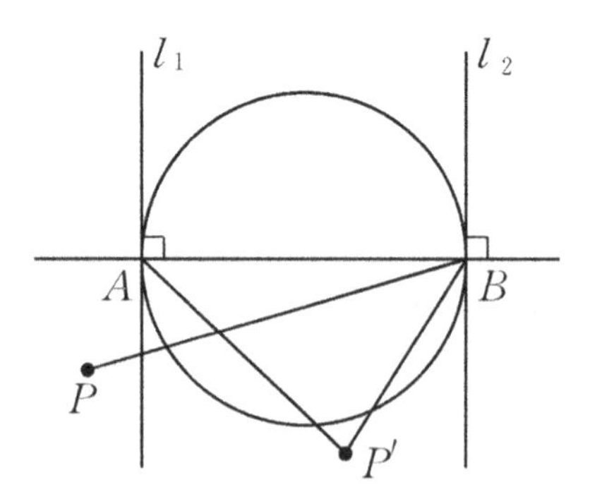

Fig. 26.4

Remark

(1) This example is a kind of covering problems in combinatorial geometry problems, which often need to be solved by the idea of an extreme principle.
(2) This example can be generalized to the following proposition.
Given n points in the plane, the distance between any two points is less than a constant a (e.g. 1), and any three points are the vertices of an obtuse triangle. Prove that there are circles with a diameter of not more than a, covering these n points.

Example 3. There are $n\,(n \geq 2)$ players taking part in the table tennis competition, and any two players have one competition. If every match is decided, prove that there is a player A so that any other player loses either to A or to a player defeated by A.

Analysis Read and understand the questions, and look for the extreme A.

Proof A is the player who won the most games.

For any other player B, if he loses to A, the proposition holds; if B does not lose to A, then B beats A.

Because B can't defeat more opponents than A, except A and B, A must defeat more opponents than B. Thus, there must be a player C, who is defeated by A but beats B, that is, B loses to the player C defeated by A.

Example 4. Given that n numbers are selected from $1, 2, \ldots, 9$ at random, in which there must be some numbers (at least one or all) whose sum can be divisible by 10, find the minimum value of n.

Solution When $n = 4$, in numbers 1, 3, 5, and 8, there are no several numbers whose sum can be divisible by 10.

When $n = 5$, Let $a_1, a_2, \ldots, a_5$ be 5 different numbers in numbers $1, 2, \ldots, 9$. If any several numbers of them whose sum cannot be divisible by 10, then in numbers $a_1, a_2, \ldots, a_5$, it is impossible to have 1 and 9, 2 and 8, 3 and 7 or 4 and 6 at the same time. Thus 5 must be a number in $a_1, a_2, \ldots, a_5$.

If 1 is included in $a_1, a_2, \ldots, a_5$, then 9 is not included. Therefore, 4 is not included $(4 + 1 + 5 = 10)$, 6 should be included. 3 is not included $(3 + 6 + 1 = 10)$, 7 should be included. 2 is not included $(2 + 1 + 7 = 10)$, 8 should be included. However, $5 + 7 + 8 = 20$ is a multiple of 10, which contradicts with the given conditions.

If 9 is included in $a_1, a_2, \ldots, a_5$, then 1 is not included. Therefore, 6 is not included $(6+9+5=20)$, 4 should be included. 7 is not included $(7+4+9=20)$, 3 should be included. 8 is not included $(8+9+3=20)$, 2 should be included. However, $5+3+2=10$ is a multiple of 10, which contradicts with the given conditions.

To sum up, the minimum value of n is 5.

Example 5. If 5 different pair-wise coprime integers a_1, a_2, a_3, a_4, a_5 are selected from 1, 2, 3, ..., n at random, in which there is always a prime integer, find the maximum value of n.

Solution If $n \geq 49$, we consider the numbers 1, 2^2, 3^2, 5^2 and 7^2. These five integers are five different pair-wise coprime integers, but none of them is a prime number. So $n \leq 48$.

5 different pair-wise coprime integers a_1, a_2, a_3, a_4, a_5 are selected from 1, 2, 3, ..., 48 at random.

Suppose that all of a_1, a_2, a_3, a_4 and a_5 are not prime numbers, then at least four of them are composite numbers. Let a_1, a_2, a_3 and a_4 be composite numbers, and p_1, p_2, p_3 and p_4 be the minimum prime factors of a_1, a_2, a_3 and a_4 respectivaly.

Because a_1, a_2, a_3 and a_4 are pair-wise coprime, p_1, p_2, p_3 and p_4 are different from each other. Let p be the maximum of p_1, p_2, p_3 and p_4. Then $p \geq 7$.

Because a_1, a_2, a_3 and a_4 are composite numbers, there must be a number a_i in a_1, a_2, a_3 and a_4 so that $a_i \geq p^2 \geq 7^2 = 49$, which contradicts with $n \leq 48$. Therefore, one of a_1, a_2, a_3, a_4 and a_5 must be a prime number.

To sum up, the maximum value of the positive integer n is 48.

Example 6. There are $n(\geq 7)$ circles, in which any three circles do not intersect pair-wise (including tangency). Prove that there must be a circle which can only intersect five circles at most.

Proof As shown in Figure 26.5, let $O_1, O_2, \ldots, O_n$ be the centers of n circles, and $\odot O_1$ be the circle with the smallest radius among n circles.

If $\odot O_1$ intersects with 6 (or more) circles $O_2, O_3, \ldots, O_7$, we can join the line segments $O_1O_2, O_1O_3, \ldots, O_1O_7$, then one angle of $\angle O_2O_1O_3$, $\angle O_3O_1O_4$, $\angle O_4O_1O_5$, $\angle O_5O_1O_6$, $\angle O_6O_1O_7$ and $\angle O_7O_1O_2$ must be not more than $\frac{360^\circ}{6} = 60^\circ$. Might as well let $\angle O_2O_1O_3 \leq 60^\circ$.

Join O_2O_3, let r_1, r_2 and r_3 be the radii of $\odot O_1$, $\odot O_2$ and $\odot O_3$. Then $r_1 \leq r_2$, $r_1 \leq r_3$.

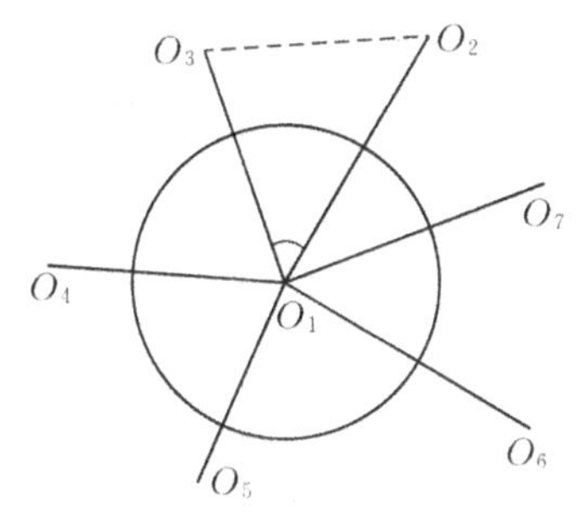

Fig. 26.5

Since $\odot O_1$ intersects with $\odot O_2$ and $\odot O_3$, we can know that

$$r_1 + r_2 \geq O_1O_2, r_1 + r_3 \geq O_1O_3.$$

Therefore, $O_1O_2 \leq r_1 + r_2 \leq r_2 + r_3$, $O_1O_3 \leq r_2 + r_3$.

From $\angle O_2O_1O_3 \leq 60°$, it can be seen that one of the other two angles must be greater than or equal to $60°$. Let $\angle O_2 \geq 60°$. Then $O_1O_3 \geq O_2O_3$, so $O_2O_3 \leq r_2 + r_3$.

Thus $\odot O_2$ and $\odot O_3$ must intersect, and then we can deduce that $\odot O_1$, $\odot O_2$ and $\odot O_3$ intersect pair-wise. This contradicts with the given conditions.

Hence, there must be a circle that can only intersect five circles at most.

Example 7. Let a, x_1, x_2 and x_3 be four integers, which satisfy

$$a = (1 + x_1)(1 + x_2)(1 + x_3) = (1 - x_1)(1 - x_2)(1 - x_3).$$

Find the value of $ax_1x_2x_3$.

Analysis By using the non-negativity of real numbers, we first consider the "extreme situation", that is, "the contradiction between positive and negative", and get the basic idea to deal with the problem.

Solution 1 Since $a = (1+x_1)(1+x_2)(1+x_3) = (1-x_1)(1-x_2)(1-x_3)$, we get

$$a^2 = (1 - x_1^2)(1 - x_2^2)(1 - x_3^2). \tag{1}$$

Because a, x_1, x_2 and x_3 are integers, we know that x_1^2, x_2^2 and x_3^2 are non-negative integers.

If all of x_1^2, x_2^2 and x_3^2 are not less than 2, then $(1-x_1^2)(1-x_2^2)(1-x_3^2) < 0$, which contradicts with $a^2 \geq 0$. So at least one of x_1^2, x_2^2 and x_3^2 is equal to 1 or 0.

If one of x_1^2, x_2^2 and x_3^2 is equal to 1, then $a = 0$. Let $x_1 = 1$, $x_2 = -1$, and x_3 be any real number, which meets the conditions. Thus the value of $ax_1x_2x_3$ is 0.

If one of x_1^2, x_2^2 and x_3^2 is equal to 0, then let $x_1 = 0$, $x_2 = -x_3$, which meets the conditions. Thus the value of $ax_1x_2x_3$ is also 0.

To sum up, the value of $ax_1x_2x_3$ is 0.

Solution 2 From $(1+x_1)(1+x_2)(1+x_3) = (1-x_1)(1-x_2)(1-x_3)$, we get

$$x_1x_2x_3 + x_1 + x_2 + x_3 = 0. \tag{2}$$

If x_1, x_2 and x_3 are all positive numbers, then $x_1x_2x_3 + x_1 + x_2 + x_3 > 0$, which contradicts with the equation .

If x_1, x_2 and x_3 are all negative numbers, then $x_1x_2x_3 + x_1 + x_2 + x_3 < 0$, which contradicts with the equation .

If one of x_1, x_2 and x_3 is equal to 0, might as well let $x_1 = 0$, $x_2 = -x_3$, which meets the conditions. At this time, the value of $ax_1x_2x_3$ is 0.

If one of x_1, x_2 and x_3 is negative and the other two are positive, might as well let $x_1 < 0$, x_2, $x_3 > 0$. Then x_2 and x_3 are more than or equal to 1.

Because a, x_1, x_2 and x_3 are integers, we know that $x_1 \leq -1$, $x_2x_3 + 1 > 1$. Then

$$\begin{aligned} 0 &= x_1x_2x_3 + x_1 + x_2 + x_3 = x_1(x_2x_3 + 1) + x_2 + x_3 \\ &\leq -(x_2x_3 + 1) + x_2 + x_3. \end{aligned}$$

So $x_2x_3 + 1 - x_2 - x_3 \leq 0$, i.e. $(x_2 - 1)(x_3 - 1) \leq 0$.

Thus at least one of x_2 and x_3 is equal to 1, then $a = 0$.

Let $x_1 = -1$, $x_2 = 1$, and x_3 be any real number, which meets the conditions. Thus the value of $ax_1x_2x_3$ is 0.

If one of x_1, x_2 and x_3 is positive and the other two are negative, similarly, it can be proved that $a = 0$. Thus $ax_1x_2x_3 = 0$.

To sum up, the value of $ax_1x_2x_3$ is 0.

Example 8. Given that $n(n \geq 3)$ points in the plane are not all on a straight line, prove that there must be a straight line in the plane, which just passes through two of these points.

Analysis Using the distance from a point to a line, consider "extreme situation" — the minimum.

Proof For a line l determined by at least two points of the given points, since the given points are not all on a straight line, there are several of the given points (finite number) outside the straight line l. Let $d(l)$ be the minimum distance between these points and line l.

There are only finite straight lines passing through any two of these given points, then $d(l)$ satisfying the above requirements must be finite. Let d be the minimum of $d(l)$. Therefore, d is the distance from the point P_0 to the line l_0, which just passes through two of these given points.

This is because, as shown in Figure 26.6, it is assumed that at least three points A, B and C among the given points are on the straight line l_0. Draw a line segment P_0Q through the point P_0 which is perpendicular to l_0 and intersects l_0 at the point Q, then at least two points of A, B and C are on the same side of point Q.

Let A and B be on the same side of the point Q, and the point B be between A and Q. Suppose that the line passing through the points A and P_0 is l_1, and the foot point of the perpendicular line BR from B to l_1 is R, then we have

$$\triangle ABR \backsim \triangle AP_0Q.$$

So $\dfrac{BR}{P_0Q} = \dfrac{AB}{AP_0}$, that is,

$$\begin{aligned} d(l_1) \leq BR = P_0Q \cdot \frac{AB}{AP_0} \\ \leq P_0Q \cdot \frac{AQ}{AP_0} \\ < P_0Q = d. \end{aligned}$$

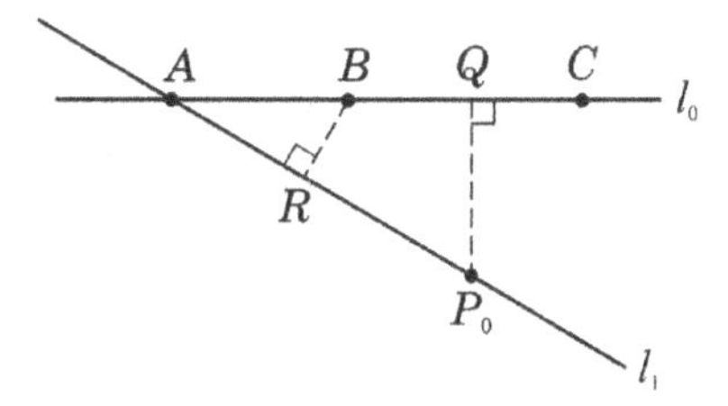

Fig. 26.6

This contradicts the minimum of d.

Hence, there are only two given points on the line l_0.

Remark An equivalent expression of Example 8 is as follow: $n(n \geq 3)$ points are given in the plane. If there are at least 3 points in each line determined by these points, then all the n points are in the same line.

Exercises

1. 1600 peanuts are distributed to 100 monkeys. Prove that no matter how to distribute, at least four monkeys get as many peanuts as they have.
2. n numbers are selected from the 2018 numbers of 1, 2, ..., 2018, so that the difference between any two of these n numbers is a composite number. Find the maximum value of n.
3. As shown in Figure 26.7, PA and PB are tangent to $\odot O$ at the points A and B respectively, PCD is a secant line, E is the midpoint of CD. If $\angle APB = 40°$, then the size of $\angle AEP$ is ( ).

   (A) 40° (B) 50° (C) 60° (D) 70°

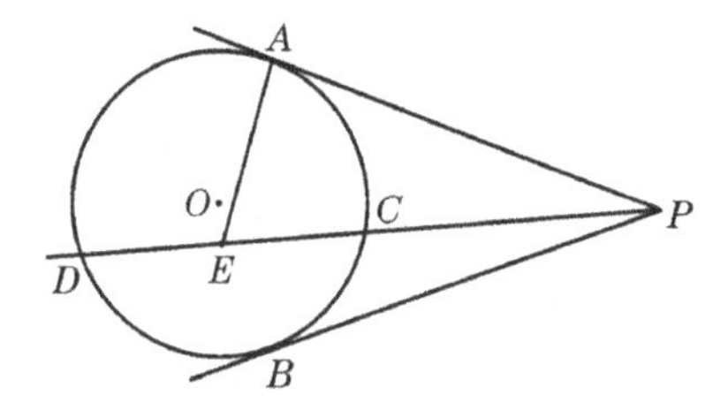

Fig. 26.7

4. Draw four semicircles on the side where the quadrilateral is located with the four sides of the quadrilateral as the diameter. Prove that these four semicircles can cover the entire quadrilateral.
5. There are 100 points in the plane, and any three points are not collinear. Prove that three points must be found so that there are no other points inside and on the edges of the triangle with these three points as its vertices.
6. Unload several boxes from the freighter, the total weight of which is 10 tons, and the weight of each box is not more than 1 ton. In order to ensure that these boxes can be transported away at one time, then how many cars with a capacity of 3 tons are needed at least?

7. There are n $(n \geq 3)$ points in the plane, and they are not all on a straight line. Prove that there are three points which are not collinear, so that there are no other given points inside the circle which passes through these three points.
8. Can n $(n \geq 2)$ line segments be made on the plane so that any end point of any line segment is in the interior of another line segment (excluding the end point)?
9. As shown in the Figure 26.8, two circles are externally tangent at the point P. Draw two secant lines APA_1 and BPB_1 which are perpendicular to each other through the point P. Let the diameter of two circles be d_1 and d_2. Prove that $AA_1^2 + BB_1^2$ is equal to a fixed value.

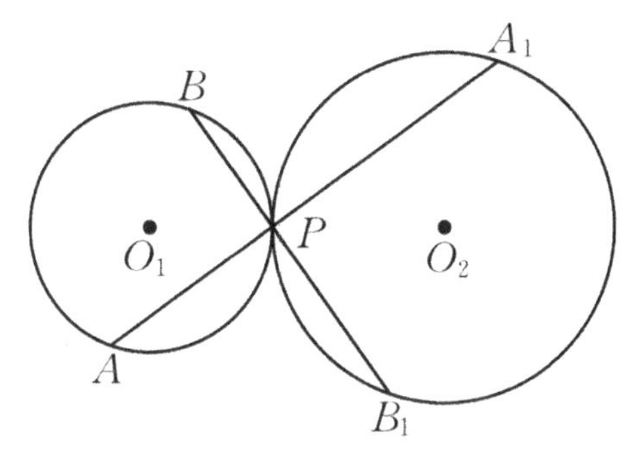

Fig. 26.8

Chapter 27

Coloring Problems

We call the problems that are required to make or prove the existence of points, grids, straight lines, quadrilateral regions and graphs satisfying some coloring properties as coloring problems.

A coloring problem is a kind of mathematical problem which is connected with drawer principle and graph theory. According to the different objects (points, line segments or regions) of coloring, we divide it into three categories: point coloring, line segment coloring and region coloring. No matter what kinds of coloring problems, they are mostly discussed around homochromatic points or homochromatic triangles.

Example 1. Color each vertex of a regular thirteen-sided polygon black or white, and each vertex is dyed in only one color. Prove that there exist three homochromatic vertices which happen to be the vertices of an isosceles triangle.

Proof Let $A_1, A_2, \ldots, A_{12}, A_{13}$ be the 13 vertices in turn. If all the 13 vertices are colored black or white, the conclusion is obviously true.

So we only need to consider the case that 13 vertices are colored some black and some white. At this time, there must be two adjacent vertices of the same color, might as well let A_1 and A_2 be homochromatic. Now we consider the five vertices A_{13}, A_1, A_2, A_3 and A_8, there must be three homochromatic vertices in these five vertices according to the drawer principle. And it can be divided into the following three situations:

(1) A_{13}, A_1, A_2 and A_3 have three points of the same color.A_1 and A_2 are known to be homochromatic, then A_{13}, A_1 and A_2 are homochromatic

or A_1, A_2 and A_3 are homochromatic. At this time, $\triangle A_{13}A_1A_2$ or $\triangle A_1A_2A_3$ is an isosceles triangle with three vertices of the same color.

(2) A_{13}, A_3 and A_8 are homochromatic. At this time, $\triangle A_{13}A_3A_8$ is an isosceles triangle with three vertices of the same color.

(3) A_1, A_2 and A_8 are homochromatic. At this time, $\triangle A_1A_2A_8$ is an isosceles triangle with three vertices of the same color.

Remark It can be seen from this example that the drawer principle and the classification thought are often accompanied by coloring problems and methods.

Example 2. Six points are given in the space, where any three points are not collinear and any four points are not coplanar. Connect them in pairs to get 15 line segments and color these line segments in red and blue (where a line segment is dyed in only one color). Prove that no matter how to color them, there always exist homochromatic triangles.

Proof Let A, B, C, D, E and F be the six given points.

Considering the lines AB, AC, AD, AE and AF with A as the end point, we know at least three of the five lines have the same color. Might as well let these three lines be AB, AC and AD and they are colored red. Then let's consider the three sides of ΔBCD. If one of them is red (such as side BC), the homochromatic triangle has appeared (red ΔABC).

If all the three sides of ΔBCD are not red, then it is a blue triangle, the homochromatic triangle has also appeared.

Hence, in either case, the homochromatic triangle exists.

Remark In Example 2, if six points are regarded as six people, two points with a red line indicate that two people know each other, and two points with a blue line indicate that two people don't know each other, then the equivalent proposition is obtained.

Of any six people, there must be three people who know each other or do not know each other.

It belongs to Ramsay problem in the graph theory. In the graph theory, the graph obtained by connecting every two points of n points with a line segment is called an n-point complete graph, which is denoted as K_n.

These points are called "vertices" and these line segments are called "sides", so we can use graph theory language to describe two important conclusions.

Theorem 1 If red and blue are dyed arbitrarily in K_6, there must exist a homochromatic triangle.

Theorem 2 If red, yellow and blue are dyed arbitrarily in K_{17}, there must exist a homochromatic triangle.

Example 3. Prove that if there are six points in a plane and any three points are the vertices of a scalene triangle, then the longest side of one of these triangles is the shortest side of another triangle.

Proof Color the shortest side of each triangle red, and color the rest sides white if they haven't dyed yet. According to Theorem 1, a homochromatic triangle must appear. Since every triangle has the shortest side, every triangle has a red side. Then the homochromatic triangle mentioned above is red. The longest side of this triangle is red, meanwhile it is also the shortest side of another triangle.

Example 4. As shown in Figure 27.1, each of the small squares of 3 rows and 7 columns is dyed red or blue. Prove that there exist a rectangle where the small squares on its four corners have the same color.

Proof According to the drawer principle, at least 4 of the 7 squares in the first row are homochromatic, which might as well be set as red (shaded) and be distributed in the columns 1, 2, 3 and 4, as shown in Figure 27.2. In columns 1, 2, 3 and 4 (there is no need to consider the columns 5, 6, and 7 below), if there are two small red squares in the rows 2 or 3, then the

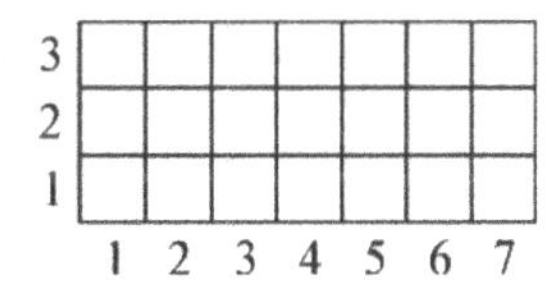

Fig. 27.1

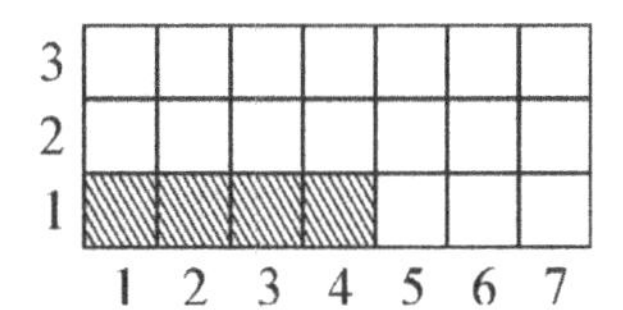

Fig. 27.2

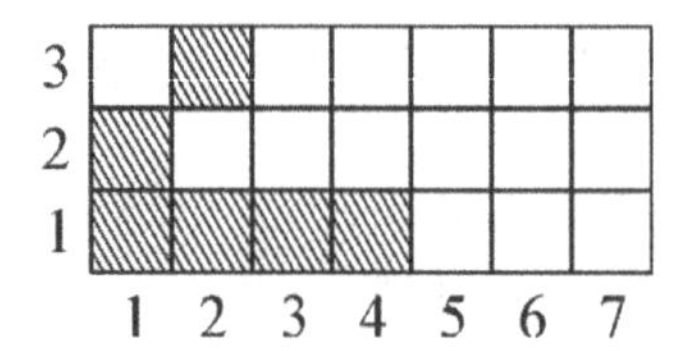

Fig. 27.3

problem has been proved. If each row in the rows 2 and 3 has at most one small red square, as shown in Figure 27.3, then there must be a rectangle with four corners in blue in these two rows.

Remark In the proof process of this example, in addition to using the drawer principle, an effective method of thinking about problems is also used, which is to gradually narrow the scope of the object to be discussed, and gradually change the complex problem into a simple problem.

Example 5. 120 people participated in the mathematics competition. There were 5 questions in the test. It was known that 96, 83, 74, 66 and 35 people did right for the first, second, third, fourth and fifth questions, respectively. If they answer at least three questions correctly, they would win the prize. How many people won the prize at least in this competition?

Solution The 120 people were numbered as $P_1, P_2, \ldots, P_{120}$, which were regarded as 120 points on the number axis. Use A_k to represent the group composed of people who did not answer question k correctly among these 120 people, and $|A_k|$ was the number of people in the group, where $k = 1, 2, 3, 4, 5$. Then we had

$$|A_1| = 24, |A_2| = 37, |A_3| = 46, |A_4| = 54, |A_5| = 85.$$

The above five groups were given five colors respectively. If someone did not do the question k correctly, then the point which represented the person was dyed the kth color, where $k = 1, 2, 3, 4, 5$. The question was transformed into finding out how many points at most were dyed with at least three colors?

Since $|A_1|+|A_2|+|A_3|+|A_4|+|A_5| = 246$, the points dyed with at least three colors were not more than $\frac{246}{3} = 82$. The above figure was the best coloring method to meet the conditions, that is, 85 points P_1, P_2, $\ldots$, P_{85} were dyed the fifth color; 37 points P_1, P_2, $\ldots$, P_{37} were dyed the second

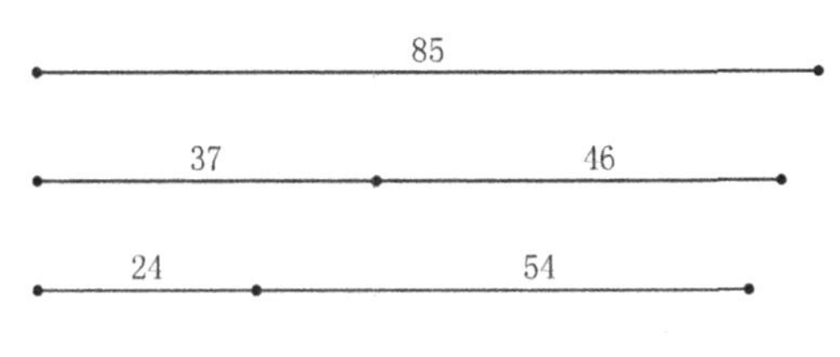

Fig. 27.4

color; 46 points P_{38}, P_{39}, ..., P_{83} were dyed the fourth color; 24 points P_1, P_2, ..., P_{24} were dyed the first color; 54 points P_{25}, P_{26}, ..., P_{78} were dyed the third color. So there were at most 78 points dyed with three colors.

Therefore, there were at least 42 points with not more than two types of coloring, that is, there were at least 42 winners (each of them could answer at most two questions wrong, but at least three questions correctly, such as the 42 peopleP_{79}, P_{80}, ..., P_{120}).

Example 6. Prove that 15 rectangular tiles with the size of 4×1 and a square tile with the size of 2×2 cannot cover the square ground with the size of 8×8.

Proof 1 As shown in Figure 27.5, dye the entire ground in two colors of black and white by using the coloring method that small grids with two spaces apart and parallel to the sub diagonal are homochromatic. Obviously, there are 32 black and 32 white grids on the ground.

No matter whether it is a horizontal cover or a vertical cover, and no matter where the cover is, each tile with the size of 4×1 always covers two black and two white grids. Because the grids on the oblique lines parallel to the sub diagonal are always homochromatic, and the adjacent two grids on the oblique lines parallel to the main diagonal are always in different colors, no matter how to place it, a tile with the size of 2×2 always covers three black and one white grids or three white and one black grids. Therefore, after 15 pieces of 4×1 tiles are covered, there are still two white girds and two black grids left, and it is impossible to cover it with a piece of 2×2 tile. So the proposition is proved.

Proof 2 Color the ground with 1, 2, 3 and 4 colors as shown in Figure 27.6, so that the grids on the oblique lines parallel to the main diagonal always keep the same color. Therefore, no matter how to place the tiles with the size of 4×1, they always cover one grid for each of four colors, while two grids on the oblique line parallel to the main diagonal are

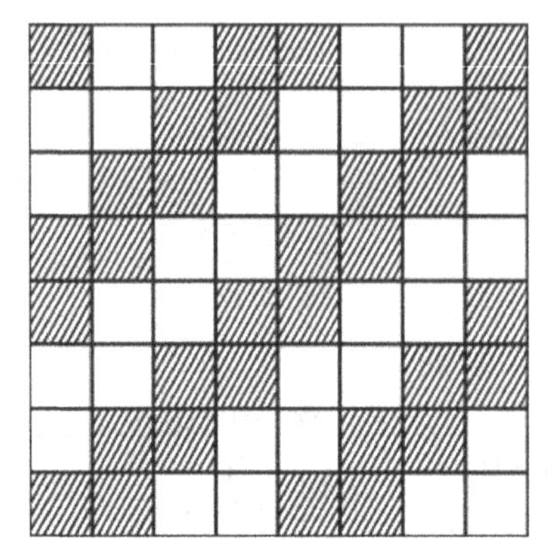

Fig. 27.5

1	2	3	4	1	2	3	4
4	1	2	3	4	1	2	3
3	4	1	2	3	4	1	2
2	3	4	1	2	3	4	1
1	2	3	4	1	2	3	4
4	1	2	3	4	1	2	3
3	4	1	2	3	4	1	2
2	3	4	1	2	3	4	1

Fig. 27.6

always homochromatic in the four grids covered by a 2×2 tile. That is, a 2×2 tile can't cover four different color grids at the same time no matter how it is placed. After 15 pieces of 4×1 tiles are covered, there is one grid of 1, 2, 3 and 4 colors left, and it is impossible to cover it with a piece of 2×2 tiles. So the proposition is proved.

Remark It can be seen from this example that the key to solve the problem of grid dyeing is to determine how many colors to dye the grids and how to dye the grids. It is common to dye the grids with two colors and use the methods of dyeing alternate intervals or alternate columns (rows).

However, to solve some problems, we should use a variety of dyeing methods to test according to the meaning of the problem, so as to find out the correct way to solve the problem.

Example 7. A 100×100 grid table is dyed black and white, all the grids on the boundary are dyed black, and there is no monochrome 2×2 sub grid table in the grid table. Prove that there is a 2×2 sub grid table in this grid table, so that the colors of two grids on one diagonal are the same, and the colors of two grids on the other diagonal are the same, but the colors of the grids on the two diagonals are different.

Proof Suppose the conclusion is not true.

Since there is no monochrome 2×2 sub grid table in the grid table, all the 2×2 sub grid tables can only be one of the three situations (or after rotation) in Figure 27.7.

If two adjacent grids are dyed black and white, they are called "black-and-white pairs".

It is easy to see that there are only two black-and-white pairs in the above three cases. The number of black-and-white pairs in the 100×100 grid table is denoted as S.

Since all the grids on the boundary are dyed black, the numbers of black-and-white pairs in each row and column are even. So S is even.

Because there are 99^2 sub grid tables with the size of 2×2 in the 100×100 grid table, each 2×2 sub grid table has exactly two black-and-white pairs, and the boundaries are black, that is, the black-and-white pairs cannot be between the marginal grids. Thus, each black-and-white pair is calculated twice, then $S = 99^2$ is obtained. This contradicts the conclusion that S is even.

Therefore, the hypothesis does not hold.

So the proposition is proved.

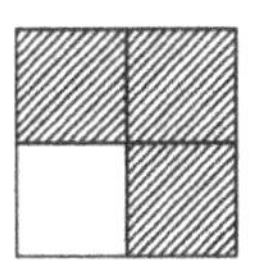
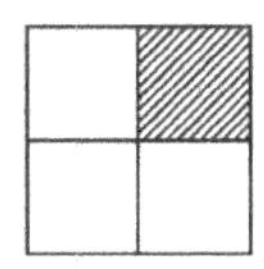
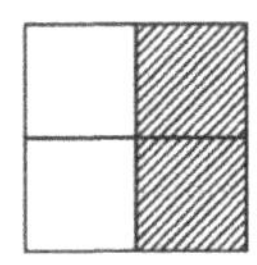

Fig. 27.7

Reading

Twenty-Four Bridge

There are many bridges in Yangzhou, and Du Mu of the Tang Dynasty has a well-known poem:

Green mountains loom afar and streams off flow,
Fall's gone but grasses in South still well grow.
When moon shines on the Twenty-Four Bridge bright,
Where are you teaching flute-blowing tonight?

Is the twenty-four bridge a bridge called the twenty-four bridge, or twenty-four bridges, or many bridges? As opinions vary, no unanimous conclusion can be drawn.

More than forty years ago, the author once saw a small bridge in Yangzhou, and there was a stone monument next to it proving that this bridge is the twenty-four bridge. But the bridge is very small, and it doesn't seem to be as famous as Fan Chuan scholar said. Now the bridge has disappeared for a long time and I don't know where it is.

Let's assume that twenty-four bridge is twenty-four bridges. Here's the problem: In the Figure, A, B, C, D, E, F, G and H are lands or islands connected by 24 Bridges. Can we start from A and cross each bridge exactly once (neither repeated nor omitted)?

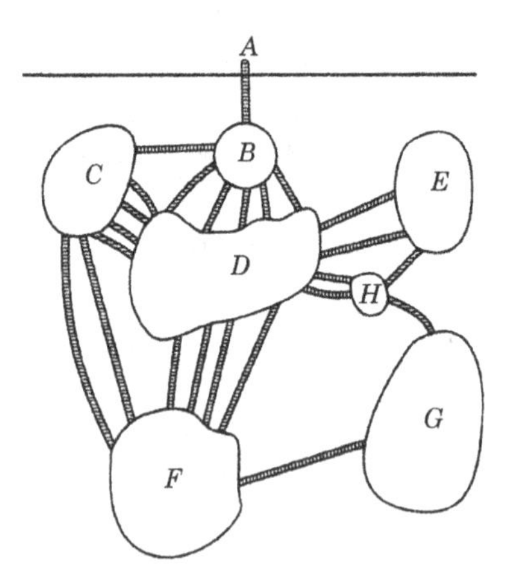

(Excerpted from: Shan Zun. *Happy mathematics*. Nanjing: Jiangsu Education Publishing House, 2006, 4.)

Exercises

1. Two grids with common sides in the grid table are called adjacent. How can we dye each grid in the 6×6 grid table into one of the three colors, so that the adjacent grids of each grid have exactly one grid in each of the other two colors.
2. Each point in the plane is dyed with one of two colors, and any equilateral triangle with the side length of 1 is known to have vertices of two colors.
   (1) Prove that there exist homochromatic triangles with the side length of $\sqrt{3}$ (i.e. the vertices are homochromatic).
   (2) Give an example of the plane that the dying method meets the given requirements.
3. Given a group of balls, each ball is dyed red or blue, each color has at least one ball, each ball weighs 1 or 2 pounds, and each weight has at least one ball. Prove that there are two balls with different weights and different colors.
4. There are 27 dots distributed at equal intervals on the circumference, which are dyed black or white respectively. It is known that any two black dots are separated by at least two dots. Prove that three white points can be found to form three vertices of an equilateral triangle.
5. Prove that only two kinds of ceramic tiles with the sizes of 2×2 and 3×3 can't cover the square ground with the size of 23×23 exactly.
6. The 36 chords connecting 9 different points on the circle are dyed red or blue. Suppose that the triangle determined by every 3 points in the 9 points contains at least one red side. Prove that there are four points, where the connection line of every two points is red.
7. Nine mathematicians met at an international conference of mathematicians. Assume that two of every three people know each other. Prove that there must exist four mathematicians who know each other.
8. There are $2n(n \geq 2)$ points in the plane, where any three points are not collinear, and n points of them are painted blue, n points are painted white. Prove that we can connect them into n line segments by one blue and one white, so that these line segments do not intersect each other.

Chapter 28

Probability

In junior high schools, the knowledge of probability mainly involves the understanding of the likelihood of an event happening. We should learn to use simple calculation and experiment to get the probability of events. We can recognize the changes of events through the understanding of statistical data and the likelihood of events happening.

Example 1. Xiao Ming and Xiao Liang play a game with two spinners A and B (as shown in Figure 28.1).

(1) Let Xiao Ming spin the spinner A and Xiao Liang spin the spinner B. If the indicators of both spinners point to black at the end, Xiao Ming wins; if one of the indicators points to black and the other points to white, Xiao Liang wins. Question: is this rule fair to both sides? Why?
(2) Let Xiao Ming spin the spinner A. When the spinner stops, the number indicated by the pointer is Xiao Ming's score. Xiao Liang doesn't spin the spinners, but he still gets 3 points (as the base score). If they do this for many times, the one with the highest score will win the game. Question: what is Xiao Ming's average score (of one rotation)? If you are allowed to participate in the game, will you choose to play the role of Xiao Ming or Xiao Liang?

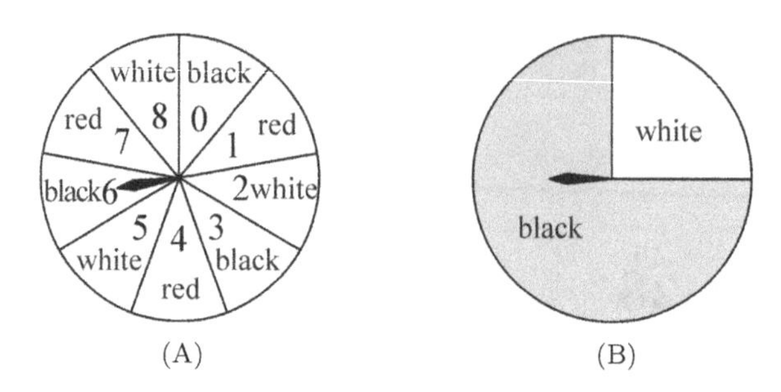

Fig. 28.1

Solution

(1) Divide the black area of spinner B into three equal parts, recorded as black_1, black_2 and black_3. The black, white and red areas in spinner A have the probability $\frac{1}{3}$ of occurring respectively. The list is as follows:

spinner B / spinner A	black_1	black_2	black_3	white
black	(black, black_1)	(black, black_2)	(black, black_3)	(black, white)
white	(white, black_1)	(white, black_2)	(white, black_3)	(white, white)
red	(red, black_1)	(red, black_2)	(red, black_3)	(red, white)

The probability of getting two black is $\dfrac{3}{12} = \dfrac{1}{4}$. The probability of getting a black and a white is $\dfrac{4}{12} = \dfrac{1}{3}$.

The game is unfair, because Xiao Liang is more likely to win.

(2) After several turns, Xiao Ming's average score is

$$\frac{1}{9}\times 0+\frac{1}{9}\times 1+\frac{1}{9}\times 2+\frac{1}{9}\times 3+\frac{1}{9}\times 4+\frac{1}{9}\times 5+\frac{1}{9}\times 6+\frac{1}{9}\times 7$$
$$+\frac{1}{9}\times 8=\frac{1}{9}\times(1+2+3+4+5+6+7+8)=4.$$

Since $4 > 3$, it is more likely to win when playing the role of Xiao Ming (who spins the spinner).

Example 2. 3 cards are randomly selected from 5 cards numbered 2, 3, 4, 5 and 6 respectively. The probability that the number written on these 3 cards can be used as the lengths of three sides of a triangle is ( ).

(A) $\frac{1}{2}$ (B) $\frac{3}{5}$ (C) $\frac{7}{10}$ (D) $\frac{4}{5}$

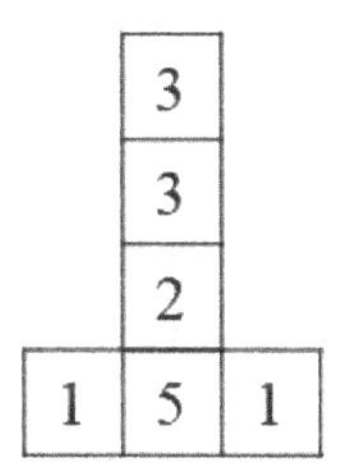

Fig. 28.2

Solution The event of selecting 3 cards randomly from 5 cards includes 10 possible outcomes in total. When the numbers of 3 cards selected are 2, 3, 5; 2, 3, 6 or 2, 4, 6, the three numbers can't be used as the lengths of three sides of the triangle. The rest 7 outcomes can be used. So the probability of the event asked is $\frac{7}{10}$.

Hence, the answer is C.

Example 3. A homogeneous cube has six numbers 1, 1, 2, 3, 3 and 5 marked on its faces, and the surface development of the cube is shown in Figure 28.2. The cube is rolled once, and the number facing up is written as the x-coordinate of a point in the plane rectangular coordinate system while the number facing down is the y-coordinate of the point. According to this rule, the coordinates of a point in the plane are obtained every time the small cube is rolled. If the two points obtained by Xiao Ming's first two throws determine a straight line l, and this straight line passes through the point $P(4, 7)$, then the probability that the point obtained by his third throw is also on the straight line l is ( ).

(A) $\frac{2}{3}$ (B) $\frac{1}{2}$ (C) $\frac{1}{3}$ (D) $\frac{1}{6}$

Solution The coordinates of six points which may be obtained for each throw (two of them are coincident) are (1, 1), (1, 1), (2, 3), (3, 2), (3, 5) and (5, 3). By tracing points and calculation, we can find the straight line determined by any two points of (1, 1), (2, 3) and (3, 5) passes through the point $P(4, 7)$. So the probability that the point obtained by his third throw is also on the straight line l is $\frac{4}{6} = \frac{2}{3}$.

Hence, the answer is A.

Example 4. A triangle is randomly selected from the triangles whose three sides are all integers and the perimeter is 24, determine the probability that the triangle is a right angled triangle.

Solution Let the lengths of three sides of a triangle be a, b and $c\,(a \geq b \geq c)$. Then $3a \geq a + b + c = 24$, $2a < a + (b + c) = 24$, so we get $8 \leq a < 12$. Thus the possible values of a are 8, 9, 10 or 11.

So the array (a, b, c) satisfying the question can be: (8, 8, 8), (9, 9, 6), (9, 8, 7), (10, 10, 4), (10, 9, 5), (10, 8, 6), (10, 7, 7), (11, 11, 2), (11, 10, 3), (11, 9, 4), (11, 8, 5) and (11, 7, 6).

There are 12 groups of the array, of which only one group can be the three sides of a right triangle.

So the probability that the triangle is a right angled triangle is $\frac{1}{12}$.

Example 5. In order to prepare for the school sports meeting, the sports team of grade 7 needs to purchase a batch of sports shoes. Given that there are 20 students in the team, and the statistical table is as follows. Because of carelessly soiling the table, two data can't be seen.

Shoe size	38	39	40	41	42
Number of students	5			3	2

Which one of the following statements is correct? ( ).

(A) The median of this group of data is 40, and the mode is 39.
(B) The median and the mode of this group of data must be equal.
(C) The average number P of this group of data satisfies $39 < P < 40$.
(D) None of the above is true.

Solution According to the meanings of the median and the mode, it can be seen from the table that when the median of this group of data is 40, the mode must be 40, so A is wrong.

When the number of students whose shoes sizes are 39 and 40 is 5, the median and mode are different, so B is wrong.

Suppose that all the remaining 10 students wear the shoes of size 39, then the average number is 39.35. Suppose that all the remaining 10 students wear the shoes of size 40, then the average number is 39.85. So C is correct. (Or suppose that there are x students wearing the shoes of size 39, it can be solved by $0 \leq x \leq 10$.)

So the answer is C.

Example 6. A math game is played between two very clever students A and B. The referee first writes the following positive integers 2, 3, 4, ..., 2006 on the blackboard, and then wipes off a number at random. Next, B and A shall wipe off one of the numbers in turn (that is, B shall wipe off one of the numbers first, and then A shall wipe off another number, and so on in turn). If the last two numbers are coprime, then A wins; otherwise, B wins. According to the rules of the game, determine the probability that A wins.

Solution Since both A and B are very clever, the key to their success depends on which number the referee wipes off. Notice that there are 1002 odd numbers and 1003 even numbers among the numbers 2, 3, 4, ..., 2006.

(i) If the referee wipes off an odd number, then B must win.
No matter what number A selects, as long as there is an odd number, B should wipe off the odd number, so that the last two numbers must be even. Therefore, the remaining two numbers are not coprime, so B wins.

(ii) If the referee wipes off an even number, then A must win.
If the number wiped off by the referee is $2m$, the remaining numbers can be matched into 1002 pairs:

$$(2,3), \ldots, (2m-2, 2m-1), (2m+1, 2m+2), \ldots, (2005, 2006).$$

In this way, no matter which number B selects, A will wipe off the other number in the matched number pair, so that the last two numbers must be coprime, so A wins.

Hence, the probability that A wins is $\frac{1003}{2005}$.

Exercises

1. There are four brand-new playing cards, including two spades and two hearts, all of which have the same back. After the cards are well shuffled with the back facing up on the table, two cards are randomly selected from them. The probability of getting different suits is ( ).

(A) $\frac{3}{4}$ (B) $\frac{2}{3}$ (C) $\frac{1}{3}$ (D) $\frac{1}{2}$

2. A homogeneous cube dice has the numbers 1, 2, 2, 3, 3 and 4 marked on its faces respectively; and another homogeneous cube dice has the numbers 1, 3, 4, 5, 6 and 8 marked on its faces respectively. If these two dices are rolled once at the same time, determine the probability that the sum of the two numbers of the upward face is 5.
3. A homogeneous cube dice has the numbers 1 to 6 marked on its faces. If it is rolled two times successively, and the number of points rolled for the first time is recorded as a while the number of points rolled for the second time is recorded as b, then the probability that the quadratic equation $x^2 - ax + b^2 = 0$ has two identical real roots is ( ).

   (A) $\frac{1}{6}$ (B) $\frac{1}{9}$ (C) $\frac{1}{12}$ (D) $\frac{1}{18}$

4. Four students A, B, C and D take part in the 4×100 metres relay race of the school track and field games. If the running sequence of four students is arranged at random, the probability that A will give the baton to B is ( ).

   (A) $\frac{1}{4}$ (B) $\frac{1}{6}$ (C) $\frac{1}{8}$ (D) $\frac{1}{12}$

5. There are 20 balls in the bag, including 9 white balls, 5 red balls and 6 black balls. 10 balls are randomly selected from the bag, so that there are not less than 2 but not more than 8 white balls, not less than 2 red balls and not more than 3 black balls. So the number of the above methods is ( ).

   (A) 14 (B) 16 (C) 18 (D) 20

6. Let P_n be the probability that the sum of the numbers on the top of two cube dice of the same shape and uniform texture is n, then $P_1 + P_2 + P_3 + P_4 + P_5 =$ ________.
7. A homogeneous cube dice has the numbers 1 to 6 marked on its faces. If it is rolled two times successively, and the numbers facing upward are denoted as m and n respectively, then the probability that the numbers m and n make the graph of quadratic function $y = x^2 + mx + n$ have two different intersections with the x-axis is ( ).

   (A) $\frac{5}{12}$ (B) $\frac{4}{9}$ (C) $\frac{17}{36}$ (D) $\frac{1}{2}$

8. When Xiao Ding, Xiao Ming and Xiao Qian are playing games together, they need to determine the sequence of playing games. They agree to

use the method of "scissors, cloth and hammer" to determine. Then the probability that they all give cloth in a round is _______.

9. As shown in the Figure 28.3, three identical coins are put into a 4×4 square grid (only one coin can be placed in each square grid). Determine the probability that any two of the three coins are in different rows and columns.

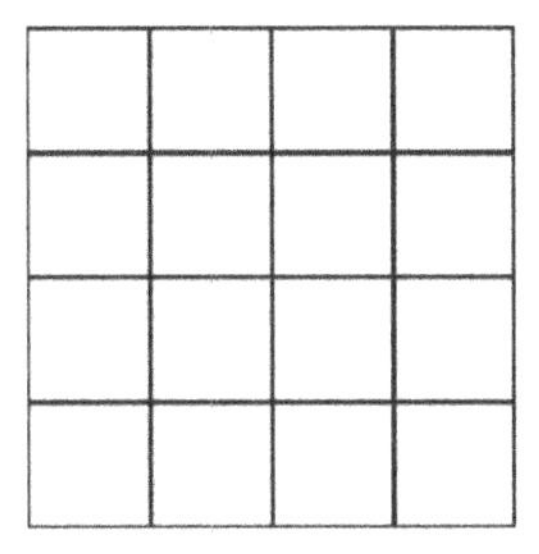

Fig. 28.3

Solutions

Solution 1

Quadratic Equations

1. From $1^2 + 1 \cdot b + c = 3$, $3^2 + 3 \cdot b + c = 5$, we get that $b = -3$, $c = 5$. At this time, $11^2 - 3 \times 11 + 5 = 93$, but $6^2 + 6 \times (-3) + 5 \neq 21$. So the answer is C.
2. Solution 1 The original equation is transformed into $(x-1)[(2x-1)-(3x+2)] = 0$, that is, $(x-1)(-x-3) = 0$. Solving the equation gives $x_1 = 1$, $x_2 = -3$.
   Solution 2 Simplify the equation directly. From $2x^2 - 2x - x + 1 = 3x^2 - 3x + 2x - 2$, we get $x^2 + 2x - 3 = 0$. Solving the equation gives $x_1 = 1$, $x_2 = -3$.
3. Solution 1 The problem is discussed in the two cases. When $x \geq 0$, the original equation is turned into $x^2 - x - 1 = 0$. Solving it gives $x = \frac{1+\sqrt{5}}{2}$ or $x = \frac{1-\sqrt{5}}{2}$ (discarded). When $x < 0$, the original equation is turned into $x^2 + x - 1 = 0$. Solving it gives $x = -\frac{1+\sqrt{5}}{2}$. Hence, the roots of the original equation are $x = \frac{1+\sqrt{5}}{2}$ or $x = -\frac{1+\sqrt{5}}{2}$.
   Solution 2 From the given condition, $|x|^2 - |x| - 1 = 0$. Solving the equation gives $|x| = \frac{1 \pm \sqrt{5}}{2}$. Since $|x| \geq 0$, we know that $|x| = \frac{1+\sqrt{5}}{2}$. Hence, $x = -\frac{1+\sqrt{5}}{2}$ or $x = \frac{1+\sqrt{5}}{2}$.
4. By substituting $x = -1$ into the original equation, we get $(-1)^2 - 3k \cdot (-1) + k^2 - k = 0$, i.e. $1 + 2k + k^2 = 0$. Solving the equation gives $k = -1$.
5. Let the common real root of two equations be x_0. Then $x_0^2 + bx_0 + 1 = 0$, $x_0^2 - x_0 - b = 0$. Subtracting these two equations gives $(b+1)(x_0+1) = 0$. Therefore, when $x_0 \neq -1$, $b = -1$; when $x_0 = -1$, $b = 2$. However, when $b = -1$, $x^2 - x + 1 = \left(x - \frac{1}{2}\right)^2 + \frac{3}{4} > 0$, the equation has no real roots, so $b = -1$ does not satisfy the given condition. Hence, $b = 2$.

6. From $\sqrt{4-2\sqrt{3}} = \sqrt{3}-1$ and the given condition, we get $(\sqrt{3}-1)^2 + a(\sqrt{3}-1) + b = 0$. Rearranging it gives $(4-a+b)+(a-2)\sqrt{3} = 0$.

   Because a and b are integers, and $\sqrt{3}$ is an irrational number, we obtain that $a-2=0$, $4-a+b=0$. Solving the equations gives $a=2$, $b=-2$.

   Hence, $a+b=0$.

7. From the given condition, $a \neq 0$, $a^2+1=5a$, and $\frac{1}{a^2}+1=\frac{5}{a}$. So we get that $(a^2+1)^2=25a^2$, $\left(\frac{1}{a^2}+1\right)^2=\frac{25}{a^2}$, that is, $a^4+1=23a^2$, $\frac{1}{a^4}+1=\frac{23}{a^2}$. Then $a^4+a^{-4}=23a^2+\frac{23}{a^2}-2=\frac{23(a^4+1)}{a^2}-2=\frac{23\times 23a^2}{a^2}-2=23^2-2=527$. So the last digit number of a^4+a^{-4} is 7.

8. From the given condition, suppose that x_0 satisfies $x_0^2-3x_0+c=0$ and $(-x_0)^2+3\cdot(-x_0)-c=0$. Subtracting these two equations gives $c=0$. Thus $x^2-3x+c=0$ is turned into $x^2-3x=0$. Solving the equation gives $x=0$ or $x=3$.

9. Solution 1 From $x=\frac{4-\sqrt{7}}{3}$, we get $\sqrt{7}=4-3x$. Squaring both sides of the equation gives $7=16-24x+9x^2$, i.e. $x^2+1=\frac{8}{3}x$. So $(x^2+1)^2=\left(\frac{8}{3}x\right)^2$, i.e. $x^4+1=\left(\frac{8}{3}x\right)^2-2x^2$. Then $x^4+x^2+1=\left(\frac{8}{3}x\right)^2-x^2$. Hence,

$$\frac{x^4+x^2+1}{x^2}=\frac{\left[\left(\frac{8}{3}\right)^2-1\right]x^2}{x^2}=\frac{55}{9}.$$

   Solution 2 From $x=\frac{4-\sqrt{7}}{3}$, we know that $\frac{1}{x}=\frac{4+\sqrt{7}}{3}$. So $\frac{x^4+x^2+1}{x^2}=x^2+1+\frac{1}{x^2}=\left(x+\frac{1}{x}\right)^2-1=\left(\frac{4-\sqrt{7}}{3}+\frac{4+\sqrt{7}}{3}\right)^2-1=\left(\frac{8}{3}\right)^2-1=\frac{55}{9}$. (This solution is given by Huang Shengping of Zhongshan City, Guangdong Province)

10. Subtracting the given two equations gives $m^2-n^2=m-n$. And since $m\neq n$, we get $m+n=1$, then

$$\begin{aligned} m^5 &= m(m^2)^2 = m(m+1)^2 = m(m^2+2m+1) = m(3m+2) \\ &= 3m^2+2m = 5m+3. \end{aligned}$$

   Similarly, $n^5=5n+3$. Then we obtain that

$$m^5+n^5=5(m+n)+6=11.$$

Solution 2

Equations that can be Transformed into Quadratic Equations

1. The original equation is converted to $x - 2 = 0$ or $x^2 - 3x - 4 = 0$. Solving the equation $x - 2 = 0$ gives $x_1 = 2$. Solving the equation $x^2 - 3x - 4 = 0$ gives $x_2 = 4$, $x_3 = -1$. Hence, the roots of the original equation are $x_1 = 2$, $x_2 = 4$, $x_3 = -1$.
2. When $x \geq \frac{1}{2}$, the original equation is converted to $x^2 - (2x - 1) - 2 = 0$. Solving the equation gives $x_1 = 1 + \sqrt{2}$, $x_2 = 1 - \sqrt{2}$. And since $x_2 = 1 - \sqrt{2} < \frac{1}{2}$, we discard it.

   When $x < \frac{1}{2}$, the original equation is converted to $x^2 - (-2x + 1) - 2 = 0$. Solving the equation gives $x_3 = -3$, $x_4 = 1$. And since $x_4 = 1 > \frac{1}{2}$, we discard it.

   Hence, the roots of the original equation are $x = 1 + \sqrt{2}$ or $x = -3$.
3. The original equation is converted to
$$\left(1 - \frac{1}{x+2}\right) + \left(1 - \frac{1}{x+9}\right) = \left(1 - \frac{1}{x+3}\right) + \left(1 - \frac{1}{x+8}\right),$$
   i.e. $\frac{1}{x+8} - \frac{1}{x+9} = \frac{1}{x+2} - \frac{1}{x+3}$.

   Rearranging the equation, we get $(x + 2)(x + 3) = (x + 8)(x + 9)$.

   Solving the equation gives $x = -\frac{11}{2}$.

   After checking, we know that the root of the original equation is $x = -\frac{11}{2}$.
4. Let $y = x - \frac{1}{x}$. Then the original equation is converted to $y^2 + 2 - \frac{7}{2}y + 1 = 0$, i.e. $2y^2 - 7y + 6 = 0$. Solving it gives $y_1 = \frac{3}{2}$, $y_2 = 2$. From $y_1 = \frac{3}{2}$, we get $x - \frac{1}{x} = \frac{3}{2}$, i.e. $2x^2 - 3x - 2 = 0$, so $x_1 = -\frac{1}{2}$, $x_2 = 2$.

From $y_2 = 2$, we get $x - \frac{1}{x} = 2$, i.e. $x^2 - 2x - 1 = 0$, so $x_3 = 1 + \sqrt{2}$, $x_4 = 1 - \sqrt{2}$.

Hence, the roots of the original equation are $x_1 = -\frac{1}{2}$, $x_2 = 2$, $x_3 = 1 + \sqrt{2}$, $x_4 = 1 - \sqrt{2}$.

5. Let $y = x + \frac{1}{x}$. Then the original equation is converted to $2y^2 - 3y - 5 = 0$. Solving it gives $y = \frac{5}{2}$ or $y = -1$. When $y = \frac{5}{2}$, $x + \frac{1}{x} = \frac{5}{2}$, i.e. $2x^2 - 5x + 2 = 0$, so $x_1 = \frac{1}{2}$, $x_2 = 2$. When $y = -1$, $x + \frac{1}{x} = -1$, i.e. $x^2 + x + 1 = 0$, the equation has no real roots. Hence, the roots of the original equation are $x_1 = \frac{1}{2}$, $x_2 = 2$.
6. Let $y = \frac{x^2-1}{x}$. Then the original equation is converted to $6y^2 - 7y - 24 = 0$. Solving it gives $y_1 = \frac{8}{3}$, $y_2 = -\frac{3}{2}$. When $y_1 = \frac{8}{3}$, $\frac{x^2-1}{x} = \frac{8}{3}$, i.e. $3x^2 - 8x - 3 = 0$, so $x_1 = -\frac{1}{3}$, $x_2 = 3$. When $y_2 = -\frac{3}{2}$, $\frac{x^2-1}{x} = -\frac{3}{2}$, i.e. $2x^2 + 3x - 2 = 0$, so $x_3 = \frac{1}{2}$, $x_4 = -2$. Hence, the roots of the original equation are $x_1 = -\frac{1}{3}$, $x_2 = 3$, $x_3 = \frac{1}{2}$, $x_4 = -2$.
7. Let $y = \frac{a+x}{b+x}$. Then the original equation is converted to $y + \frac{1}{y} = 2 + \frac{1}{2}$. Solving the equation gives $y = 2$ or $y = \frac{1}{2}$. From $y = 2$, we get $\frac{a+x}{b+x} = 2$, i.e. $x_1 = -2b + a$. From $y = \frac{1}{2}$, we get $\frac{a+x}{b+x} = \frac{1}{2}$, i.e. $x_2 = b - 2a$. After checking, we have as follows: when $a \neq b$, the roots of the equation are $x_1 = a - 2b$, $x_2 = b - 2a$; when $a = b$, the equation has no roots.
8. The original equation is converted to $\sqrt{x+8} + 2 = \sqrt{5x+20}$. Squaring both sides of the equation, we get $x + 2 = \sqrt{x+8}$. Then squaring both sides again, we obtain $x^2 + 3x - 4 = 0$. Solving it gives $x_1 = -4$, $x_2 = 1$. After checking, we know that $x = 1$ is the root of the original equation.
9. The original equation is converted to $x^2 + 2x\sqrt{x+2} + (\sqrt{x+2})^2 = 16$, i.e. $x + \sqrt{x+2} = 4$ or $x + \sqrt{x+2} = -4$. Solving the equation $x + \sqrt{x+2} = 4$ gives $x = 2$. The equation $x + \sqrt{x+2} = -4$ has no real roots. After checking, we know that the root of the original equation is $x = 2$.
10. Let $y = \sqrt{x^2 + 18x + 45}$. Then $y \geq 0$, and the original equation is converted to $y^2 - 2y - 15 = 0$. Solving it gives $y = -3$(discarded) or $y = 5$. From $y = 5$, we get $x^2 + 18x + 45 = 25$, i.e. $x^2 + 18x + 20 = 0$. Solving the equation gives $x_1 = -9 - \sqrt{61}$, $x_2 = -9 + \sqrt{61}$. Hence, the roots of the original equation are $x_1 = -9 - \sqrt{61}$, $x_2 = -9 + \sqrt{61}$.
11. From the given condition, $\frac{1}{a} - \frac{1}{b} + \frac{1}{a-b} = 0$, we get that $a \neq 0$, $b \neq 0$, and $\frac{b-a}{ab} + \frac{1}{a-b} = 0$. Simplifying it gives $a^2 - 3ab + b^2 = 0$. That is, $\left(\frac{b}{a}\right)^2 - 3 \cdot \frac{b}{a} + 1 = 0$. Solving the equation gives $\frac{b}{a} = \frac{3 \pm \sqrt{5}}{2}$.

12. The original equation is converted to $(x + \sqrt{3x^2 + x})^2 = 9$, i.e. $x + \sqrt{3x^2 + x} = 3$ or $x + \sqrt{3x^2 + x} = -3$. From $x + \sqrt{3x^2 + x} = 3$, we get $2x^2 + 7x - 9 = 0$, then $x_1 = 1$, $x_2 = -\frac{9}{2}$. From $x + \sqrt{3x^2 + x} = -3$, we obatian that the equation has no real roots. After checking, we know that the roots of the original equation are $x_1 = 1$, $x_2 = -\frac{9}{2}$.
13. The original equation is converted to $(x - 1)(x^2 + 3x - 2) = 0$. Solving the equation gives $x_1 = 1$, $x_2 = \frac{-3-\sqrt{17}}{2}$, $x_3 = \frac{-3+\sqrt{17}}{2}$.

Solution 3

Discriminant of Quadratic Equation

1. From the given condition, $\Delta = (k+6)^2 - 4 \times 9 \times (k+1) = 0$, that is, $k^2 - 24k = 0$. Solving it gives $k_1 = 0$, $k_2 = 24$.
2. From the given condition, $\Delta = 1^2 - 4 \times a \times 2 < 0$, so $a > \frac{1}{8}$.
3. From the given condition, we know that this equation is a quadratic equation. So $k \neq 0$, and the discriminant is

$$\Delta = (2k+1)^2 - 4k^2 = 4k + 1.$$

   (1) From $\Delta > 0$, we get $4k + 1 > 0$, so $k > -\frac{1}{4}$ and $k \neq 0$.
   (2) From $\Delta = 0$, we get $4k + 1 = 0$, that is, $k = -\frac{1}{4}$.
   (3) From $\Delta < 0$, we get $4k + 1 < 0$, that is, $k < -\frac{1}{4}$.

4. Since the equation $x^2 + 2x = n - 1$ has no real roots, we get $2^2 + 4(n-1) < 0$, that is, $n < 0$. Then the discriminant of the equation $x^2 + nx + (2n-1) = 0$ is $\Delta = n^2 - 4(2n-1) = n^2 - 8n + 4 > 0$. So we can see that the conclusion is true.
5. When $a = 1$, the original equation is converted to $x = \frac{1}{4}$, the equation has exactly one real root. When $a = -1$, the original equation has no roots. When $a^2 \neq 1$, if the original equation has exactly one real root, then $\Delta = 4(a+1)^2 - 4(a^2 - 1) = 0$, that is, $a = -1$, which contradicts with $a^2 \neq 1$. So $a = 1$ satisfies the requirement.
6. The original equation is converted to $(2k-1)x^2 - 8x + 6 = 0$. When $k = \frac{1}{2}$, the equation has a real root $x = \frac{3}{4}$. When $k \neq \frac{1}{2}$, from $\Delta = 64 - 4 \times 6(2k-1) < 0$, we get $k > \frac{11}{6}$. Hence, the value range of k is $k > \frac{11}{6}$.

7. Because the equation has a real root x_0, we get that $\Delta = (-2\sqrt{-a})^2 - 4 \times \frac{(a-1)^2}{4} = -a^2 - 2a - 1 \geq 0$, that is $(a+1)^2 \leq 0$. And since $(a+1)^2 \geq 0$, we obtain $a = -1$. Thus $x_0 = 1$, $a^5 - x_0^5 = (-1)^5 - 1^5 = -2$.
8. From the given condition,

$$\begin{cases} 2(m+1) \neq 0, \\ \Delta = (2\sqrt{6}m)^2 - 4 \times 2(m+1)(3m-2) > 0. \end{cases}$$

Then we get that $m < 2$ and $m \neq -1$.
9. We can directly transform the equation into a quadratic equation about x, and then use the discriminant to solve the problem. Or we can observe the characteristics of the equation, and transform the equation into $(x-a)^2 - b(x-a) - 1 = 0$. The discriminant of the equation about $(x-a)$ is $b^2 + 4 > 0$. Therefore, the equation has two distinct real roots, so the original equation has two distinct real roots.
10. From $|x^2 + 3x| = a$, it can be seen that $a \geq 0$.

From the given condition, the equation $|x^2 + 3x| = a$ can be converted to the equation

$$x^2 + 3x - a = 0 \tag{1}$$

and the equation

$$x^2 + 3x + a = 0. \tag{2}$$

Since the discriminant of the equation (1) is $\Delta_1 = 9 + 4a > 0$, the equation (1) has two distinct real roots $x_1 = \frac{-3-\sqrt{9+4a}}{2}$, $x_2 = \frac{-3+\sqrt{9+4a}}{2}$. Because the equation $|x^2 + 3x| = a$ has exactly three distinct real roots, the equation (2) must have real roots, and the two roots are identical but different from x_1 and x_2. Thus the discriminant of the equation (2) is $\Delta_2 = 9 - 4a = 0$, then $a = \frac{9}{4}$, the roots of the equation (2) are $x_{3,4} = -\frac{3}{2}$. Hence, $a = \frac{9}{4}$.
11. The discriminant of the equation $ax^2 + 2bx + c = 0$ is $\Delta_1 = 4(b^2 - ac)$, the discriminant of the equation $bx^2 + 2cx + a = 0$ is $\Delta_2 = 4(c^2 - ab)$, and the discriminant of the equation $cx^2 + 2ax + b = 0$ is $\Delta_3 = 4(a^2 - bc)$. Suppose that $\Delta_1 = 0$, $\Delta_2 = 0$, $\Delta_3 = 0$. Then $a = b = c$, which contradicts the condition that a, b and c are unequal real numbers. Therefore, Δ_1, Δ_2 and Δ_3 can not be equal to 0 at the same time, that is to say, these three equations can not have two identical roots at the same time.

12. The original equation is converted to

$$5x^2 - (3y+2)x + \frac{2y^2+2y+1}{4} = 0.$$

From $\Delta \geq 0$, we get $(y-1)^2 \leq 0$. And since $(y-1)^2 \geq 0$, it can be deduced that $y = 1$. Then we obtain $x = \frac{1}{2}$. So the real roots of the equation are $x = \frac{1}{2}$, $y = 1$.

Remark It can also be solved by completing the square directly.

13. The original equation is converted to

$$\frac{2x^2+2x+a+4}{x^2-1} = 0,$$

that is, $2x^2 + 2x + a + 4 = 0$, and $x^2 - 1 \neq 0$.

If the equation $2x^2 + 2x + a + 4 = 0$ has real roots, there must be

$$\Delta = 4 - 4(a+4) \times 2 \geq 0.$$

Solving it gives $a \leq -\frac{7}{2}$.

When $\Delta = 0$, i.e. $a = -\frac{7}{2}$, the equation $2x^2 + 2x + a + 4 = 0$ only has one real root $x = -\frac{1}{2}$, and $x^2 - 1 \neq 0$.

When $\Delta > 0$, i.e. $a < -\frac{7}{2}$, the equation $2x^2 + 2x + a + 4 = 0$ has two distinct real roots. However, the original equation only has one real root, it can be seen that the equation $2x^2 + 2x + a + 4 = 0$ must have one root be 1 or -1.

When $x = 1$, $2 \times 1^2 + 2 \times 1 + a + 4 = 0$, so $a = -8 < -\frac{7}{2}$, which meets the given requirements.

When $x = -1$, $2 \times (-1)^2 + 2 \times (-1) + a + 4 = 0$, so $a = -4 < -\frac{7}{2}$, which meets the given requirements.

Hence, the sum of all the values of a satisfying the requirements is

$$-\frac{7}{2} + (-8) + (-4) = -\frac{31}{2}.$$

14. Since the equation $x^2 + 2ax + 7a - 10 = 0$ has no real roots, its discriminant is $(2a)^2 - 4(7a - 10) < 0$. Then we get $2 < a < 5$. The discriminants of the equations A, B, C and D are

$$\Delta_{\mathrm{A}} = 4(a-1)(a-2), \quad \Delta_{\mathrm{B}} = 4(a-2)(a-3),$$
$$\Delta_{\mathrm{C}} = 4(a-3)(a-7), \quad \Delta_{\mathrm{D}} = 4(a+1)(a-3).$$

Obviously, for each value a satisfying $2 < a < 5$, we can ensure that $\Delta_{\mathrm{A}} > 0$, but we can't guarantee that Δ_{B}, Δ_{C} and Δ_{D} are not negative (that is, the interval of a in which the equations B, C and D have no real roots and the interval $(2, 5)$ have an overlap, and the interval $(1, 2)$ in which the equation A has no real roots and the interval $(2, 5)$ have no overlap). Thus the equation A must have real roots, and the other equations don't necessarily have real roots. So the answer is A.

Solution 4

Relationship Between Roots and Coefficients and Application

1. From the given condition and Vieta's Formula, $ab = -1$, $a + b = 1$, and $a^2 = a + 1$, $b^2 = b + 1$. So $3a^3 + 4b + \frac{2}{a^2} = 3a^3 + 4b + 2b^2 = 3a(a+1) + 4b + 2(b+1) = 3a^2 + 3a + 6b + 2 = 3(a+1) + 3a + 6b + 2 = 6(a+b) + 5 = 11$.
2. From the given condition, a and b are two roots of the equation $x^2 - 3x + 1 = 0$, then $a + b = 3$, $ab = 1$. Therefore, we get that
$$\frac{1}{a^2} + \frac{1}{b^2} = \frac{a^2 + b^2}{a^2b^2} = \frac{(a+b)^2 - 2ab}{a^2b^2} = \frac{3^2 - 2 \times 1}{1^2} = 7.$$
3. From the given condition, $a \neq 0$, then $\left(\frac{1}{a}\right)^2 + 99\left(\frac{1}{a}\right) + 19 = 0$. Since $ab \neq 1$, $\frac{1}{a}$ and b are two distinct real roots of the equation $x^2 + 99x + 19 = 0$. From Vieta's Formula, $\frac{1}{a} + b = -99$, $\frac{b}{a} = 19$, that is, $1 + ab = -99a$, $b = 19a$. Thus we get that
$$\frac{ab + 4a + 1}{b} = \frac{-99a + 4a}{b} = \frac{-95a}{19a} = -5.$$
4. Let the required equation be $x^2 + bx + c = 0$, and the roots of the equation $x^2 - 5x + 2 = 0$ be x_1 and x_2. Then $x_1 + x_2 = 5$, $x_1x_2 = 2$. From the given condition, $-x_1^2 + (-x_2^2) = -b$, $(-x_1^2) \cdot (-x_2^2) = c$. Then we get that $b = x_1^2 + x_2^2 = (x_1 + x_2)^2 - 2x_1x_2 = 5^2 - 2 \times 2 = 21$, $c = (x_1x_2)^2 = 2^2 = 4$. Hence, the quadratic equation meeting the requirements is $x^2 + 21x + 4 = 0$.

5. From $(2x+7)(x-3) = 1$, we obtain that $2x^2+x-22 = 0$. Let x_1 and x_2 be the roots of this equation. Then $x_1 + x_2 = -\frac{1}{2}$, $x_1x_2 = -11$. Thus we get that $(x_1+x_2)+(x_1x_2) = -\frac{1}{2}-11 = -\frac{23}{2}$, $(x_1+x_2)(x_1x_2) = \frac{11}{2}$. Hence, the quadratic equation meeting the requirements is $x^2 + \frac{23}{2}x + \frac{11}{2} = 0$, that is, $2x^2 + 23x + 11 = 0$.
6. Let x_1 and x_2 be two roots of the equation $ax^2 + bx + c = 0$. Then from Vieta's Formula, $x_1+x_2 = -\frac{b}{a}$, $x_1x_2 = \frac{c}{a}$, and $c \neq 0$. Thus we get that

$$\left(\frac{1}{x_1}+1\right) + \left(\frac{1}{x_2}+1\right) = \frac{x_1+x_2}{x_1x_2} + 2 = \frac{-\frac{b}{a}}{\frac{c}{a}} + 2 = -\frac{b}{c} + 2,$$

$$\left(\frac{1}{x_1}+1\right)\left(\frac{1}{x_2}+1\right) = \frac{1}{x_1x_2} + \left(\frac{1}{x_1} + \frac{1}{x_2}\right) + 1 = \frac{1+x_1+x_2}{x_1x_2} + 1$$

$$= \frac{1-\frac{b}{a}}{\frac{c}{a}} + 1 = \frac{a-b+c}{c}.$$

Hence, the quadratic equation meeting the requirements is

$$x^2 - \left(-\frac{b}{c}+2\right)x + \frac{a-b+c}{c} = 0,$$

i.e. $cx^2 + (b-2c)x + (a-b+c) = 0$.
7. Let α and β be two roots of the equation, where $\alpha = 2\beta$. From Vieta's Formula, $\alpha + \beta = \frac{3}{2}$, $\alpha \cdot \beta = \frac{m}{2}$. Thus we get that $\beta = \frac{1}{2}$, $\alpha = 1$, and $\frac{1}{2} \times 1 = \frac{m}{2}$, i.e. $m = 1$. When $m = 1$, the discriminant of the original equatios is $\Delta = (-3)^2 - 4 \times 2 \times 1 > 0$, which implies that the equation has real roots. Hence, $m = 1$.
8. Let x_1 and x_2 be two roots of the equation. Then $x_1 + x_2 = \frac{1}{2}(a-1)$, $x_1x_2 = \frac{1}{2}(a+1)$. Since $(x_1 - x_2)^2 = (x_1+x_2)^2 - 4x_1x_2 = 1$, we get

$$\left[\frac{1}{2}(a-1)\right]^2 - 4 \times \frac{1}{2}(a+1) = 1,$$

i.e. $a^2 - 10a - 11 = 0$. Solving this equation gives $a = 11$ or $a = -1$. When $a = -1$ or 11, the discriminant of the original equatios is $\Delta = (a-1)^2 - 8(a+1) = a^2 - 10a - 7 = (a^2 - 10a - 11) + 4 > 0$. So $a = -1$ or 11.
9. Let x_1 and x_2 be two roots of the equation. Then $x_1 + x_2 = -2a$, $x_1x_2 = -3$, $x_1^2 + x_2^2 = 10$. Since $(x_1+x_2)^2 = (x_1^2+x_2^2) + 2x_1x_2$, that is, $4a^2 = 10+2\times(-3) = 4$, we get $a^2 = 1$. At this time, $\Delta = 4a^2+3\times4 > 0$. So $|a| = 1$.

10. Observing the relationship between the terms in the equation, we find that the sum and the product of the left two terms of the equation are constant. Suppose that

$$\frac{x^2+3x}{2x^2+2x-8}=A,\ \frac{x^2+x-4}{3x^2+9x}=B,$$

then $\begin{cases} A+B=\frac{11}{12}, \\ AB=\frac{1}{6}. \end{cases}$

So A and B are two roots of the equation $t^2-\frac{11}{12}t+\frac{1}{6}=0$. Solving the equation gives $t_1=\frac{2}{3}$, $t_2=\frac{1}{4}$.

When $A=\frac{2}{3}$, that is, $\frac{x^2+3x}{2x^2+2x-8}=\frac{2}{3}$, we get that $x_1=\frac{5+\sqrt{89}}{2}$, $x_2=\frac{5-\sqrt{89}}{2}$.

When $A=\frac{1}{4}$, that is, $\frac{x^2+3x}{2x^2+2x-8}=\frac{1}{4}$, we obtain that $x_3=-1$, $x_4=-4$.

It's easy to know that x_1, x_2, x_3 and x_4 satisfy the given equation. So the roots of the original equation are

$$x_1=\frac{5+\sqrt{89}}{2},\ x_2=\frac{5-\sqrt{89}}{2},\ x_3=-1,\ x_4=-4.$$

11. Student A misread a as a', then from Vieta's Formula, $-\frac{b}{a'}=6$, $\frac{c}{a'}=8$. So we get $\frac{b}{c}=-\frac{3}{4}$.

Because changing the sign of the coefficient of the linear term does not change the value of the discriminant, Student B misread a or c. Then we obtain $\frac{c}{a}=4$. It can be deduced that $\frac{b}{a}=-3$. Hence, $\frac{2b+3c}{a}=6$.
12. Let x_1 and x_2 be two roots of the equation. Then $\Delta=(-4a)^2-4\times(5a^2-6a)=4(6a-a^2)>0$, that is, $0<a<6$. Since $x_1+x_2=4a$, $x_1\cdot x_2=5a^2-6a$, $|x_1-x_2|=6$, and $(x_1-x_2)^2=(x_1+x_2)^2-4x_1x_2$, we obtain $16a^2=36+4\times(5a^2-6a)$. Solving the equation about a gives $a=3$, which satisfies the requirements. Hence, $a=3$.
13. Since the equation has real roots, we get $\Delta=(2m)^2-4(m^2+2m+3)=-8m-12\geq 0$, that is, $m\leq -\frac{3}{2}$. From $x_1+x_2=2m$, $x_1x_2=m^2+2m+3$, we obtain that $x_1^2+x_2^2=(x_1+x_2)^2-2x_1x_2=2[(m-1)^2-4]$, where $m\leq -\frac{3}{2}$. Hence, when $m=-\frac{3}{2}$, the minimum value of $x_1^2+x_2^2$ is $\frac{9}{2}$.
14. From the given condition, $a+c=2+\sqrt{3}$, $ac=2\sqrt{3}$. Since $(c-a)^2=(c+a)^2-4ca=(2+\sqrt{3})^2-8\sqrt{3}=(2-\sqrt{3})^2$, $c>a$, we get $c-a=2-\sqrt{3}$. Therefore, $b^2=c^2-a^2=(c-a)(c+a)=4-(\sqrt{3})^2=1$, that is, $b=1$.

15. From $\Delta \geq 0$, we get $k^2 - 4k - 4 \geq 0$. Since $x_2^2 = -kx_2 - k - 1$, $x_1 + 2x_2^2 = k$, we obtain $x_1 - 2kx_2 = 3k + 2$, then

$$(x_1 + x_2) - (2k+1)x_2 = 3k + 2, -k - (2k+1) \cdot x_2 = 3k + 2,$$

i.e. $(2k+1)(x_2 + 2) = 0$. If $2k + 1 = 0$, i.e. $k = -\frac{1}{2}$, then $k^2 - 4k - 4 < 0$, which contradicts the fact $\Delta \geq 0$. So there must be $x_2 = -2$. Substituting it into the original equation gives $k = 5$.

16. From Vieta's Formula, $x_1 + x_2 = -a$, $x_1 x_2 = b$. From the given condition,

$$(x_1 + x_2)[(x_1 + x_2)^2 - 3x_1 x_2] = (x_1 + x_2)^2 - 2x_1 x_2 = x_1 + x_2,$$

i.e. $-a(a^2 - 3b) = a^2 - 2b = -a$. If $a = 0$, then $b = 0$. If $a \neq 0$, from $a^2 - 3b = 1$, $a^2 - 2b + a = 0$, we get $a + b = -1$. Then $(1+b)^2 - 3b - 1 = 0$, $b(b-1) = 0$, so $b = 0$ or $b = 1$.

Hence, there are three groups of solutions: $\begin{cases} a = 0, \\ b = 0, \end{cases}$ $\begin{cases} a = -1, \\ b = 0 \end{cases}$ or $\begin{cases} a = -2, \\ b = 1, \end{cases}$ and the two corresponding roots are $\begin{cases} x_1 = 0, \\ x_2 = 0, \end{cases}$ $\begin{cases} x_1 = 0, \\ x_2 = 1 \end{cases}$ or $\begin{cases} x_1 = 1, \\ x_2 = 1. \end{cases}$

Solution 5

Simultaneous Quadratic Equations with Two Unknowns

1. If x and y are regarded as two roots of the equation $t^2 - 5t + 4 = 0$, then

$$\begin{cases} x = 1, \\ y = 4 \end{cases} \quad \text{or} \quad \begin{cases} x = 4, \\ y = 1. \end{cases}$$

2. From $a^2 + b^2 - 11 = 0$ and $a^2 - 5b - 5 = 0$, we get $b^2 + 5b - 6 = 0$. Solving the equation gives $b = -6$ or $b = 1$. Since $a^2 = 5b + 5 \geq 0$, we discard $b = -6$. So $b = 1$.

   When $b = 1$, $a = \pm\sqrt{10}$ meets the requirements. Hence, the value of b is 1.

3. The original simultaneous equations are converted to

$$\begin{cases} x + y = 4 - a, \\ xy = 5 - a(4 - a), \end{cases}$$

   so the roots x and y are two roots of the quadratic equation $X^2 - (4 - a)X + [5 - a(4 - a)] = 0$. Then from $\Delta = (4 - a)^2 - 4 \times [5 - a(4 - a)] \geq 0$, we obtain $\frac{2}{3} \leq a \leq 2$. So the answer is D.

4. From the given condition, $x^2 - 4x - 2(kx + 2) + 1 = 0$, i.e. $x^2 - (4 + 2k)x - 3 = 0 \ldots (*)$. The discriminant of the equation $(*)$ is $\Delta = (4 + 2k)^2 + 3 \times 4 > 0$, so for any value of k, there are two groups of distinct real roots of the original simultaneous equations.

5. Since $x^2 + 2xy - 10x = x(x + 2y - 10)$, $y^2 + 2xy - 10y = y(y + 2x - 10)$, the original simultaneous equations are converted to

$$\begin{cases} x = 0, \\ y = 0, \end{cases} \begin{cases} x = 0, \\ y + 2x - 10 = 0, \end{cases} \begin{cases} x + 2y - 10 = 0, \\ y = 0, \end{cases} \text{or} \begin{cases} x + 2y - 10 = 0, \\ y + 2x - 10 = 0. \end{cases}$$

So we obtain that

$$\begin{cases} x_1 = 0, \\ y_1 = 0, \end{cases} \begin{cases} x_2 = 0, \\ y_2 = 10, \end{cases} \begin{cases} x_3 = 10, \\ y_3 = 0, \end{cases} \begin{cases} x_4 = \dfrac{10}{3}, \\ y_4 = \dfrac{10}{3}. \end{cases}$$

6. From $2x^2 - 3xy - 2y^2 = 0$, we get $(2x + y)(x - 2y) = 0$. Then the original simultaneous equations are converted to $\begin{cases} x^2 + y^2 = 5, \\ 2x + y = 0 \end{cases}$ or $\begin{cases} x^2 + y^2 = 5, \\ x - 2y = 0. \end{cases}$ To solve these two systems of equations, the solutions of the original simultaneous equations are

$$\begin{cases} x_1 = 1, \\ y_1 = -2, \end{cases} \begin{cases} x_2 = -1, \\ y_2 = 2, \end{cases} \begin{cases} x_3 = 2, \\ y_3 = 1, \end{cases} \begin{cases} x_4 = -2, \\ y_4 = -1. \end{cases}$$

7. Complete the square or transform the simultaneous equations into a quadratic equation (including z) about x, and then solve it by the discriminant.

   From the given condition, $x(2-x)-z^2 = 1$, that is, $(x-1)^2+z^2 = 0$. Then we get that $x = 1$, $z = 0$. So the solutions of the simultaneous equations are $x = 1$, $y = 1$, $z = 0$.
8. From $x = y + \sqrt{2}$, we get $(x - y)^2 = 2$, i.e. $(x + y)^2 - 4xy = 2$, so $2xy = \frac{1}{2}(x + y)^2 - 1$. From $2xy + 2\sqrt{2}z^2 + 1 = 0$, we obtain $\frac{1}{2}(x + y)^2 + 2\sqrt{2}z^2 = 0$. Therefore, $x + y = 0$, $z = 0$, then $x + y + z = 0$.
9. Let the speed of the team's advance be v_1, and the speed of the messenger be v_2. Then the time taken by the messenger from the end of the line to the front of the line is $\frac{40}{v_2-v_1}$, and the time taken by the messenger from the front of the line to the end of the line is $\frac{40}{v_2+v_1}$. And the team advanced 30 meters in this period, which took the time of $\frac{30}{v_1}$. Thus we have $\frac{40}{v_2-v_1} + \frac{40}{v_2+v_1} = \frac{30}{v_1}$. Then $v_2 = -\frac{1}{3}v_1$ (discarded) or $v_2 = 3v_1$. So the messenger advanced 60 meters more than the team.
10. Suppose that the unit prices of goods of A, B and C are x yuan, y yuan and z yuan respectively, then $\begin{cases} 3x + 7y + z = 3.15, \\ 4x + 10y + z = 4.20. \end{cases}$ Let $x + y + z = a(3x + 7y + z) + b(4x + 10y + z)$. Then we have $a = 3$, $b = -2$. Hence, $x + y + z = 1.05$(yuan).
11. Let $x - 2y = t$. Then $\begin{cases} x^2 - 2x - 4y = 5, \\ x - 2y = t \end{cases}$ has real roots.

    From $x^2 - 4x + 2t - 5 = 0$, we know that $\Delta = 4^2 - 4(2t - 5) = 4 \times (9 - 2t) \geq 0$, so $t \leq \frac{9}{2}$, i.e. $x - 2y \leq \frac{9}{2}$.

12. Let a km/h be the speed of water flow and b km/h be Xiaoming's usual rowing speed in still water. Then

$$\begin{cases} \dfrac{15}{a+b} = \dfrac{15}{b-a} - 5, \\ \dfrac{15}{2b+a} = \dfrac{15}{2b-a} - 1. \end{cases}$$

Solving the simultaneous equations gives $a = 2$. So the speed of water flow is 2 kilometers per hour.

13. From the given condition, $3y^2 - 6y + z^2 + 3 = 0$. Since y is a real number, the discriminant of the equation about y is $\Delta_y = 36 - 4 \times 3(z^2 + 3) = -12z^2 \geq 0$. So we get $z = 0$, and then $y = 1$, $x = 3$. Hence, $x^{2y+z} = 9$.

Solution 6

Integer Roots of Quadratic Equation

1. From $\sqrt{27-10\sqrt{2}} = 5-\sqrt{2} = a+b$, we get that $a = 3$, $b = 2-\sqrt{2}$, then $a-b = \sqrt{2}+1$. Hence, $\frac{a+b}{a-b} = (5-\sqrt{2})(\sqrt{2}-1) = 6\sqrt{2}-7$.
2. Let $n-1$, n and $n+1$ be three consecutive positive integers respectively (n is an integer greater than 1). When the coefficient of the linear term is $n-1$ or n, the discriminant Δ of the equation is less than zero, and the equation has no real roots. When the coefficient of the linear term is $n+1$, the discriminant of the equation is $\Delta = (n+1)^2 - 4n(n-1) = -3(n-1)^2+4$. In order to make $\Delta \geq 0$, since n is an integer greater than 1, n can only be taken as 2. When $n = 2$, the equations $x^2+3x+2 = 0$ and $2x^2+3x+1 = 0$ both have real roots. So there are only two groups of a, b and c that meet the requirements: $(1,3,2)$ and $(2,3,1)$.
3. When $r = 0$, the original equation is converted to $x = 1$, which satisfies the requirements. When $r \neq 0$, let the roots of the original equation be x_1 and x_2. Then $x_1+x_2 = 2+\frac{7}{r}$, $x_1x_2 = 1+\frac{7}{r}$. So $r = 1$ or 7, which meet the requirements. Hence, $r = 0$, 1, 7.
4. From the given condition, $a \neq 0$. Because the two roots of the equation $ax^2+bx+c = 0$ are exactly a and b, from Vieta's Formula, we get $a+b = -\frac{b}{a}$, then $b = -\frac{a^2}{a+1} = 1-a-\frac{1}{a+1}$. Since a and b are integers, $a+1 = 1$ or -1. When $a+1 = 1$, $a = 0$, which contradicts $a \neq 0$. So $a+1 = -1$, that is, $a = -2$. Therefore, $b = 4$, $c = 16$.
5. From the given condition, $|ab| = 0$ and $|a+b| = 1$, or $|ab| = 1$ and $|a+b| = 0$. Therefore, $a = 0$, $b = 1$; $a = 1$, $b = 0$; $a = 0$, $b = -1$; $a = -1$, $b = 0$; $a = 1$, $b = -1$; $a = -1$, $b = 1$. There are 6 pairs of (a, b). So the answer is C.

6. From $\frac{1}{x} - \frac{1}{y} = \frac{1}{100}$, we get that $x < y$ and $(100 + y)(100 - x) = 100^2$. To maximize y, $100 - x = 1$, then $100 + y = 100^2$. So $y = 9900$.
7. Let x_1 and x_2 be the two integer roots of the equation. Then $(x - a)(x - 8) - 1 = (x - x_1)(x - x_2)$. Let $x = 8$. Then $(8 - x_1)(8 - x_2) = -1$. So $8 - x_1 = -1$ and $8 - x_2 = 1$, or $8 - x_1 = 1$ and $8 - x_2 = -1$. Then we obtain that $x_1 = 7$, $x_2 = 9$ or $x_1 = 9$, $x_2 = 7$. Hence, $a = 8$.
8. From the given condition, $\triangle AEF \backsim \triangle ABC$, then $\frac{EF}{BC} = \frac{AD-EF}{AD}$, i.e. $\frac{c}{10a+b} = \frac{d-c}{d}$. And since $b = a + 1$, $c = a + 2$, $d = a + 3$, we get $a^2 - 6a + 5 = 0$. Solving the equation gives $a_1 = 1$, $a_2 = 5$, that is, $BC = 12$, $d = 4$ or $BC = 56$, $d = 8$. Hence, $S_{\triangle_{ABC}} = 24$ or 224.
9. Let a and $b(a < b)$ be the lengths of the legs of a right triangle. Then from the given condition, $a + b = k + 2$, $ab = 4k$. From the fact that a and b are integers, we know that the discriminant of the equation is a perfect square number, that is, $(k + 2)^2 - 4 \times 4k = n^2$ (n is a non-negative integer), so

$$(k - 6 + n)(k - 6 - n) = 1 \times 32 = 2 \times 16 = 4 \times 8.$$

Because $k + n - 6$ and $k - n - 6$ are both odd or even numbers, and $k - 6 + n > k - 6 - n$, we get that $\begin{cases} k - 6 + n = 16, \\ k - 6 - n = 2 \end{cases}$ or $\begin{cases} k - 6 + n = 8, \\ k - 6 - n = 4. \end{cases}$

Solving the systems of equations gives $k = 15$ or $k = 12$.

When $k = 15$, $a + b = 17$, $ab = 60$, we obtain that $a = 5$, $b = 12$, $c = 13$;

When $k = 12$, $a + b = 14$, $ab = 48$, we obtain that $a = 6$, $b = 8$, $c = 10$.

10. From the quadratic formula, $x_{1,2} = -5m \pm \sqrt{25m^2 + 5n - 3}$. Since the last digit of $25m^2 + 5n - 3$ can only be 2 or 7, but the last digit of a perfect square number can't be 2 or 7, we obtain that $25m^2 + 5n - 3$ can't be a perfect square number. So the original equation has no integer roots.
11. Suppose that the equation has an integer root α, then $a\alpha^2 + b\alpha + c = 0$. If α is even, all of a, b and c are odd, then $\alpha^2 a + \alpha b$ is even, so $a\alpha^2 + b\alpha + c$ is odd, not 0. If α is odd, all of a, b and c are odd, then $a\alpha^2 + b\alpha + c$ is also odd, not 0. The above results contradict the condition $a\alpha^2 + b\alpha + c = 0$. Therefore, the conclusion is true.

Solution 7

Perfect Square Numbers

1. (1) $\underbrace{11\ldots1}_{n\text{ digits}}\underbrace{55\ldots5}_{(n-1)\text{ digits}}6 = \underbrace{11\ldots1}_{n\text{ digits}}\underbrace{00\ldots0}_{n\text{ digits}} + \underbrace{55\ldots5}_{(n-1)\text{ digits}}0 + 6$

$$= \underbrace{11\ldots1}_{n\text{ digits}} \times 10^n + \underbrace{55\ldots5}_{(n-1)\text{ digits}} \times 10 + 6$$

$$= \frac{1}{9} \times \underbrace{99\ldots9}_{n\text{ digits}} \times 10^n + \frac{5}{9} \times \underbrace{99\ldots9}_{(n-1)\text{ digits}} \times 10 + 6$$

$$= \frac{1}{9} \times (10^n - 1) \times 10^n + \frac{5}{9} \times (10^{n-1} - 1) \times 10$$

$$+ \frac{6}{9} \times 9 = \frac{1}{9} \times 10^{2n} - \frac{1}{9} \times 10^n + \frac{5}{9} \times 10^n$$

$$- \frac{5}{9} \times 10 + \frac{6}{9} \times 9 = \frac{1}{9}[10^{2^n} + 4 \times 10^n + 4]$$

$$= \left(\frac{10^n + 2}{3}\right)^2.$$

Because the sum of all the digits of the sum number $10^n + 2$ is 3, $\frac{10^n+2}{3}$ is an integer, that is, $\underbrace{11\ldots1}_{n\text{ digits}}\underbrace{55\ldots5}_{(n-1)\text{ digits}}6$ is a perfect square number.

(2) Because $8\underbrace{99\ldots9}_{(n-1)\text{ digits}}4\underbrace{00\ldots0}_{(n-1)\text{ digits}}1 = 9\underbrace{00\ldots0}_{2n\text{ digits}} - 5\underbrace{00\ldots0}_{n\text{ digits}} - \underbrace{99\ldots9}_{n\text{ digits}} = 9 \times 10^{2n} - 5 \times 10^n - (10^n - 1) = 9 \times 10^{2n} - 6 \times 10^n + 1 = (3 \times 10^n - 1)^2$, we get that $8\underbrace{99\ldots9}_{(n-1)\text{ digits}}4\underbrace{00\ldots0}_{(n-1)\text{ digits}}1$ is a perfect square number.

2. (1) Observing the largest several two-digit numbers in the perfect square numbers: 81, 64, 49, 36, ..., we know easily that the minimum three-digit number P is $49 + 64 = 113$.
   (2) The last digit of the squares of the integers whose units digits are from 0 to 9 are 0, 1, 4, 9, 6, 5, 6, 9, 4, 1 in turn. Among them, the units digit numbers of the sum of the two adjacent numbers are 1, 5, 3, 5, 1, 1, 5, 3, 5, 1 in turn. That is to say, the last digit of P can only be 1, 3, 5.
3. Since $a > 0$ and $41 - 2$ is a perfect square number and an odd number, it can be deduced that $41 - 2a = 1^2, 3^2, 5^2$. So $a = 20, 16, 8$.
4. If the last digit numbers of integer n are 1, 2, 3, ..., 8, 9, 0 in turn, then the last digit numbers of n^2 are 1, 4, 9, 6, 5, 6, 9, 4, 1, 0. Suppose that $a = n - 1$, $b = n$, $c = n + 1$, then the last digit numbers of $a^2 + b^2 + c^2$ are 5 (i.e. $0 + 1 + 4$), 4, 9, 0, 7, 0, 9, 4, 5, 2. That is, only when the last digit of b is 0, the last digit of $a^2 + b^2 + c^2$ is 2.
5. Since $(10a + b)^2 = 100a^2 + 10 \cdot (2ab) + b^2$, when $b = 1$ or 3, the tens digit of $(10a + b)^2$ has the same parity as $2ab$, that is, it is an even number. So the sum of the last two digits is odd. When $b = 5, 7, 9$, its tens digit has the same parity as $2ab + 2$, $2ab + 4$ and $2ab + 8$, that is, it is an even number. So the sum of the last two digits is odd. Therefore, the sum of the last two digits of an odd perfect square number is odd, that is, (1) is correct. Since $4^2 = 16$, $6^2 = 36$ and so on, we know that (2) is incorrect. So the answer is A.
6. If the middle number of these 75 numbers is n, then the required number (i.e. the first number) is $n - 37$, and their sum is $75n = 25 \times 3 \times n$. When $3n$ is a perfect square number (that is, n equals to the product of 3 and a perfect square number), $75n$ is also a perfect square number. Since $n - 37 > 0$, we know that $n > 37 > 3 \times 12$. Then $n = 3 \times 16 = 48$, $n - 37 = 48 - 37 = 11$. That is to say, the minimum value of this number is 11.
7. Let $B = (10m + n)^2 = 100m^2 + 20mn + n^2$, where m is a natural number and n is a number among 0, 1, 2, ..., 9. It is easy to know that the last digit of B is the same as that of n^2. If the last digit of n^2 is even, the conclusion holds. Now suppose that the last digit of n^2 is odd, then n itself must be odd (1, 3, 5, 7 or 9), thus the tens digit of n^2 is either zero or even. Then the tens digit of B equals to the tens digit of n^2 plus an even number, and the sum must be even. Hence, the conclusion holds.

8. Because $\overline{ab} + \overline{ba} = 10a + b + 10b + a = 11(a + b)$, where a and b are both one-digit number, we know that only when $a+b = 11$, $\overline{ab}+\overline{ba}$ is a perfect square number. That is to say, there are 8 double-digit numbers $\overline{ab}$ as follows: 29, 38, 47, 56, 65, 74, 83, 92.
9. Let the sum and difference of a positive integer x and an odd number a be perfect square numbers. That is, $x+a = m^2$, $x-a = n^2$. Subtracting these two equations gives $m^2 - n^2 = 2a$, i.e. $(m+n)(m-n) = 2a$. Because $m+n$ and $m-n$ have the same parity, but a is an odd number, there is only one even prime factor 2 in the prime factorizations, that is, two factors of $2a$ must be one odd and one even. So if a positive integer is added and subtracted by the same odd number, the sum and difference cannot both be perfect square numbers.
10. From the given condition, $5 \times 2^m + 1 \geq 5 \times 2^1 + 1 = 11$, $5 \times 2^m + 1$ is odd and a perfect square number. Then let $5 \times 2^m + 1 = (2k-1)^2$, where $k \geq 3$ and k is a positive integer. So we get $5 \times 2^m = 4k(k-1)$, then $m \geq 3$, and $5 \times 2^{m-2} = k(k-1)$.

    Since $k \geq 3$, k and $k-1$ are coprime, we obtain that $\begin{cases} k = 5, \\ k - 1 = 2^{m-2}, \end{cases}$ or $\begin{cases} k = 2^{m-2}, \\ k - 1 = 5. \end{cases}$ Solving them gives $k = 5$, $m = 4$. Hence, $m = 4$.
11. Among 10 figures, the figures whose last digit after squaring is itself can only be 0, 1, 5 and 6. The discussion is as follows:

    (1) Let the double-digit number be $\overline{a0}$. Then $(10a)^2 = 100a^2$. From the given condition, the last two digits of $100a^2$ is $\overline{00} = \overline{a0}$, i.e. $a = 0$ (which does not satisfy the question).
    (2) Let the double-digit number be $\overline{b1}$. Then $(10b+1)^2 = 100b^2 + 20b + 1$. From the given condition, the tens digit of $20b$ (that is, the units digit of $2b$) is b, i.e. $b = 0$ (which does not satisfy the question).
    (3) Let the double-digit number be $\overline{c5}$. Then $(10c+5)^2 = 100c^2 + 100c + 25$. From the given condition, the last two digits of it is $25 = \overline{c5}$, i.e. $c = 2$.
    (4) Let the double-digit number be $\overline{d6}$. Then $(10d+6)^2 = 100d^2 + 120d + 36 = 100d^2 + 100d + 20d + 36$. From the given condition, the tens digit of $20d + 36$ (that is, the units digit of $2d + 3$) is d, i.e. $d = 7$.

    To sum up, there are 2 double-digit numbers (i.e. 25 and 76).
12. Let A and B be two groups, where 1 is in the group A. From $1+3 = 4$, we get that 3 is in the group B. From $3+6 = 9$, $6+10 = 16$, we obtain

that 6 is in the group A and 10 is in the group B. From $10 + 15 = 25$, we can see that 15 is in the group A. However, $15 + 1 = 16$, so 15 is not in the group A. Hence, $n \leq 14$.

After trying, we know that $A = \{1, 2, 4, 6, 9, 11, 13\}$, $B = \{3, 5, 7, 8, 10, 12, 14\}$ meet the grouping requirements. So the maximum value of n is 14.

13. From the same idea as the above exercise, it can be seen that $n \leq 28$, and when $A = \{2, 4, 6, 9, 11, 13, 15, 17, 18, 20, 22, 24, 26, 28\}$, $B = \{3, 5, 7, 8, 10, 12, 14, 16, 19, 21, 23, 25, 27\}$, it meets the grouping requirements. So the maximum value of n is 28.
14. Start with simple situations. When n =1, 2, 3, 4, 5, 6, 7, 8, the minimum values can be obtained as follows: 1, 3, 4, 2, 3, 1, 0, 0. When $n = 9$, 10, 11, 12, 13, 14, 15, 16, the minimum values can be obtained as follows: 1, 1, 0, 0, 1, 1, 0, 0.

   Notice that $-(m+1)^2 + (m+2)^2 + (m+3)^2 - (m+4)^2 + (m+5)^2 - (m+6)^2 - (m+7)^2 + (m+8)^2 = 0$. Therefore, if the minimum value is $f(n)$, we get that

$$f(8k) = 0, \quad k = 1, 2, \ldots;$$

$$f(8k+1) = 1, \quad k = 0, 1, 2, \ldots;$$

$$f(8k+2) = \begin{cases} 3, k = 0, \\ 1, k \geq 1; \end{cases}$$

$$f(8k+3) = \begin{cases} 4, k = 0, \\ 0, k \geq 1; \end{cases}$$

$$f(8k+4) = \begin{cases} 2, k = 0, \\ 0, k \geq 1; \end{cases}$$

$$f(8k+5) = \begin{cases} 3, k = 0, \\ 1, k \geq 1; \end{cases}$$

$$f(8k+6) = 1, \quad k = 0, 1, 2, \ldots;$$

$$f(8k+7) = 0, \quad k = 0, 1, 2, \ldots.$$

15. From $a^2 \equiv 0$ or 1 (mod4), we get that $x^2 + y^2 \equiv 0$, 1 or 2 (mod4). And since $2019 \equiv 3 (\text{mod} 4)$, this equation has no integer roots.
16. Considering $44^2 + 117^2 = 125^2$, $117^2 + 240^2 = 267^2$, $240^2 + 44^2 = 244^2$, we know that in 44^2, 117^2 and 240^2, there must be two numbers in a group. So the proposition is proved.

Solution 8

Quadratic Functions

1. Since $0 = -0^2 + 2m \cdot 0 + 1 - m^2$, we get $m^2 - 1 = 0$, i.e. $m = \pm 1$. Thus, when $m = 1$, the quadratic function is $y = -x^2 + 2x = -(x-1)^2 + 1$, and the vertex of the graph is (1, 1); when $m = -1$, the quadratic function is $y = -x^2 - 2x = -(x+1)^2 + 1$, and the vertex of the graph is $(-1, 1)$.
2. From the graph, it can be seen that $a < 0$, $-\frac{b}{2a} > 0$, $0 \cdot a + b \cdot 0 + c < 0$, $a \cdot 1^2 + b \cdot 1 + c > 0$, $a \cdot (-1)^2 + b \cdot (-1) + c < 0$, $-\frac{b}{2a} < 1$. So we get that $a < 0$, $b > 0$, $c < 0$, $a + b + c > 0$, $a - b + c < 0$, $2a + b < 0$, then $ab < 0$, $ac > 0$, $2a - b < 0$. So the answer is A.
3. From the graph, it can be seen that $a > 0$, $-\frac{b}{2a} < 0$, $a \cdot 0^2 + b \cdot 0 + c < 0$, then $b > 0$, $c < 0$. Thus we get that $a + b > 0$, $ac < 0$. Point $(a + b, ac)$ is in the fourth quadrant, so the answer is D.
4. From the given condition, $\begin{cases} 0.64a - 0.8b + c = 4.132, \\ 1.44a + 1.2b + c = -1.948, \\ 7.84a + 2.8b + c = -3.932. \end{cases}$ Solving the system of equations gives $a = 0.5, b = -3.24, c = 1.22$. Hence, $y = 0.5 \times 3.24 - 3.24 \times 1.8 + 1.22 = -2.992$.
5. The graph of the quadratic function $y = -x^2 + 6x - \frac{27}{4}$ intersects the x-axis at the points $\left(\frac{3}{2}, 0\right)$ and $\left(\frac{9}{2}, 0\right)$. There are three integers between $x = \frac{3}{2}$ and $x = \frac{9}{2}$, which are 2, 3 and 4.

   When $x = 2$ or 4, $y = \frac{5}{4}$, the integers satisfying $0 \le y \le \frac{5}{4}$ are 0 and 1, so there are 4 integral points.

   When $x = 3$, $y = \frac{9}{4}$, the integers satisfying $0 \le y \le \frac{9}{4}$ are 0, 1 and 2, so there are 3 integral points.

   In summary, there are 7 integral points, so the answer is C.

6. The vertex of the graph of the quadratic function is $\left(-\frac{b}{2a}, \frac{4ac-b^2}{4a}\right)$. In the option (A), it can be seen from the graph of the linear function that $a > 0$ and $b > 0$, then $-\frac{b}{2a} < 0$ should be obtained, which contradicts the graph of the quadratic function.

   In the option (B), it can be seen from the graph of the linear function that $a < 0$, then the graph of the quadratic function opens downwards, which doesn't match too. In the option (D), it can be seen from the graph of the linear function that $a > 0$, then the graph of the quadratic function opens upwards, which doesn't match too. So the answer is C.
7. As shown in Figure 8.1, $AB = 1$. Let the coordinates of point C be $(1.1, h)$. Then the coordinates of point D are $(2.1, h)$.

   From the symmetry, we know that the x-coordinate of the point P is

$$1.1 + \frac{2.1 - 1.1}{2} = 1.6.$$

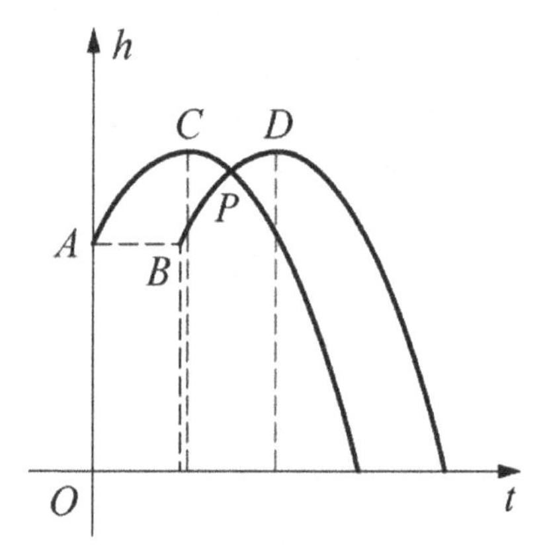

Fig. 8.1

8. Since the graph of the linear function $y_2 = dx + e (d \neq 0)$ passes through the point $(x_1, 0)$, we get that

$$y_2 = d(x - x_1),$$

$$y = y_1 + y_2 = (x - x_1)[a(x - x_2) + d].$$

Since the graph of function $y = y_1 + y_2$ has only one intersection point with x-axis, it is obtained that $y = y_1 + y_2$ is a quadratic function, and its vertex is on the x-axis, that is, $y = y_1 + y_2 = a(x - x_1)^2$. Thus we get

$$a(x - x_2) + d = a(x - x_1).$$

Let $x = x_1$. Then $a(x_2 - x_1) = d$.

Hence, $x_2 - x_1 = \frac{d}{a}$. So the answer is D.

9. Since $|m| + |n| - 2\sqrt{|mn|} = (\sqrt{|m|} - \sqrt{|n|})^2 \geq 0$, it is obtained that $|m| + |n| \geq 2\sqrt{|mn|}$. Then from $|m| + |n| \leq 1$, we get $2\sqrt{|mn|} \leq 1$.

   So $|mn| \leq \frac{1}{4}$, and the equal sign holds if and only if $|m| = |n|$. That is, $-\frac{1}{4} \leq mn \leq \frac{1}{4}$.

   Since m and n are the x-intercepts of the graph of the function $y = x^2 + ax + b$, m and n are two roots of the equation $x^2 + ax + b = 0$. Then $mn = b$ is obtained from the relationship between the roots and the coefficients.

   Therefore, $-\frac{1}{4} \leq b \leq \frac{1}{4}$. Then we get that $p \leq \frac{1}{4}$, $q \geq -\frac{1}{4}$.

   When $m = n = \frac{1}{2}$, it is obtained that $p = \frac{1}{4}$, and $y = x^2 - x + \frac{1}{4}$.

   When $m = -\frac{1}{2}$, $n = \frac{1}{2}$, it is obtained that $q = -\frac{1}{4}$, and $y = x^2 - \frac{1}{4}$.

   Hence, $|p| + |q| = \frac{1}{2}$.

10. (1) It can be seen from the graph that $a < 0$, $-\frac{b}{2a} > 0$, $\frac{c}{a} < 0$, $b^2 - 4ac > 0$, then $a < 0$, $b > 0$, $c > 0$, $b^2 - 4ac > 0$.

    (2) When $x = 0$, $y = a \cdot 0^2 + b \cdot 0 + c = c; y = ax^2 + bx + c = a\left(x + \frac{b}{2a}\right)^2 + \frac{4ac-b^2}{4a}$; the roots of equation $ax^2 + bx + c = 0$ are $x_{1,2} = \frac{-b\pm\sqrt{b^2-4ac}}{2a}$. So the coordinates of the points A, B, D and M are $(\frac{-b\pm\sqrt{b^2-4ac}}{2a}, 0)$, $(\frac{-b\pm\sqrt{b^2-4ac}}{2a}, 0)$, $(0, c)$ and $\left(-\frac{b}{2a}, \frac{4ac-b^2}{4a}\right)$ respectively.

    (3) $|AB| = \left|\frac{-b+\sqrt{b^2-4ac}}{2a} - \frac{-b+\sqrt{b^2-4ac}}{2a}\right| = \frac{\sqrt{b^2-4ac}}{|a|}$

    (4) $|OA| \cdot |OB| = -\frac{c}{a}$.

    (5) From $|OA| = |OD|$, it is obtained that $\left|\frac{-b+\sqrt{b^2-4ac}}{2a}\right| = c$, that is, $ac - b + 1 = 0$.

11. Let A be $(x_1, 0)$ and B be $(x_2, 0)$. From the fact that ΔABC is a right triangle, we can know that x_1 and x_2 must be of different signs, then $x_1x_2 = \frac{c}{a} < 0$. From the knowledge of similar triangles, we get that $|OC|^2 = |AO| \cdot |BO|$, that is, $c^2 = |x_1| \cdot |x_2| = |\frac{c}{a}|$. So from $|ac| = 1$, it is obtained that $ac = -1$.

12. (1) Let $y_1 = mx + n$, $y_2 = ax^2 + bx + c$. From the given condition, $\begin{cases} 4 = 1 \times m + n, \\ -2 = -2m + n, \end{cases}$ then $m = 2$, $n = 2$.

From $\begin{cases} 4 = a + b + c, \\ -2 = a \times (-2)^2 + b \times (-2) + c, \\ 3 = a \cdot 0^2 + b \cdot 0 + c, \end{cases}$ it is obtained that $a = -\frac{1}{2}$, $b = \frac{3}{2}$, $c = 3$. So the expression of these two functions are $y_1 = 2x + 2$, $y_2 = -\frac{1}{2}x^2 + \frac{3}{2}x + 3$.

(2) From $y_1 > y_2$, it is obtained that $2x + 2 > -\frac{1}{2}x^2 + \frac{3}{2}x + 3$, that is, $x^2 + x - 2 > 0$. Solving the inequality gives $x < -2$ or $x > 1$.

(3) It can be seen from the graph of the function $y_2 = -\frac{1}{2}(x^2 - 3x - 6) = -\frac{1}{2}\left(x - \frac{3}{2}\right)^2 + \frac{33}{8}$ that y_2 increases as x increases when $x \leq \frac{3}{2}$. It's known that y_1 increases as x increases on the real number set R. Hence, both y_1 and y_2 increase as x increases when $x \leq \frac{3}{2}$.

13. It's known from the condition (2) that $m \neq 0$. Suppose that the point (x_0, y_0) meets the conditions, then

$$y_0 = -2x_0 + 3. \tag{1}$$

And for any non-zero real number m, we have

$$y_0 \neq mx_0^2 + \left(m - \frac{2}{3}\right) x_0 - \left(2m - \frac{3}{8}\right). \tag{2}$$

Substituting the formula (1) into the formula (2) gives $(x_0 - 1)(x_0 + 2)m \neq -\frac{4}{3}x_0 + \frac{21}{8}$. Hence, $x_0 = 1$, -2 or $\frac{63}{32}$. Substituing these values into the formula (1) gives that the coordinates of points satisfying these two conditions are $(1, 1)$, $(-2, 7)$ and $\left(\frac{63}{32}, -\frac{15}{16}\right)$.

14. (1) From $B(0, 4)$, we get $c = 4$. The x-intercept of G is $A\left(-\frac{b}{2a}, 0\right)$. From $ac = b$, it is obtained that $\frac{b}{a} = c$, so $-\frac{b}{2a} = -\frac{c}{2} = -2$, that is $A(-2, 0)$.

Then we obtain $\begin{cases} b = 4a, \\ 4a - 2b + 4 = 0. \end{cases}$ Solving the system of equations gives $\begin{cases} a = 1, \\ b = 4. \end{cases}$

So the expression of the quadratic function is $y = x^2 + 4x + 4$.

(2) Let the analytic expression of the graph L be $y = -3x + b$. Since the graph L passes through the point $A(-2, 0)$, it is obtained that $b = -6$, that is, the analytic expression of the linear function obtained after translation is $y = -3x - 6$. Let $-3x - 6 = x^2 + 4x + 4$. Then we get that $x_1 = -2$, $x_2 = -5$. Substituting them into $y = -3x - 6$ gives $y_1 = 0$, $y_2 = 9$. So the other intersection of L and G is $C(-5, 9)$. As shown in Figure 8.2, construct a line segment CD

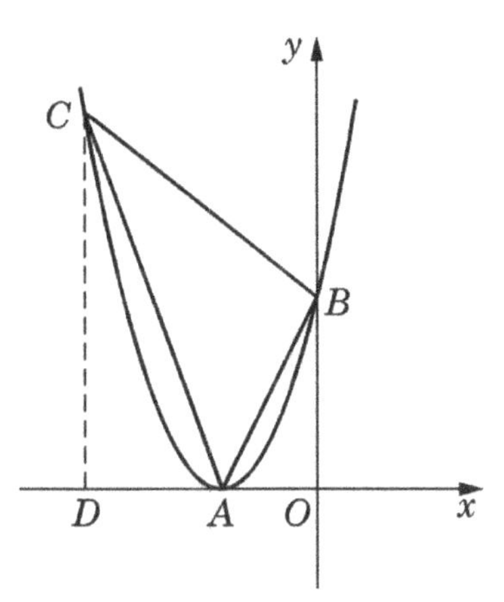

Fig. 8.2

perpendicular to the x-axis through the point C which intersects the x-axis at the point D, then $S_{\Delta_{ABC}} = S_{\text{trapezoid}BCDO} - S_{\Delta_{ACD}} - S_{\Delta_{ABO}} = \frac{1}{2}(4+9)\times 5 - \frac{1}{2}\times 3\times 9 - \frac{1}{2}\times 2\times 4 = 15.$

15. (1) From

$$y = ax^2 - 2amx + am^2 + 2m + 1 = a(x-m)^2 + 2m + 1,$$

it is obtained that the vertex is $A(m, 2m+1)$. Let the analytic expression of AB be $y = kx + b$. Substituting the coordinates of the points A and $P(1, 3)$ into the expression gives

$$\begin{cases} 2m + 1 = km + b, & (1) \\ 3 = k + b. & (2) \end{cases}$$

From (1)–(2), it is obtained that $2m - 2 = (m-1)k$. Since $m \neq 1$(if $m = 1$, then three points A, B and P coincide, which doesn't meet the requirements), we get that $k = 2$, $b = 1$, that is, the analytic expression of AB is $y = 2x + 1$. Because the point B is also the intersection of the line AB and the y-axis, it is obtained that $y = 2\times 0 + 1$, and the vertex of S_2 is $B(0, 1)$.

Because S_2 and S_1 are centrosymmetric about the point P, the parabolas have the same size of its opening and the directions are opposite, then the analytical expression of S_2 is $y = -ax^2 + 1$.

Since the points A and B are centrosymmetric about the point $P(1, 3)$ (as shown in Figure 8.3(1)), construct PE perpendicular to the y-axis at the point E and AF perpendicular to the y-axis at

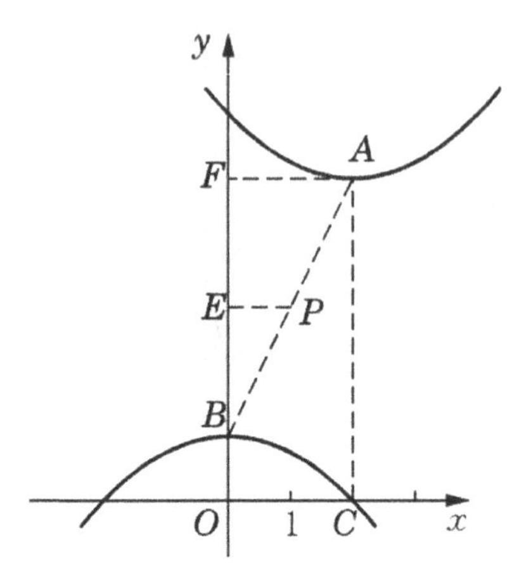

Fig. 8.3(1)

the point F, then $\triangle BPE \backsim \triangle BAF$. So we get that $AF = 2PE$, i.e. $m = 2$, then we obtain $A(2, 5)$.

Hence, when $a = 1$, the analytic expression of S_2 is $y = -x^2 + 1$, $m = 2$.

(2) In RtΔABF, $AB = \sqrt{(2-0)^2 + (5-1)^2} = 2\sqrt{5} < 5$, so when ΔABC is an isosceles triangle, there are only two cases as follows: (1) As shown in Figure 8.3(2), if $BC = AB = 2\sqrt{5}$, then $OC = \sqrt{BC^2 - OB^2} = \sqrt{(2\sqrt{5})^2 - 1^2} = \sqrt{19}$, so $C(\sqrt{19}, 0)$. Since the point $C(\sqrt{19}, 0)$ is on $y = -ax^2 + 1$, it is obtained that $a = \frac{1}{19}$. (2) As shown in Figure 8.3(3), if $AC = BC$, let C be $(x, 0)$. Construct AD perpendicular to the x-axis through the point A which intersects the x-axis at the point D. In RtΔOBC, $BC^2 = x^2 + 1$, and in RtΔADC, $AC^2 = (x-2)^2 + 25$. From $x^2 + 1 = (x-2)^2 + 25$, we get $x = 7$. Since the point $C(7,\ 0)$ is on $y = -ax^2 + 1$, it is obtained that $a = \frac{1}{49}$.

To sum up, it can be concluded that there are two values of a to make ΔABC be an isosceles triangle, that is,

$$a_1 = \frac{1}{19}, a_2 = \frac{1}{49}.$$

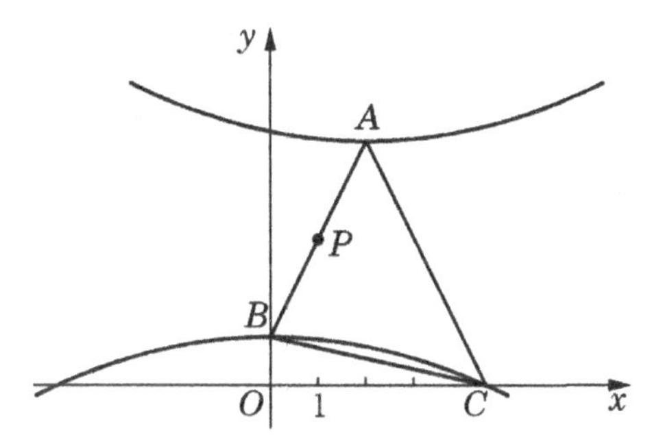

Fig. 8.3(2)

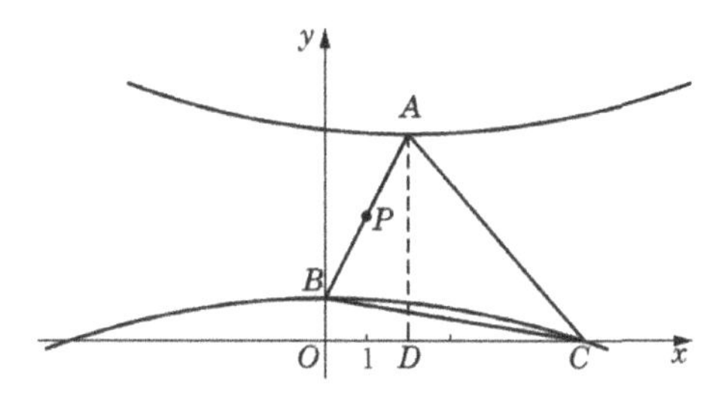

Fig. 8.3(3)

16. (1) When $x = c$, $y = 0$, that is, $ac^2 + bc + c = 0$, $c(ac + b + 1) = 0$. Since $c > 1$, it is obtained that $ac + b + 1 = 0$.

Let x_1 and x_2 be two roots of the quadratic equation $ax^2 + bx + c = 0$ $(x_1 \le x_2)$.

From $x_1 x_2 = \frac{c}{a} > 0$ and $x = c > 1$, we get that $x_1 > 0$, $x_2 > 0$. And because $y > 0$ when $0 < x < c$, we obtain that $x_1 = c$. So the axis of symmetry for the quadratic function $y = ax^2 + bx + c$ is $x = -\frac{b}{2a} \ge c$, i.e. $b \le -2ac$.

Hence, $b = -ac - 1 \le -2ac$, i.e. $ac \le 1$.

(2) Because $y > 0$ when $0 < x = 1 < c$, we get $a + b + c > 0$.

From $ac \le 1$, $a > 0$ and $c > 1$, it is obtained that $0 < a < 1$. Since

$$\begin{aligned}\frac{a}{x+2} + \frac{b}{x+1} + \frac{c}{x} &= \frac{(a+b+c)x^2 + (a+2b+3c)x + 2c}{x(x+1)(x+2)} \\ &= \frac{(a+b+c)x^2 + (a-2ac-2+3c)x + 2c}{x(x+1)(x+2)},\end{aligned}$$

from $a+b+c>0$, $0<a<1$, $c>1$, it is known that

$$a-2ac-2+3c=(1-a)(2c-1)+(c-1)>0.$$

Hence, when $x>0$,

$$\frac{(a+b+c)x^2+(a-2ac+3c-2)x+2c}{x(x+1)(x+2)}>0,$$

that is, $\frac{a}{x+2}+\frac{b}{x+1}+\frac{c}{x}>0$.

Solution 9

Quadratic Inequalities

1. From $-1 < x < 0$, it is obtained that $x + 1 > 0$, then $x + x^2 = x(x+1) < 0$, so $x < -x^2$. Since $\frac{1}{x} - x = \frac{1-x^2}{x} = \frac{(1-x)(1+x)}{x} < 0$, we get $\frac{1}{x} < x$. So the answer is D.
2. The original inequality is converted to $x^2 - 3x + \frac{9}{4} < 0$, that is, $\left(x - \frac{3}{2}\right)^2 < 0$. So the original inequality has no solution.
3. When $a < 0$, the solution of the inequality is $x < \frac{a}{8}$ or $x > -\frac{a}{7}$. When $a > 0$, the solution of the inequality is $x < -\frac{a}{7}$ or $x > \frac{a}{8}$. When $a = 0$, x is any real number not equal to 0.
4. If $x^2 + ax + 1 \leq 0$ has only one value of x, then it must be obtained that $a^2 - 4 = 0$, i.e. $a = \pm 2$. When $a = 2$, $x = -1$; when $a = -2$, $x = 1$, which satisfy the inequality $x^2 + ax + 5 > 0$. So the value of a is ± 2.
5. When $x \geq 1$, the original inequality is converted to $x^2 - 5x > 0$, then $x > 5$. When $x < 1$, the original inequality is converted to $x^2 + x - 6 > 0$, then $x < -3$. So the solution of the inequality is $x < -3$ or $x > 5$.
6. The fractional equation is converted to $x^2 + (k-5)x + (2k-2) = 0$, and $x \neq \pm 2$. Then we have that $\Delta = (k-5)^2 - 4(2k-2) = k^2 - 18k + 33 > 0$, $2^2 + 2(k-5) + 2k - 2 \neq 0$, $(-2)^2 - 2(k-5) + 2k - 2 \neq 0$. That is, $k^2 - 18k + 33 > 0$, and $k \neq 2$. So the answer is B.
7. Solving the inequality $x^2 - 2 + x > 0$ gives $x > 1$ or $x < -2$. Solving the inequality $x^2 - x - 2 < 0$ gives $-1 < x < 2$. So the solution of the original inequality is $1 < x < 2$.
8. From $\alpha < \beta < \gamma$, it can be discussed in different situations:

   (1) When $x < \alpha$, it's obtained that $x - \alpha < 0$, $x - \beta < 0$, $x - \gamma < 0$. So the original inequality holds.

(2) When $\alpha < x < \beta$, it's obtained that $x - \alpha > 0$, $x - \beta < 0$, $x - \gamma < 0$. So $(x-\alpha)(x-\beta)(x-\gamma) > 0$, the original inequality doesn't hold.

(3) When $\beta < x < \gamma$, it's obtained that $x - \alpha > 0$, $x - \beta > 0$, $x - \gamma < 0$. So the original inequality holds.

(4) When $x > \gamma$, it's obtained that $x - \alpha > x - \beta > x - \gamma > 0$. So the original inequality doesn't hold.

Hence, the solution of the original inequality is $x < \alpha$ or $\beta < x < \gamma$.

9. Because $f(x) = x^2 - 2ax + 6 = (x-a)^2 + 6 - a^2$, when $a < -2$, only $f(-2) \geq a$ is needed, then we obtain $-\frac{10}{3} \leq a < -2$. When $-2 \leq a \leq 2$, only $6 - a^2 \geq a$ is needed, then we get $-2 \leq a \leq 2$. When $a > 2$, only $f(2) \geq a$ is needed, but the inequality has no solution. Hence, the value range of a is $-\frac{10}{3} \leq a \leq 2$.
10. Because $1 < x < 2$ is equivalent to $x^2 - 3x + 2 < 0$, it is obtained that $a < 0$ and $x^2 + bx + \frac{b}{a} < 0$. Thus we get that $b = -3$, $\frac{b}{a} = 2$, i.e. $a = -\frac{3}{2}$.
11. Let $\frac{x+y}{2} = \frac{y+z}{3} = \frac{z+x}{7} = k$. Then $x = 3k$, $y = -k$, $z = 4k$. It can be concluded that for any value of k, $26k^2 + 6ak + 1 > 0$. Therefore, $\Delta = (6a)^2 - 4 \times 26 < 0$, then $-\frac{\sqrt{26}}{3} < a < \frac{\sqrt{26}}{3}$.
12. From the given condition, $b + c = \frac{1}{2}(a^2 - a) > a$, so $a > 3$. And from $a + 2(b-c) + 3 = 0$, it is obtained that $c - b = \frac{1}{2}(a+3) > 0$, so $c > b$. Adding the two formulas in the question gives $4c = a^2 + 3$. Then we get $4c - 4a = a^2 - 4a + 3 = (a-3)(a-1) > 0$, that is, $c > a$. Hence, side c is longest.

Solution 10

Distribution of Roots of Quadratic Equation

1. Solution 1 The conditions for the quadratic equation with real coefficients to have two negative roots can be expressed as $\Delta \geq 0$ and $x_1 + x_2 < 0$, $x_1 x_2 > 0$. Then from the relationship between the roots and the coefficients, we can get that

$$\begin{cases} (m-2)^2 - 16(m-5) \geq 0, \\ \dfrac{m-2}{4} > 0, \\ \dfrac{m-5}{4} > 0, \end{cases} \quad \text{that is,} \quad \begin{cases} m \geq 14 \text{ or } m \leq 6, \\ m > 2, \\ m > 5. \end{cases}$$

So the range of m is the solution set of the system of inequalities, which is $5 < m \leq 6$ or $m \geq 14$.

Solution 2 Let $f(x) = 4x^2 + (m-2)x + m - 5$, $a = 4$, $b = m - 2$, $c = m - 5$. The condition that the roots of the quadratic equation $f(x) = 0$ are all negative is equivalent to the following requirments: the axis of symmetry for the parabola $y = f(x)$ is on the left of the y-axis, the vertex is not above the x-axis, and the y-intercept of parabola is positive, that is,

$$\begin{cases} -\dfrac{b}{2a} = -\dfrac{m-2}{8} < 0, \\ \dfrac{4ac - b^2}{4a} = \dfrac{16(m-5) - (m-2)^2}{16} \leq 0, \\ f(0) = m - 5 > 0. \end{cases}$$

Solving the system of inequalities gives $5 < m \leq 6$ or $m \geq 14$.

2. In order to make the requirement of the problem hold, as long as the equation $kx^2 + (k-3)x + 1 = 0$ has at least one positive root. For this reason, first we need that $\Delta = (k-3)^2 - 4k = k^2 - 10k + 9 \geq 0$, then $k \geq 9$ or $k \leq 1$. When $\Delta \geq 0$, there are two negative roots of the equation (it can be seen from the equation that $x = 0$ is not the root) if $\frac{3-k}{k} < 0$ and $\frac{1}{k} > 0$, then $k > 3$. Hence, when $\Delta \geq 0$, the equation has at least one positive root if $k \leq 3$. Combining with the conditions of $\Delta \geq 0$, we obtain that $k \leq 1$ and $k \neq 0$.
3. (1) In order to make the original parabola and the x-axis have two intersections, we should have that $m-1 \neq 0$, and the discriminant is $\Delta = 64-4(m-1)(m-7) > 0$. That is, $m \neq 1$ and $m^2-8m-9 < 0$. So the solution is $-1 < m < 1$ or $1 < m < 9$.
   (2) In addition to the condition of (1), as long as $\frac{m-7}{m-1} > 0$ is satisfied (i.e. the x-coordinates of the two intersection points have the same sign), we get that $m > 7$ or $m < 1$. Combined with (1), it is obtained that $7 < m < 9$ or $-1 < m < 1$.
4. It can be seen from Figure 10.1 that the parabola $y = x^2 - 2mx - (m-12)$ opens upwards. To make the two intersections of the parabola and the x-axis lie on the right side of the line $x = 2$, the axis of symmetry $x = m$ should be on the right side of the line $x = 2$, the vertex should be on the x-axis or below the x-axis, and the value of the function is greater than 0 when $x = 2$. From these conditions, we can get

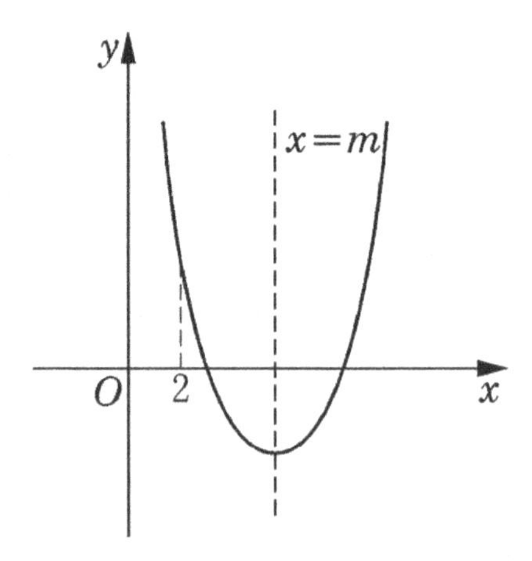

Fig. 10.1

the system of inequalities

$$\begin{cases} m > 2, \\ m^2 - 2m^2 - m + 12 \leq 0, \\ 4 - 4m - m + 12 > 0. \end{cases}$$

Solving it gives $3 \leq m < \frac{16}{5}$.

5. The parabola $y = 3x^2 - 5x + a$ opens upwards. To make the intersections of the parabola and the x-axis lie in two ribbon areas as shown in Figure 10.2, the following conditions should be met:

$$\begin{cases} 3 \times (-2)^2 - 5 \times (-2) + a > 0, \\ 3 \times 0^2 - 5 \times 0 + a < 0, \\ 3 \times 1^2 - 5 \times 1 + a < 0, \\ 3 \times 3^2 - 5 \times 3 + a > 0. \end{cases}$$

Solving this system of inequalities gives $-12 < a < 0$.

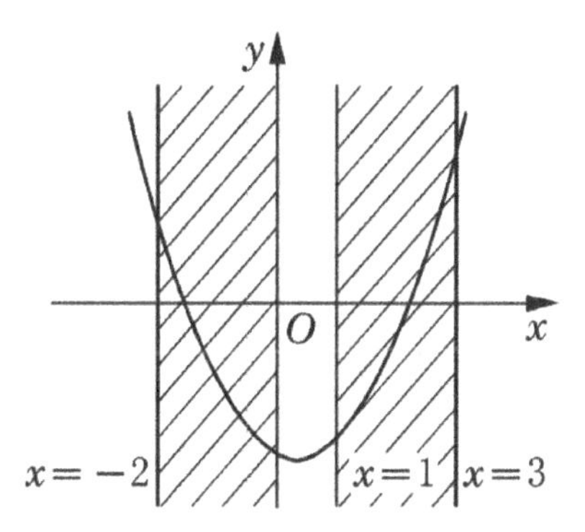

Fig. 10.2

6. The two intersections of the parabola $y = (m-1)x^2 + (3m+2)x + 2m - 1$ and the x-axis are located on both sides of the straight line $x = 1$.

   If the parabola opens upwards, as shown in Figure 10.3(1), then the intersection of the parabola and the straight line $x = 1$ is below the x-axis, we can get that

$$\begin{cases} m - 1 > 0, \\ m - 1 + 3m + 2 + 2m - 1 < 0. \end{cases}$$

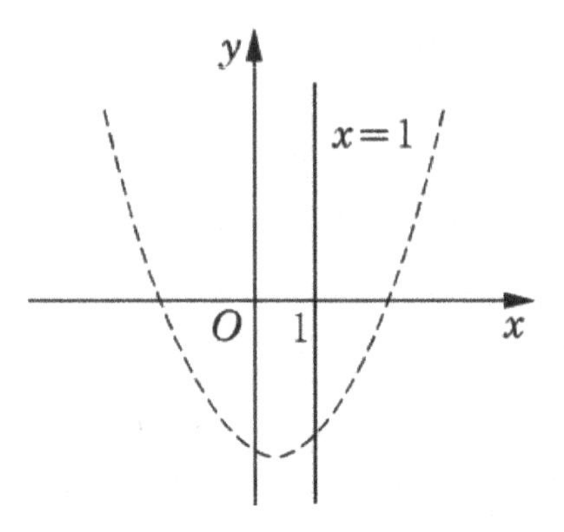

Fig. 10.3(1)

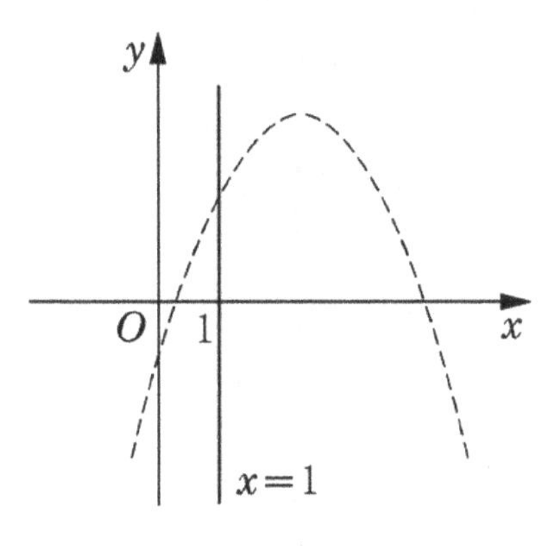

Fig. 10.3(2)

If the parabola opens downwards, as shown in Figure 10.3(2), the same principle can be used to obtain the conditions that

$$\begin{cases} m-1<0, \\ m-1+3m+2+2m-1>0. \end{cases}$$

Solving these two systems of inequalities respectively, we get that the value range of m is $0<m<1$.

7. Because two roots have oppsite signs, it is obtained that

$$\frac{2m-1}{m+3}<0. \quad (1)$$

Since the positive root is less than the absolute value of negative root, it is obtained that

$$\frac{4m}{m+3} < 0. \tag{2}$$

From (1) and (2), we can get $-3 < m < 0$. Hence, when $-3 < m < 0$, the equation has two roots of opposite signs, and the absolute value of negative root is greater.

8. $af(p) = a(ap^2 + bp + c) = a^2\left[\left(p + \frac{b}{2a}\right)^2 - \frac{b^2-4ac}{4a^2}\right]$. Because $af(p) < 0$, it is obtained that $\frac{b^2-4ac}{4a^2} > \left(p + \frac{b}{2a}\right)^2 \geq 0$, that is, $b^2 - 4ac > 0$. Suppose that the equation $f(x) = 0$ has two distinct real roots α and β, then $f(x) = a(x - \alpha)(x - \beta)$, $af(p) = a^2(p - \alpha)(p - \beta)$. From $af(p) < 0$, we get $(p - \alpha)(p - \beta) < 0$. Hence, $\alpha < p < \beta$ or $\beta < p < \alpha$, that is, one root of the equation $f(x) = 0$ is less than p, and the other root is greater than p.
9. Let $f(x) = 4x^2 - 2mx + n$. To make the two roots of the equation be both greater than 1 and less than 2, it is obtain from the graph of the function that

$$\begin{cases} \Delta \geq 0, \\ 1 < \dfrac{m}{4} < 2, \\ f(1) = 4 - 2m + n > 0, \\ f(2) = 16 - 4m + n > 0, \end{cases}$$

i.e.

$$\begin{cases} m^2 \geq 4n, & (1) \\ 4 < m < 8, & (2) \\ 4 + n > 2m, & (3) \\ 16 + n > 4m. & (4) \end{cases}$$

Since m and n are both positive integers, we can obtain that $m = 5$, 6 and 7 from the inequality (2). When $m = 5$, solving the inequality (1) gives $n \leq \frac{25}{4}$, i.e. $n \leq 6$. From the inequality (3), we get $n > 6$, which contradicts $n \leq 6$. When $m = 6$, solving the inequality (1) gives $n \leq 9$. From (3) or (4), we obtain $n > 8$, so $n = 9$. When $m = 7$, solving the inequality (1) gives $n \leq \frac{49}{4}$, i.e. $n \leq 12$. From the inequality (4), we have that $n > 12$, which contradicts $n \leq 12$.

In conclusion, $m = 6$, $n = 9$.

10. Let $t = |x|$. Then the equation $t^2 - 4t + 5 - m = 0$ must have two distinct positive roots. So we have

$$\begin{cases} \Delta = (-4)^2 - 4 \times (5 - m) > 0, \\ 5 - m > 0. \end{cases}$$

Solving the system of inequatilies gives $1 < m < 5$.

Hence, when $1 < m < 5$, the equation $x^2 - 4|x| + 5 = m$ has four distinct real roots.

11. Let $y = f(x) = x^2 - mx + 4$. Then from the give condition, we get $f(-1) \cdot f(1) \leq 0$, so $[(-1)^2 - m \cdot (-1) + 4] \cdot (1^2 - m \cdot 1 + 4) \leq 0$, that is, $(5 + m)(5 - m) \leq 0$. Solving the inequatily gives $m \leq -5$ or $m \geq 5$.

12. Draw a sketch (as shown in Figure 10.4). It can be seen from the Figure 10.4 that the condition 'two intersections of the line $y = \frac{1}{2}x + b$ and the parabola $y = x^2 + 2x - 3$ are in the first and third quadrants' can be transformed into that one root of the equation $x^2 + 2x - 3 = \frac{1}{2}x + b$ must be between -3 and 0 and the other root must be greater than 1. Then we have

$$\begin{cases} [(-3)^2 + 2 \times (-3) - 3 - \frac{1}{2} \times (-3) - b][0^2 + 2 \times 0 - 3 - \dfrac{1}{2} \times 0 - b] < 0, \\ 1^2 + 2 - 3 - \dfrac{1}{2} - b < 0. \end{cases}$$

So it is obtained that $-\frac{1}{2} < b < \frac{3}{2}$.

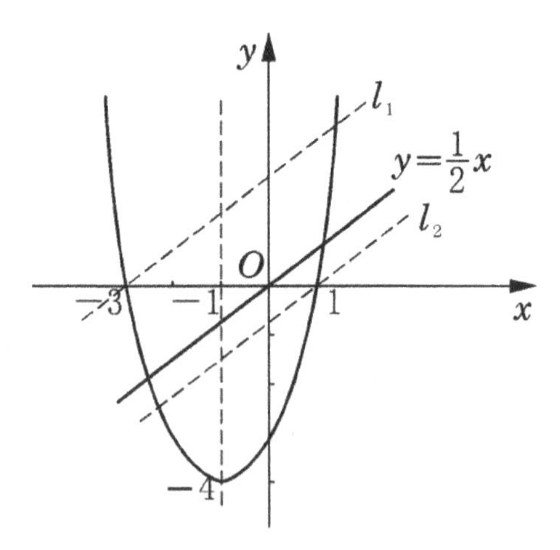

Fig. 10.4

13. Draw the graph of $y = |x^2 - 4x + 3|$ and a sketch of $y = x - b$.

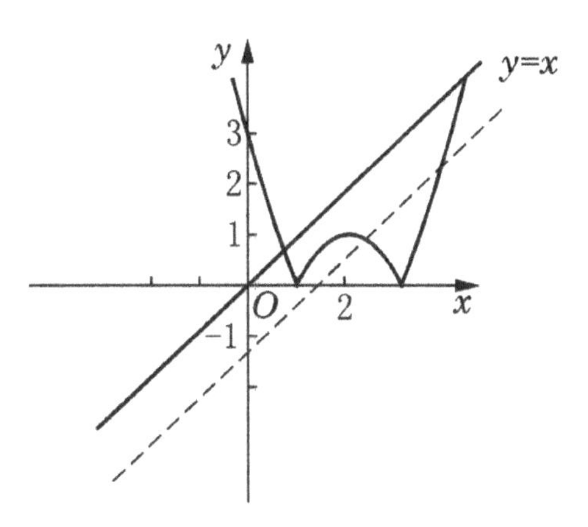

Fig. 10.5

(1) When the line $y = x - b$ passes through the point (3, 0), that is, when $b = 3$, these two curves have one intersection. When the line $y = x - b$ passes through the point (1, 0) and the line $y = x - b$ is tangent to the parabola $y = -x^2 + 4x - 3$, that is, the discriminant of the equation $-x^2 + 4x - 3 = x - b$ about x is equal to 0, we get that when $b = 1$ or $\frac{3}{4}$, these two curves have three intersections.

(2) From the idea of (1), we get that when $b < \frac{3}{4}$ or $1 < b < 3$, these two curves have two intersections.

(3) From the idea of (1), we get that when $\frac{3}{4} < b < 1$, these two curves have four intersections.

14. Let a be the coefficients of the quadratic terms for both $f_1(x)$ and $f_2(x)$. Then according to the given condition, we can get that

$$f_1(x) = a(x - \alpha_1)(x - \beta_1),$$
$$f_2(x) = a(x - \alpha_2)(x - \beta_2).$$

Therefore,

$$\begin{aligned} f_1(\alpha_2)f_1(\beta_2) &= a(\alpha_2 - \alpha_1)(\alpha_2 - \beta_1) \cdot a(\beta_2 - \alpha_1)(\beta_2 - \beta_1) \\ &= a(\alpha_1 - \alpha_2)(\alpha_1 - \beta_2) \cdot a(\beta_1 - \alpha_2)(\beta_1 - \beta_2) \\ &= f_2(\alpha_1)f_2(\beta_1). \end{aligned}$$

Solution 11

Maximum and Minimum Values of Quadratic Functions

1. Because $y = (x-2)^2 - 1$, $0 \le x \le 5$, the minimum value of the function is -1. When $x = 0$, $y = 3$; and $x = 5$, $y = 8$, so the maximum value of y is 8 when $x = 5$.
2. Because $-x^2 + 6x - 5 = -(x-3)^2 + 4 \le 4$, the equal sign holds if and only if $x = 3$. Therefore, the maximum value of $\sqrt{-x^2 + 6x - 5}$ is 2, then the minimum value of the function is 1.
3. From the given condition, we have $2y = \frac{1}{2}(x^2 - 2x - 5)$. Then
$$t = x - 2y = \frac{1}{2}(9 - (2-x)^2) \le \frac{9}{2}.$$
So the value range of t is $t \le \frac{9}{2}$.
4. From the given condition, we get that
$$\begin{cases} -\dfrac{-c}{2\left(a - \dfrac{b}{2}\right)} = 1, \\ a - \dfrac{b}{2} - c - a - \dfrac{b}{2} = -\dfrac{8}{5}b, \end{cases}$$
i.e. $\begin{cases} b + c = 2a, \\ c = \frac{3}{5}b, \end{cases}$ So $c = \frac{3}{5}b$, $a = \frac{4}{5}b$, it is obtained that $a^2 + c^2 = b^2$, then $\triangle ABC$ is a right triangle. Hence, the answer is D.
5. $y = (x-a)^2 - a^2$. When $a \ge 1$, the maximum value of y is 0, and the minimum value is $1 - 2a$. When $\frac{1}{2} < a < 1$, the maximum value of y is 0, and the minimum value is $-a^2$. When $0 \le a \le \frac{1}{2}$, the maximum

value of y is $1 - 2a$, and the minimum value is $-a^2$. When $a < 0$, the minimum value of y is 0, and the maximum value is $1 - 2a$.

6. (1) From $PF \parallel AC$ and $PE \parallel AB$, it is known that quadrilateral $AEPF$ is a parallelogram. Since $BP = x$, it is obtained that $\frac{S_{\triangle BPF}}{S_{\triangle ABC}} = \left(\frac{BP}{BC}\right)^2 = \frac{x^2}{4}$, $\frac{S_{\triangle PCE}}{S_{\triangle ABC}} = \frac{(2-x)^2}{4}$, then $2S_{\triangle PEF} = S_{\square PEAF} = S_{\triangle ABC} - S_{\triangle PBF} - S_{\triangle PCE} = -\frac{1}{2}(x-1)^2 + \frac{1}{2}$.

   Hence, $S_{\triangle PEF} = -\frac{1}{4}(x-1)^2 + \frac{1}{4}(0 < x < 2)$.

   (2) When $x = 1$, $S_{\triangle PEF}$ has the maximum value of $\frac{1}{4}$. At this time, P is the midpoint of BC.

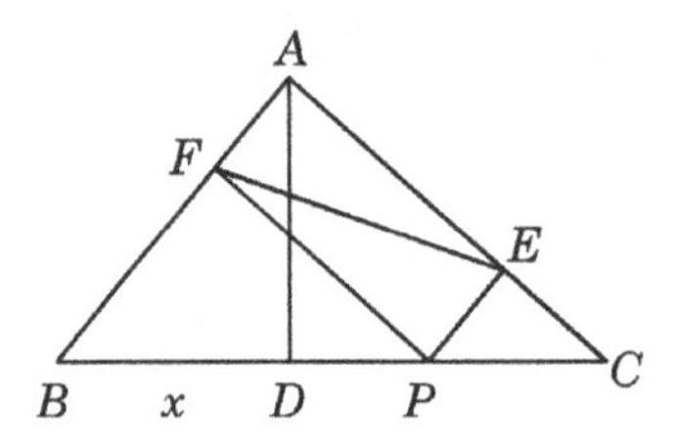

Fig. 11.1

7. From the given condition, $a + \frac{150}{b} = 40$. So the profit is

$$
\begin{aligned}
y &= x\left(a + \frac{x}{b}\right) - \left(\frac{1}{10}x^2 + 5x + 1000\right) \\
&= \left(\frac{1}{b} - \frac{1}{10}\right)x^2 + (a-5)x - 1000 \\
&= \left(\frac{40-a}{150} - \frac{1}{10}\right)x^2 + (a-5)x - 1000 \\
&= -\frac{a-25}{150}\left(x - \frac{75(a-5)}{a-25}\right)^2 + \frac{75(a-5)^2}{2(a-25)} - 1000.
\end{aligned}
$$

Then we have $\begin{cases} -\frac{a-25}{150} < 0, \\ \frac{75(a-5)}{a-25} = 150. \end{cases}$ Solving it gives $a = 45$, and then $b = -30$.

8. (1) From the given condition, $\begin{cases} -\frac{b}{2a} = \frac{3}{2}, \\ 9a + 3b + c = 3, \\ -\frac{b}{a} = 2 \cdot \frac{c}{a}, \end{cases}$ that is, $\begin{cases} b = -3a, \\ 9a + 3b + c = -3, \\ b = -2c. \end{cases}$

   Solving the simultaneous equations gives $c = -3$, $a = -2$, $b = 6$. So $y = -2x^2 + 6x - 3$.

(2) From $-2x^2 + 6x - 3 > 1$, it is obtained that $1 < x < 2$. From $-2x^2 + 6x - 3 < -3$, it is obtained that $x < 0$ or $x > 3$. Hence, when $1 < x < 2$, $y > 1$; when $x < 0$ or $x > 3$, $y < -3$.

(3) Since $y = -2\left(x - \frac{3}{2}\right)^2 + \frac{3}{2} \leq \frac{3}{2}$, we know that the maximum value of y is $\frac{3}{2}$ when $x = \frac{3}{2}$.

(4) The graph of the function is shown in the Figure 11.2.

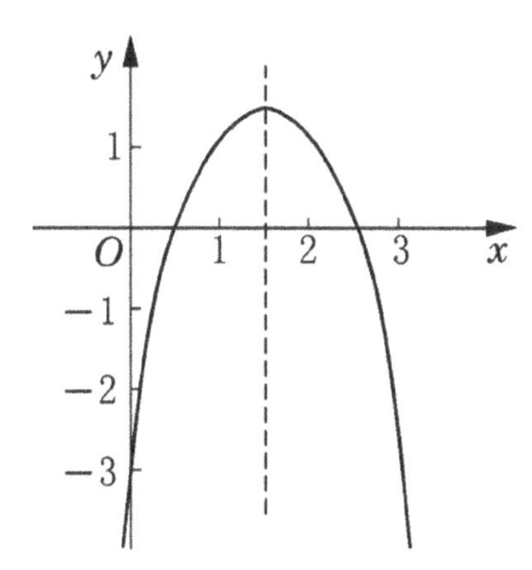

Fig. 11.2

9. $y = (m+1)(x-1)^2 - 2m - 1$. So when $m < -1$, $x = 1$, the maximum value of the function y is $-2m-1$; when $m > -1$, $x = 1$, the minimum value of the function y is $-2m - 1$.

10. Because $y = f(x) = \begin{cases} x^2 - 3x, x > 0, \\ -x^2 - x, x \leq 0, \end{cases}$ we know that when $-1 \leq x \leq 0$, the minimum of $f(x)$ is $f(-1) = f(0) = 0$; when $0 < x \leq \frac{3}{2}$, the minimum value of $f(x)$ is $f\left(\frac{3}{2}\right) = -\frac{9}{4}$. So the minimum value of $f(x)$ on the domain $-1 \leq x \leq \frac{3}{2}$ is $-\frac{9}{4}$.

11. Let $x = 0$. Then $y = b+2$, $b > -2$. Let $y = 0$. Then $x = -\frac{b+2}{k} > 0$, $k < 0$. So the coordinates of the points A and B are $A\left(-\frac{b+2}{k}, 0\right)$, $B(0, b+2)$. Then the area of $\triangle OAB$ is $S = \frac{1}{2}(b+2) \cdot \left(-\frac{b+2}{k}\right)$. From the given condition, $\frac{1}{2}(b+2) \cdot \left(-\frac{b+2}{k}\right) = -\frac{b+2}{k} + b + 2 + 3$. Solving the equation gives $k = -\frac{b^2+2b}{2(b+5)}$, $b > 0$. Thus $S = \frac{(b+2)(b+5)}{b} = \left(\sqrt{b} - \sqrt{\frac{10}{b}}\right)^2 + 7 + 2\sqrt{10}$. When $\sqrt{b} - \sqrt{\frac{10}{b}} = 0$, that is, $b = \sqrt{10}$, the minimum of S is $7 + 2\sqrt{10}$. Hence, the minimum value of the area of $\triangle OAB$ is $7 + 2\sqrt{10}$.

12. (1) Solving the system of the equations

$$\begin{cases} y = -x^2 - 3x + 4, \\ y = x^2 - 3x - 4, \end{cases}$$

gives $\begin{cases} x_1 = -2, \\ y_1 = 6, \end{cases}$ $\begin{cases} x_2 = 2, \\ y_2 = -6. \end{cases}$

So the coordinates of the points A and B are $(-2, 6)$ and $(2, -6)$ respectively.

Then we get

$$AB = \sqrt{(2+2)^2 + (-6-6)^2} = 4\sqrt{10}.$$

(2) As shown in the Figure 11.3, when PQ is parallel to the y-axis, let the coordinates of the points P and Q be

$$(t, -t^2 - 3t + 4) \quad \text{and} \quad (t, t^2 - 3t - 4), \quad -2 < t < 2.$$

Therefore,

$$PQ = (-t^2 - 3t + 4) - (t^2 - 3t - 4) = -2t^2 + 8 \leq 8.$$

The equal sign holds if and only if $t = 0$. Hence, the maximum value of the length of PQ is 8.

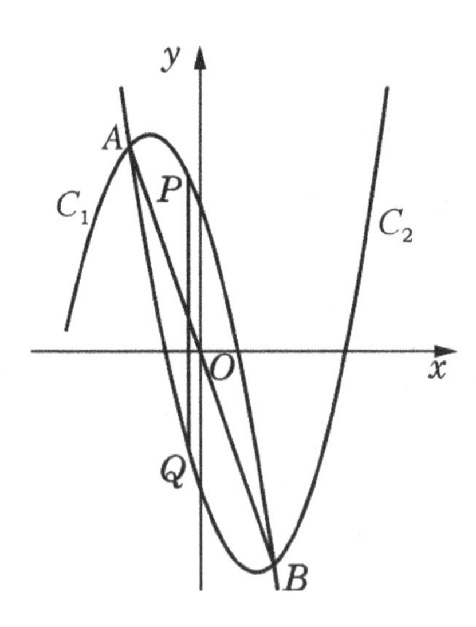

Fig. 11.3

13. Suppose that

$$\begin{aligned} f(x) &= a(1-x)(1-x-ax) - bx(b-x-bx), \\ \text{then } f(x) &= a(1-x)^2 - a^2x(1-x) + bx^2 - b^2x(1-x) \\ &= a(1-x)^2 + bx^2 - (a^2+b^2)x(1-x) \\ &= a(1-x)^2 + bx^2 - x(1-x). \end{aligned}$$

It is known from the given condition that $f(0) = a \geq 0$ and $f(1) = b \geq 0$. So $a \geq 0$, $b \geq 0$.

If $a = 0$, then $b = 1$, $f(x) = 2x^2 - x$ is not always greater than or equal to 0. Therefore, $a \neq 0$, that is, $a > 0$. For the same reason, it is obtained that $b > 0$.

When $0 < x < 1$, $f(x) = [\sqrt{a}(1-x) - \sqrt{b}x]^2 + (2\sqrt{ab}-1)x(1-x)$.

So when $\sqrt{a}(1-x) = \sqrt{b}x$, i.e. $x = \frac{\sqrt{a}}{\sqrt{a}+\sqrt{b}} \in (0,1)$,

$$f(x) = (2\sqrt{ab}-1)x(1-x) \geq 0.$$

Therefore, $2\sqrt{ab} - 1 \geq 0$, i.e. $ab \geq \frac{1}{4}$.

Conversely, when $ab \geq \frac{1}{4}$, i.e. $2\sqrt{ab}-1 \geq 0$, from $0 \leq x \leq 1$, we have

$$f(x) = [\sqrt{a}(1-x) - \sqrt{b}x]^2 + (2\sqrt{ab}-1)x(1-x) \geq 0.$$

To sum up, the minimum value of ab is $\frac{1}{4}$. Then from $ab = \frac{1}{4}$ and $a^2 + b^2 = 1$, it is obtained that

$$\begin{cases} a = \dfrac{\sqrt{6}+\sqrt{2}}{4}, \\ b = \dfrac{\sqrt{6}-\sqrt{2}}{4} \end{cases} \quad \text{or} \quad \begin{cases} a = \dfrac{\sqrt{6}-\sqrt{2}}{4}, \\ b = \dfrac{\sqrt{6}+\sqrt{2}}{4}. \end{cases}$$

Solution 12

Maximum and Minimum Values of Simple Fractional Functions

1. Let $f_1(x) = k_1 x$, $f_2(x) = \frac{k_2}{x}$. Then from the given condition, $\begin{cases} \frac{k_1}{k_2} = 2, \\ 2k_1 + \frac{4k_2}{2} = 6, \end{cases}$ that is, $\begin{cases} k_1 = 2k_2, \\ k_1 + k_2 = 3. \end{cases}$ Solving the simultaneous equations gives $k_1 = 2$, $k_2 = 1$. So $f_1(x) = 2x$, $f_2(x) = \frac{1}{x}$.
2. From $c = m - \frac{1}{m} = -\left(\frac{1}{m} - m\right) = -d$, $0 < m < 1$, it is obtained that $c = m - \frac{1}{m} < 1 - 1 = 0$, $d > 0$, thus $d > c$. From $a = 1 - \frac{1}{m} = -\left(\frac{1}{m} - 1\right) = -b$, $b = \frac{1}{m} - 1 > 1 - 1 = 0$, it is obtained that $a < 0$, $b > a$. Considering the difference between $d = \frac{1}{m} - m$ and $b = \frac{1}{m} - 1$, we get that $d - b = \left(\frac{1}{m} - m\right) - \left(\frac{1}{m} - 1\right) = 1 - m > 1 - 1 = 0$, so $d > b$, i.e. $-d < -b$, then $c < a$. Hence, we obtain that $c < a < 0 < b < d$. So the answer is C.
3. Adding up the three formulas in the question gives $\frac{1}{a} + \frac{2}{b} + \frac{2}{c} + \frac{1}{d} = \frac{1}{c} + \frac{1}{d} + \frac{1}{a}$, i.e. $\frac{2}{b} + \frac{1}{c} = 0$, so $c = -\frac{b}{2}$. By substituting it into the given formula, we get that $a = -\frac{b}{3}$, $d = -b$. To make all of a, c and d be integers, b should be a common multiple of 2 and 3. And we know that $\frac{1}{a} + \frac{1}{b} + \frac{1}{c} + \frac{1}{d} = -\frac{5}{b}$.Therefore, to make $-\frac{5}{b}$ as small as possible, just make $\frac{5}{b}$ as large as possible, so b should take the minimum common multiple of 2 and 3 which is 6. (At this time, $a = -2$, $c = -3$, $d = -6$.) So the answer is C.
4. $y = \frac{3(x-2)+11}{x-2} = 3 + \frac{11}{x-2}$. From $1 \leq x - 2 \leq 3$, it is obtained that $\frac{11}{3} \leq \frac{11}{x-2} \leq 11$, so $\frac{20}{3} \leq 3 + \frac{11}{x-2} \leq 14$, i.e. $\frac{20}{3} \leq y \leq 14$. When $x = 3$, y takes the maximum value 14; when $x = 5$, y takes the minimum value $\frac{20}{3}$.

5. From the given condition, $\begin{cases} 3 = \frac{a\cdot 0+b}{0+1}, \\ 1 = \frac{a+b}{1+1} \end{cases}$ or $\begin{cases} 1 = \frac{a\cdot 0+b}{0+1}, \\ 3 = \frac{a+b}{1+1}. \end{cases}$ Solving the simultaneous equations gives $a = -1$, $b = 3$ or $a = 5$, $b = 1$.
6. By removing the denominator, the original function can be transformed into $(2y-1)x^2 - (y+1)x + y - 2 = 0$. When $y \neq \frac{1}{2}$, this is a quadratic equation about x. Because both x and y are real numbers, $\Delta = (y+1)^2 - 4(2y-1)(y-2) \geq 0$, that is, $7y^2 - 22y + 7 \leq 0$. Solving the inequality gives $\frac{11-6\sqrt{2}}{7} \leq y \leq \frac{11+6\sqrt{2}}{7}$. When $x = -1-\sqrt{2}$, $y = \frac{11-6\sqrt{2}}{7}$; when $x = -1+\sqrt{2}$, $y = \frac{11+6\sqrt{2}}{7}$. So $y_{\min} = \frac{11-6\sqrt{2}}{7}$, $y_{\max} = \frac{11+6\sqrt{2}}{7}$.
7. From the given condition, for any real number x, $x^2 - 2bx + c^2$ is not equal to 0, that is to say, the quadratic function $f(x) = x^2 - 2bx + c^2$ never intersects the x-axis. Therefore, $\Delta = 4b^2 - 4c^2 < 0$, it is obtained that $b^2 < c^2$. So the answer is D.
8. To find the maximum value of $\frac{xy+2yz}{x^2+y^2+z^2}$, we only need to find a minimum constant k, so that $xy + 2yz \leq k(x^2+y^2+z^2)$. For the real number α satisfying $0 < \alpha < 1$, since $x^2 + \alpha y^2 \geq 2\sqrt{\alpha}xy$, $(1-\alpha)y^2 + z^2 \geq 2\sqrt{1-\alpha}yz$, it is obtained that $x^2+y^2+z^2 \geq 2\sqrt{\alpha}xy + 2\sqrt{1-\alpha}yz$. Let $2\sqrt{\alpha} = \sqrt{1-\alpha}$. Then $4\alpha = 1-\alpha$, that is, $\alpha = \frac{1}{5}$. Thus we get that $x^2+y^2+z^2 \geq \frac{2}{\sqrt{5}}(xy+2yz)$, i.e. $\frac{xy+2yz}{x^2+y^2+z^2} \leq \frac{\sqrt{5}}{2}$. When $x = 1$, $y = \sqrt{5}$ and $z = 2$, the equal sign of the above inequality holds. So the maximum value of $\frac{xy+2yz}{x^2+y^2+z^2}$ is $\frac{\sqrt{5}}{2}$.

Solution 13

Trigonometric Functions of Acute Angle

1. From $\sin\alpha > \sin\beta > 0$, we get $0 < \cos\alpha < \cos\beta$, then $\tan\alpha > \tan\beta$. So the answer is B.
2. Since $\sin^2 A + \cos^2 B = \frac{5}{4}t$ and $\cos^2 A + \sin^2 B = \frac{3}{4}t^2$, adding these two formulas gives $\frac{3}{4}t^2 + \frac{5}{4}t = 2$. Solve the equation and get $t = -\frac{8}{3}$ or $t = 1$.

   Since $\sin^2 A + \cos^2 B \geq 0$, we know that $t \geq 0$, then we discard $t = -\frac{8}{3}$. Thus $t = 1$, and the answer is C.
3. As $\frac{4}{3}\sqrt{3} > 2$, the qualified $\triangle ABC_1$ and $\triangle ABC_2$ are shown in the Figure 13.1.

   Construct the line segment BD perpendicular to AC through the point B, and the perpendicular foot is the point D.

   In the right triangle ADB,

$$BD = AB \cdot \sin\angle A = 4 \times \sin 30^\circ = 2 \quad \text{and} \quad \angle ABD = 60^\circ.$$

   In the right triangle $\triangle C_2DB$, $\sin\angle BC_2D = \frac{BD}{BC_2} = \frac{2}{\frac{4}{3}\sqrt{3}} = \frac{\sqrt{3}}{2}$, then we get that $\angle BC_2D = 600^\circ$, so $\angle DBC_2 = 30^\circ$.

   Similarly, in $\triangle BC_1D$, it is obtained that $\angle C_1BD = 30^\circ$.

   Hence, $\angle ABC_1 = \angle ABD - \angle C_1BD = 30^\circ$, $\angle ABC_2 = \angle ABD + \angle C_2BD = 90^\circ$, that is, the size of $\angle B$ is 30° or 90°.

   So the answer is D.
4. As shown in Figure 13.2, extend AD to the point E, so that $DE = AD = 2$. Join BE, then we get $\triangle BDE \cong \triangle CDA$, so $BE = AC = 3$. And since $AE = 4$, $AB = 5$, $\triangle AEB$ is a right triangle where $\angle E = 90^\circ$. Therefore, $\tan\angle BAD = \frac{BE}{AE} = \frac{3}{4}$.

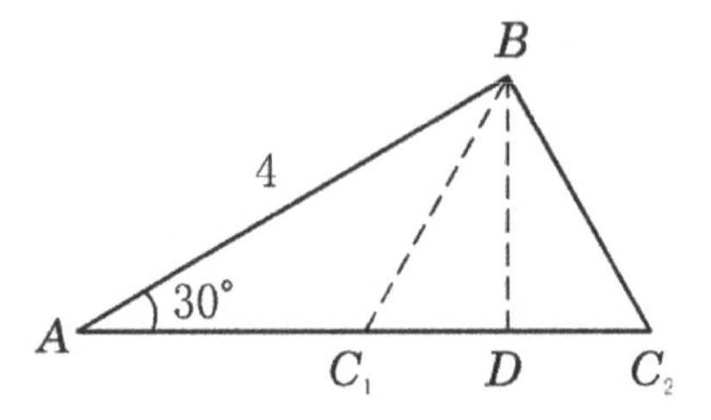

Fig. 13.1

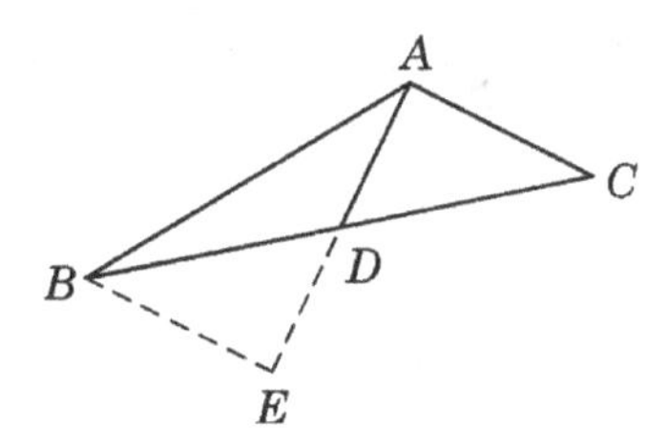

Fig. 13.2

5. From the given condition, $\tan B = \frac{AD}{BD}$, $\cos\angle DAC = \frac{AD}{AC}$. From $\tan B = \cos\angle DAC$, we get that $\frac{AD}{BD} = \frac{AD}{AC}$, i.e. $BD = AC$. In $\mathrm{Rt}\triangle ADC$, $AD^2 = AC^2 - CD^2$. Thus in $\mathrm{Rt}\triangle ADB$, $AB^2 = AD^2 + BD^2 = AC^2 - CD^2 + BD^2 = 2BD^2 - CD^2 = 2 \times 2^2 - 1^2 = 7$, that is, $AB = \sqrt{7}$. So the answer is B.

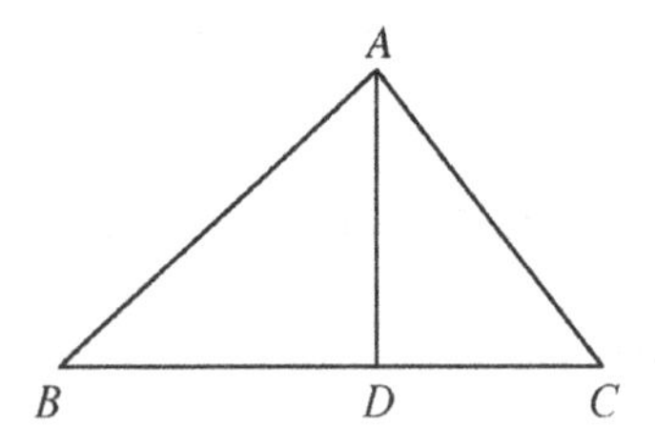

Fig. 13.3

6. From $\sin A = 1$, we get $A = 90°$. From $\cos B = \frac{\sqrt{3}}{2}$, we obtain $B = 30°$. So $C = 180° - A - B = 60°$, then $\tan C = \sqrt{3}$.

7. As shown in Figure 13.4, $S_{\triangle PAB} = \frac{1}{2} \cdot AB \cdot PH = \frac{1}{2} \cdot AP \cdot PB \cdot \sin\angle APB$, then we get $AP \cdot PB \cdot \sin\angle APB = 30$.

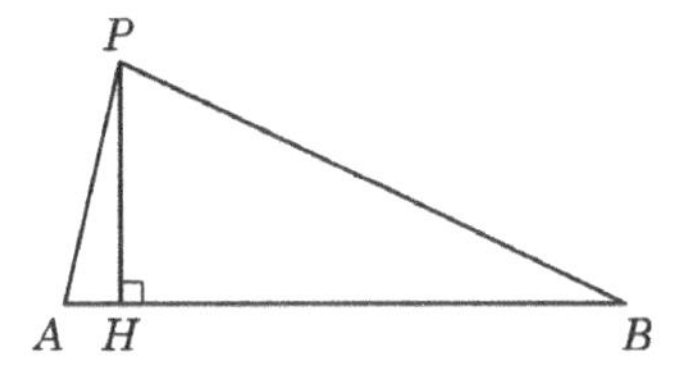

Fig. 13.4

So $AP \cdot PB \geq 30$, the equal sign holds if and only if $AP \perp BP$, at this time, $AH \cdot BH = PH^2 = 9$. And since $AH + BH = 10$, it is obtained that $AH = 1$, $BH = 9$. Hence, $AP + BP = 4\sqrt{10}$.

8. $y = 4\cos x \cdot \sin x + 2\cos x - 2\sin x - 1 = (2\sin x + 1) \cdot (2\cos x - 1)$.

From $0 \leq x \leq \frac{\pi}{2}$, we know that $1 \geq \sin x \geq 0$, $1 \geq \cos x \geq 0$, then $2\sin x + 1 > 0$. Since $y \geq 0$, it is obtained that $1 \geq \cos x \geq \frac{1}{2}$. Thus we get $0 \leq x \leq \frac{\pi}{3}$.

9. Let D be the midpoint of BC. Make $\angle CAE = 20°$ outside $\triangle ABC$, then $\angle BAE = 60°$.

Construct $CE \perp AE$ produced at the point E, and make $PF \perp AE$ produced at F, then it is easy to prove that $\triangle ACE \cong \triangle ACD$, so $CE = CD = \frac{1}{2}BC$.

And since $PF = PA\sin\angle BAE = PA\sin 60° = \frac{\sqrt{3}}{2}AP$, $PF = CE$, it is obtained that $\frac{\sqrt{3}}{2}AP = \frac{1}{2}BC$. Hence, $\frac{BC}{AP} = \sqrt{3}$.

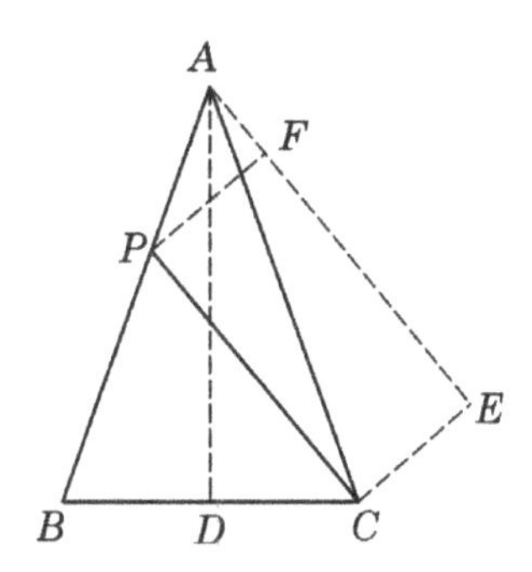

Fig. 13.5

10. From the given condition, the graph of the function opens upwards, that is, $\cos\theta > 0$, and the vertex is above the x-axis, so we have that $(-4\sin\theta)^2 - 4 \times 6\cos\theta < 0$. From $\sin^2\theta = 1 - \cos^2\theta$, we get $(2\cos\theta - 1)(\cos\theta + 2) > 0$, then $\cos\theta > \frac{1}{2}$ or $\cos\theta < -2$. Since $0 < \cos\theta < 1$, the value range of $\cos\theta$ is $\frac{1}{2} < \cos\theta < 1$.

Solution 14

Solving Right Triangles

1. From the given condition, $x^2 - 2ax + b^2 + ac - bc = 0$. Because the real number x satisfying the equation has only one value, there must be $\Delta = 0$, i.e. $b^2 + ac - bc = a^2$, then $(a-b)(c-a-b) = 0$. From $c < a + b$, we know that $a = b$, that is, this triangle is an isosceles triangle.
2. Let $AB = a$. Then $AD = \frac{a}{2}$, $BD = \frac{\sqrt{3}}{2}a$, $AE = \frac{1}{2}AD = \frac{a}{4}$. It is obtained that
$$AE : BD = \frac{a}{4} : \frac{\sqrt{3}}{2}a = 1 : 2\sqrt{3}.$$
3. As shown in Figure 14.1, join AC, it's easy to prove that $\triangle CDE \cong \triangle CBA$(SAS), $\angle ACE = 90°$. Because $CA = CE = 4\sqrt{2}$ cm, we get $AE = 8$ cm. So $AD = 5$ cm.

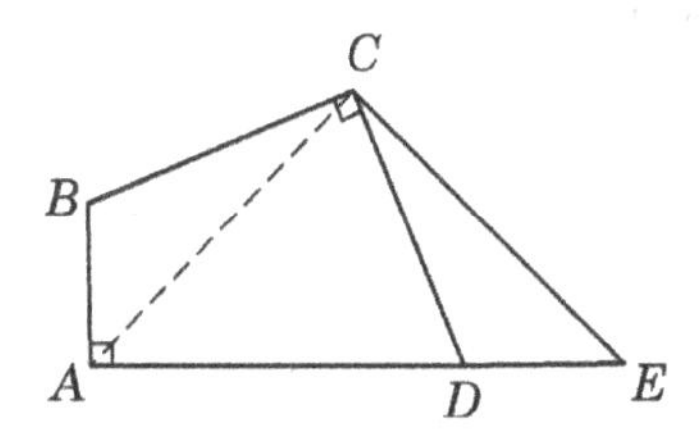

Fig. 14.1

4. It's easy to know that $\triangle AEF \backsim \triangle ABC$, then $AE = \frac{18}{5}$. So from Pythagorean Theorem, we can get $BE = \sqrt{AB^2 - AE^2} = \frac{24}{5}$.
5. From $\tan A = \sqrt{3}$, it is seen that $\angle A = 60°$, $\angle B = 30°$, $\sin A = \frac{\sqrt{3}}{2}$. From $a + c = 6$ and $a = c \sin A = \frac{\sqrt{3}}{2}c$, it is obtained that $a = 12\sqrt{3} - 18$, $c = 24 - 12\sqrt{3}$, then $b = c \sin B = 12 - 6\sqrt{3}$.

Hence, $\angle A = 60°$, $\angle B = 30°$, $a = 12\sqrt{3} - 18$, $b = 12 - 6\sqrt{3}$, $c = 24 - 12\sqrt{3}$.

6. Construct a perpendicular BD from the point B to AC which intersects AC at the point D.

   Let $BD = a$. If point D is on side AC. $CD = \sqrt{6^2 - a^2}$, $AD = a \cdot \cot 30° = 6\sqrt{3} - CD$. So we get $a = 3$, then $AB = 6$.

   If the point D is on the extended line of AC, $AD = \sqrt{3}a = 6\sqrt{3} + CD$. So we obtain $a = 6$, then $AB = 12$.

   So the length of AB is 6 or 12.

7. From $AB : BC : CD : DA = 2 : 2 : 3 : 1$, we get that $AB = BC = 2a$, $CD = 3a$, $DA = a$. In the right triangle ABC, since $\angle B = 90°$, it is obtained that $AB^2 + BC^2 = AC^2$, then $AC^2 = 8a^2$. Because $8a^2 + a^2 = 9a^2$, that is, $AC^2 + AD^2 = CD^2$, we know that $\angle DAC = 90°$. And from $\angle BAC = 45°$, we get $\angle DAB = 135°$.

8. As shown in Figure 14.2, join AC. From the given condition, $BA = BC = 6$, $\angle ABC = 60°$, so $\triangle ABC$ is a equilateral triangle. Then $AC = 6$, and $\angle BAC = 60°$. Ship P is located at the point D which is in the direction 75° north by west of Lighthouse A and $4\sqrt{2}$ nautical miles away from A, so $\angle DAB = 105°$, then $\angle DAC = \angle DAB - \angle BAC = 45°$. In $\triangle DAC$, construct a line CE perpendicular to AD through the point C and the perpendicular foot is the point E, then $CE = AE = 3\sqrt{2}$, so $DE = AD - AE = \sqrt{2}$. In the right triangle DEC,

$$CD = CE^2 + DE^2 = \sqrt{(3\sqrt{2})^2 + (\sqrt{2})^2} = 2\sqrt{5}.$$

Hence, the distance between the two ships is $2\sqrt{5}$ nautical miles.

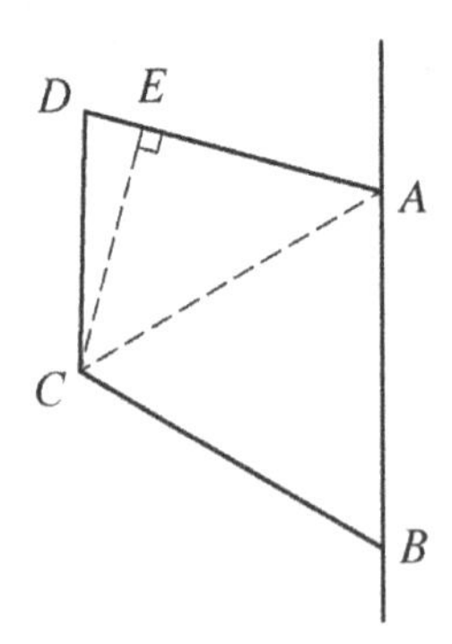

Fig. 14.2

9. Extend AD to intersect BC at the point E, then $AE = 2AB = 14$, $BE = AE \cdot \sin 60^\circ = 7\sqrt{3}$, $DE = AE - AD = 6$, $CD = DE \cdot \tan 30^\circ = 2\sqrt{3}$, $CE = 4\sqrt{3}$. Thus we get that

$$BC + CD = (BE - CE) + CD = 7\sqrt{3} - 4\sqrt{3} + 2\sqrt{3} = 5\sqrt{3}.$$

10. Since $CM = \frac{1}{2}AB$, and CM is the median on the side AB, we get $\angle ACB = 90^\circ$, then $AC + BC = 14$, $AC^2 + BC^2 = 10^2$.

   Hence, we obtain that

$$2AC \cdot BC = (AC + BC)^2 - (AC^2 + BC^2) = 4 \times 24,$$

$$S_{\triangle ABC} = \frac{1}{2}AC \cdot BC = 24.$$

11. Take a point D on the side AC, so that $BD = AD$, then we have $\angle CDB = 30^\circ$. Let $AD = a$. Then $BD = a$, $BC = \frac{1}{2}a$, and $CD = \frac{\sqrt{3}}{2}a$.

   Thus we have that

$$\left(\frac{1}{2}a\right)^2 + \left(\frac{\sqrt{3}}{2}a + a\right)^2 = 10^2,$$

i.e. $a^2 = \frac{100}{2+\sqrt{3}}$.

   Hence, $S_{\triangle ABC} = \frac{1}{2}BC \cdot CA = \frac{1}{2} \times \frac{1}{2}a \times \left(\frac{\sqrt{3}}{2}a + a\right) = \frac{1}{8}(2+\sqrt{3})a^2 = \frac{25}{2}$.

12. As shown in Figure 14.3, construct a line CD perpendicular to AB through the point C and the perpendicular foot is the point D. Take the point E as the midpoint of AB, and join CE.

   Because $\angle ACB = 90^\circ$, $S_{\triangle ABC} = \frac{1}{2}AC \cdot BC = \frac{1}{2}AB \cdot CD$, from $AC \cdot BC = \frac{1}{4}AB^2$, we obtain $CD = \frac{1}{4}AB$. And since E is the midpoint of AB, we know that $CE = \frac{1}{2}AB$. In the right triangle CDE, $CD = \frac{1}{2}CE$, so $\angle DEC = 30^\circ$.

   It is obtained that $\angle EAC = \angle ECA$ from $AE = EC$. Therefore, from $\angle DEC = 2\angle EAC$, we get $\angle EAC = 15^\circ$.

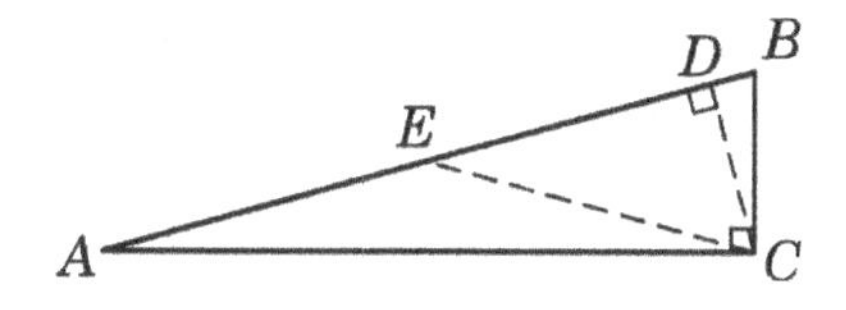

Fig. 14.3

13. From the given condition, $CE^2 + CB^2 = BE^2$, $CD^2 + AC^2 = AD^2$, D and E are the midpoints of BC and AC respectively, we get that $\left(\frac{1}{2}AC\right)^2 + CB^2 = 4^2$, $\left(\frac{1}{2}CB\right)^2 + AC^2 = 7^2$. Adding these two equalities gives $\frac{5}{4}(AC^2 + CB^2) = 65$. Then

$$AB = \sqrt{AC^2 + CB^2} = 2\sqrt{13}.$$

So the answer is C.

14. Solution 1 Construct a line CE perpendicular to AB through the point C and the perpendicular foot is the point E. Let $AE = x$, $BE = y$. Then $AE = EC$, thus we have that $x > \frac{BC}{2} = \frac{5}{2}$, and $AC = \sqrt{2}x$. Therefore, $(x+y)^2 - 3^2 = (\sqrt{2}x)^2 - 2^2$, and $x(x+y) = 5 \cdot \sqrt{(\sqrt{2}x)^2 - 2^2}$, then we get that $x + y = \sqrt{2x^2 + 5}$, and $x(x+y) = 5\sqrt{2x^2 - 4}$. Solving the equations gives $x^2 = 20$, $x^2 = \frac{5}{2}$ (discarded).

Hence, $S_{\triangle ABC} = \frac{1}{2} \cdot (3+2) \cdot \sqrt{2x^2 - 4} = 15$.

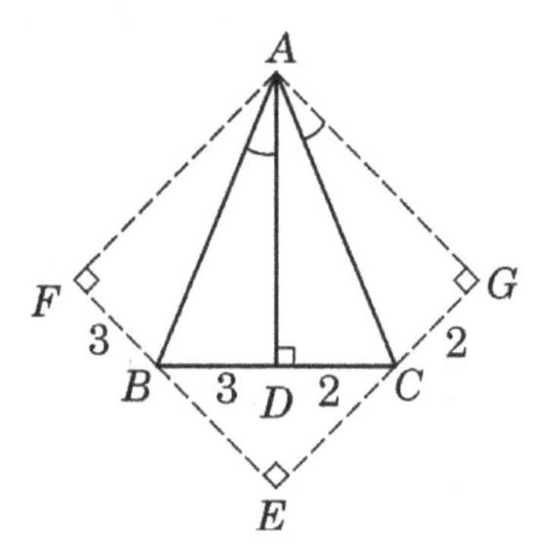

Fig. 14.4

Solution 2 Draw the symmetrical graph Rt$\triangle AFB$ for Rt$\triangle ADB$ with AB as the axis of symmetry, and draw the symmetrical graph Rt$\triangle AGC$ for Rt$\triangle ADC$ with AC as the axis of symmetry.

Since $\angle BAC = 45°$, it is obtained that $\angle FAG = 2\angle BAC = 90°$.

Extend FB to intersect the extended line of GC at the point E. It is easy to prove that the quadrilateral $AGEF$ is a square.

Let $AD = x$. Then $AF = AG = EF = EG = x$, $BE = x - 3$, $EC = x - 2$.

In Rt$\triangle BEC$, from $BE^2 + EC^2 = BC^2$, i.e.

$$(x-3)^2 + (x-2)^2 = 25,$$

we get that $x = 6$ or $x = -1$ (discarded), so $AD = 6$.

Hence, $S_{\triangle ABC} = \frac{1}{2}BC \cdot AD = \frac{1}{2} \times 5 \times 6 = 15$.

(This solution is given by Mr. Dai Chiwen in Guangzhou)

15. From the given condition, $\Delta = 4b^2 - 4(c+a)(c-a) = 0$, that is, $b^2 + a^2 = c^2$. Since all of a, b and c are positive numbers, a, b and c can be the three sides of a triangle, and this triangle is a right triangle.

Solution 15

Rotations

1. From the given condition, $CF = 3, CD = BC = 5$, $\angle CFD = 90°$, then $DF = \sqrt{DC^2 - FC^2} = 4$. And from $\angle ECF = 90°$, we know that $\angle ECF + \angle CFD = 180°$, then $EC \parallel DF$. Therefore, it is obtained that $\frac{DM}{MC} = \frac{DF}{EC} = \frac{4}{3}$. So the answer is C.
2. Because $\text{Rt}\triangle ABC \cong \text{Rt}\triangle DEC$, $\angle E = 30°$, we get $\angle B = 30°$. And since $AC = 1$, it is obtained that $AB = 2$, $BC = \sqrt{3}$. When $\triangle DMN$ is an equilateral triangle, the length of AM is $\frac{2\sqrt{3}}{3}$. So the answer is B.
3. When the board is flipped along the horizontal line for the first time, the path taken by the point A in Figure 15.18 of Chapter 15 is an arc with the center of the point C, the radius of 1 and the central angle of 120° which has the length of $\frac{2\pi}{3}$. Similarly, the path length of the second flip is $\frac{2\pi}{3}$, so the total path length is $\frac{4\pi}{3}$. Hence, the answer is D.
4. Suppose that $C'B'$ intersects AB at the point D, then it is known from the given condition that $\angle C'AD = 30°$. Since $\angle AC'B' = 90°$, $AC' = 5$, we get $C'D = \frac{5\sqrt{3}}{3}$. Therefore,
$$\begin{aligned} S_{\text{shadow part}} &= S_{\triangle AC'D} = \frac{1}{2}AC' \cdot C'D \\ &= \frac{1}{2} \times 5 \times \frac{5\sqrt{3}}{3} = \frac{25\sqrt{3}}{6}. \end{aligned}$$
5. (1) $\triangle APB$ is rotated anticlockwise through 60° about the point A to the position of $\triangle AP'C$ (obviously, P' is outside $\triangle ABC$), then $PB = P'C$, $PA = P'A$. And since $\angle PAP' = 60°$, $\triangle PAP'$ is a regular triangle, then $PA = PP'$. Hence, the three sides of $\triangle P'PC$

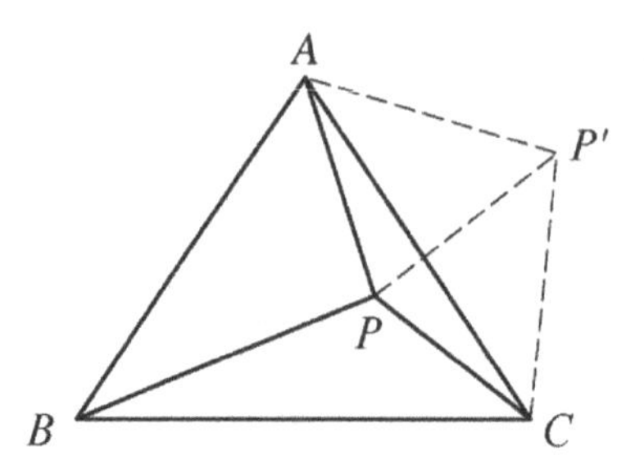

Fig. 15.1

are exactly PA, PB and PC. So a triangle can be formed with PA, PB and PC as its sides.

(2) $\angle P'PC = \angle CPA - \angle P'PA = 140° - 60° = 80°$. And from $\angle AP'C = \angle APB = 100°$, we get $\angle PP'C = \angle AP'C - \angle AP'P = 100° - 60° = 40°$, then $\angle P'CP = 60°$. That is to say, the degrees of the three angles of the triangle are $40°$, $60°$ and $80°$ respectively.

6. As shown in Figure 15.2, take A as the center of the rotation, and rotate $\triangle ABD$ anticlockwise about $\angle BAC$ to the position of $\triangle ACD'$. Because $AD = AD'$, we get $\angle ADD' = \angle AD'D$. And from $\angle AD'C = \angle ADB < \angle ADC$, we know that $\angle CD'D < \angle CDD'$. In $\triangle CDD'$, from the fact that the side opposite the greater angle is longer, we obtain that $CD < CD'$, that is, $CD < BD$.

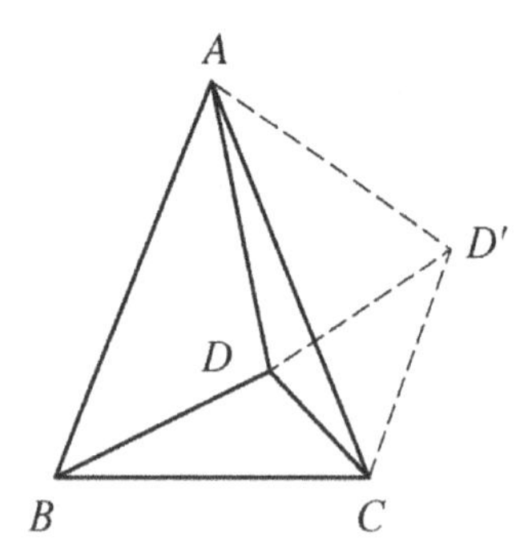

Fig. 15.2

7. The area of a polygon is usually transformed into several triangles by drawing diagonals, but the area of three triangles obtained by connecting AC and AD is not easy to find directly.

Considering $AB = AE$, $\angle B = \angle AED = 90°$, we can rotate $\triangle$ ABC to the position below $\triangle ADE$, so that the original pentagonal can be assembled into a new figure.

As shown in Figure 15.3, join AC, AD, and rotate $\triangle ABC$ about the point A to $\triangle AEF$. Since $AB = AE$, $\angle B = \angle AED = 90°$, the points D, E and F are on the same line, and $AB = CD = AE = BC + DE = 1 = DF$. And because $AF = AC$, we get that $S_{\triangle ACD} = S_{\triangle AFD} = \frac{1}{2}DF \cdot AE = \frac{1}{2}$. So the area of the pentagon $ABCDE$ is $\frac{1}{2} + \frac{1}{2} = 1$.

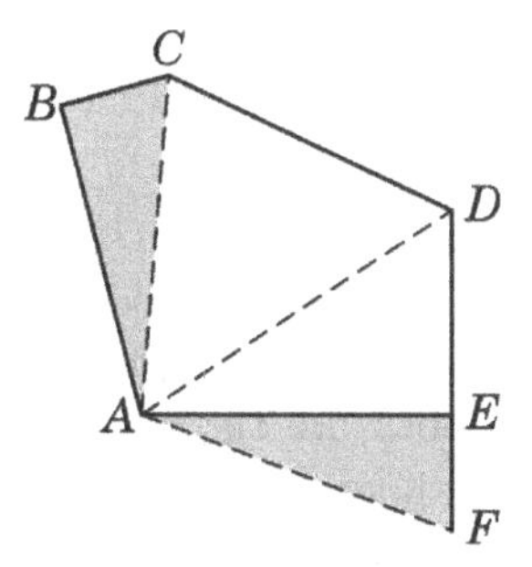

Fig. 15.3

8. Because $\angle QAP = \angle BAC$, we get that $\angle QAP + \angle PAB = \angle PAB + \angle BAC$, that is, $\angle QAB = \angle PAC$. From $AQ = AP$, $AB = AC$, we obtain $\triangle ABQ \cong \triangle ACP$. Hence, $BQ = CP$.
9. (1) Because $\angle A = \angle A'$, $AC = A'C$, $\angle ACM = 90° - \angle MCN = \angle A'CN$, it is obtained that $\triangle ACM \cong \triangle A'CN$.

   (2) In Rt$\triangle ABC$, $\angle B = 30°$, so $\angle A = 90° - 30° = 60°$.

   And since $\alpha = 30°$, we get that $\angle MCN = 30°$, then

$$\angle ACM = 90° - \angle MCN = 60°.$$

   Therefore, $\angle EMB' = \angle AMC = \angle B + \angle MCN = 60°$.

   Since $\angle B' = \angle B = 30°$, we obtain that $\triangle MEB'$ is a right triangle. And from $\angle B' = 30°$, we get that $MB' = 2ME$.
10. As shown in Figure 15.4, line segment BG is rotated anticlockwise about the point A through $90°$. Because the quadrilaterals $ACFG$ and $ABDE$ are squares, G coincides with C, B coincides with E, and BG coincides with EC, that is, $BG = EC$.

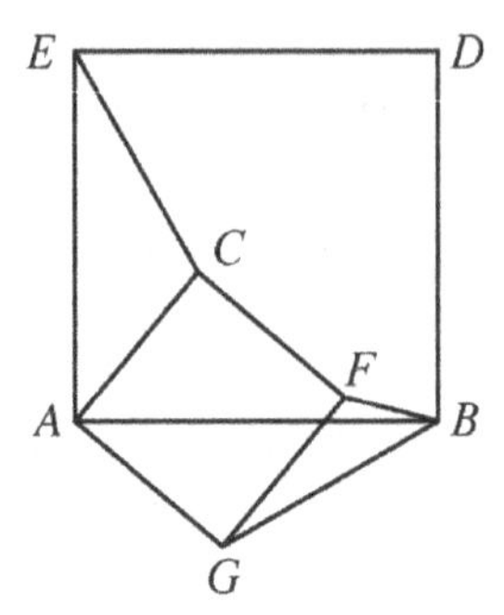

Fig. 15.4

11. (1) As shown in Figure 15.5, join AD. Since D is the midpoint of the hypotenuse BC of the isosceles right triangle, we get that $AD = BD = DC$, and $AD \perp BC$. And from $DE \perp DF$, we know that $\angle ADE = \angle CDF$ (both are the remainder of $\angle ADF$). And since $\angle EAD = \angle C = 45°$, it's seen that $\triangle DCF \cong \triangle DAE$. Thus, $\triangle DAE$ can be obtained by rotating $\triangle DCF$ counterclockwise through 90° about the point D.

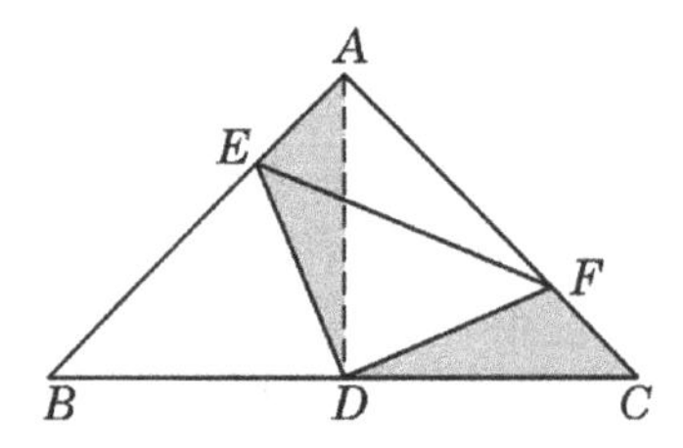

Fig. 15.5

Then $DE = DF$, $AE = CF = 5$. From $AB = AC$, we have that $AF = BE = 12$, $AB = AC = 5 + 12 = 17$.

Hence, the sum of the areas of $\triangle BDE$ and $\triangle DCF$ = the sum of the areas of $\triangle BDE$ and $\triangle DAE$ = the area of ABD = half of the area of $\triangle ABC = \frac{1}{2} \times \frac{1}{2} \times 17 \times 17 = \frac{289}{4}$ (square units).

(2) From (1), we get $EF = \sqrt{AE^2 + AF^2} = \sqrt{5^2 + 12^2} = 13$, then the height on the hypotenuse EF of the isosceles right triangle DEF is $\frac{1}{2}EF = \frac{13}{2}$. Hence, the area of $\triangle DEF = \frac{1}{2} \cdot EF \cdot \frac{1}{2}EF = \frac{13^2}{4} = \frac{169}{4}$ (square units).

12. (1) As shown in Figure 15.6, join OC and OC', which intersects PQ and NP at points D and E respectively, then we know that $\angle COC' = 45°$.

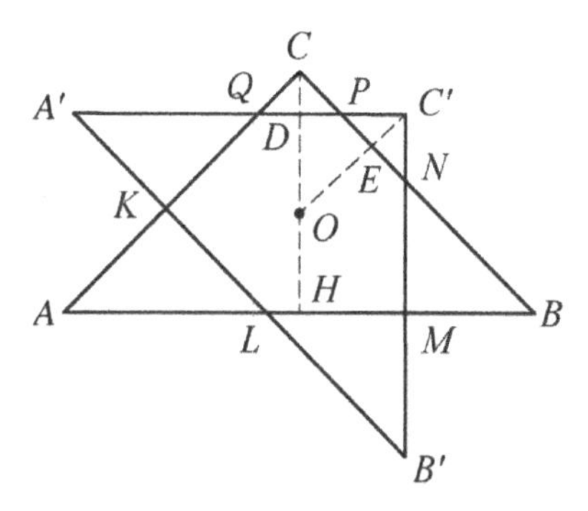

Fig. 15.6

Since the distances from the point O to AC and BC are equal to 1, OC is the bisector of $\angle ACB$. From $\angle ACB = 90°$, we get $\angle OCE = \angle OCQ = 45°$. Similarly, $\angle OC'D = \angle OC'N = 45°$. Hence, $\angle OEC = \angle ODC' = 90°$, and then $\angle CQP = \angle CPQ = \angle C'PN = \angle C'NP = 45°$. So $\triangle CPQ$ and $\triangle C'NP$ are both isosceles right triangles. It is obtained that $\angle BNM = \angle C'NP = 45°$, $\angle A'QK = \angle CQP = 45°$. From $\angle B = 45°$, $\angle A' = 45°$, we obtain that $\triangle BMN$ and $\triangle A'KQ$ are both isosceles right triangles. Then we know that $\angle B'ML = \angle BMN = 90°$, $\angle AKL = \angle A'KQ = 90°$. And $\angle B' = 45°$, $\angle A = 45°$, so $\triangle B'ML$ and $\triangle AKL$ are both isosceles right triangles too.

(2) In Rt$\triangle ODC'$ and Rt$\triangle OEC$, $OD = OE = 1$, $\angle COC' = 45°$, then we get $OC = OC' = \sqrt{2}$. Hence, $CD = C'E = \sqrt{2} - 1$.

Then

$$PQ = NP = 2(\sqrt{2} - 1) = 2\sqrt{2} - 2,$$

$$CQ = CP = C'P = C'N = 2 - \sqrt{2}.$$

Thus we know that $S_{\triangle CPQ} = \frac{1}{2} \times (2 - \sqrt{2})^2 = 3 - 2\sqrt{2}$.

Extend CO to intersect AB at the point H. Because CO is the bisector of $\angle ACB$, and $AC = BC$, we get $CH \perp AB$, then $CH = CO + OH = \sqrt{2} + 1$.

Hence,

$$AC = BC = A'C' = B'C' = \sqrt{2}(\sqrt{2}+1) = 2+\sqrt{2}.$$

So $S_{\triangle ABC} = \frac{1}{2} \times (2+\sqrt{2})^2 = 3+2\sqrt{2}$.

From $A'Q = BN = (2+\sqrt{2}) - (2\sqrt{2}-2) - (2-\sqrt{2}) = 2$, we know that

$$KQ = MN = \sqrt{2}.$$

So $S_{\triangle BMN} = \frac{1}{2} \times (\sqrt{2})^2 = 1$.

Because $AK = (2+\sqrt{2}) - (2-\sqrt{2}) - \sqrt{2} = \sqrt{2}$, we get

$$S_{\triangle AKL} = \frac{1}{2} \times (\sqrt{2})^2 = 1.$$

Hence, $S_{\text{polygon } KLMNPQ} = S_{\triangle ABC} - S_{\triangle CPQ} - S_{\triangle BMN} - S_{\triangle AKL} = 4\sqrt{2} - 2$.

Solution 16

Basic Properties of Circles

1. Let the center of the circle be O, and the radius be r. Then it is known from the given condition that $r^2 = (AD - r)^2 + BD^2$. Because $AD = 4$, $AB = 4\sqrt{3}$, we get $BD = \sqrt{AB^2 - AD^2} = 4\sqrt{2}$, that is, $r^2 = (4 - r)^2 + 32$. Solving the equation gives $r = 6$. So the answer is D.
2. Let the radius of the circle be r. As shown in Figure 16.1, construct a line segment FG perpendicular to AB through the point F which intersects AB at the point G, then $FG = AD = 4, EG = DF - AE = 5 - 3 = 2$, OG=OE-EG=R-2. Joining OF, in Rt$\triangle OGF$, from $R^2 = (R - 2)^2 + 4^2$, we get $R = 5$. Thus $2R = 10$.

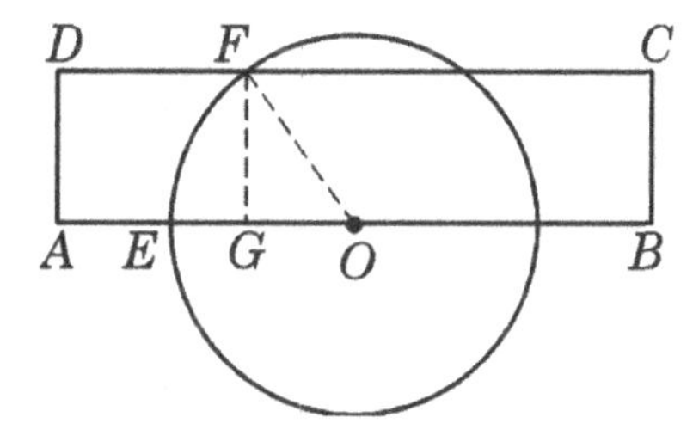

Fig. 16.1

3. Construct a line segment OH perpendicular to CD through the point F which intersects CD at the point H, then OH is the distance from the chord to the centre and $OH = \sqrt{10^2 - 8^2} = 6$. From $OH \parallel AE \parallel BF$, we get that $\frac{AE}{OH} = \frac{10+OG}{OG}$, $\frac{BF}{OH} = \frac{10-OG}{OG}$. Hence, $AE - BF = 2OH = 12$.
4. Construct a line segment DE perpendicular to BC through the point D with the perpendicular foot of the point G, which intersects the circle at the point E. Construct EF perpendicular to AB through the point

E with the perpendicular foot of F, construct GH perpendicular to AB through the point G with the perpendicular foot of H, and join AE, EB.

From the given condition and $DE \perp BC$, $BD = 5$, we obtain that $BE = BD = 5$.

Because AB is the diameter of the circle, $\angle AEB = 90°$, we get $AE^2 = AB^2 - BE^2$. From $AB = AD + DB = 9$, it is obtained that $AE = \sqrt{9^2 - 5^2} = \sqrt{56}$.

From $S_{\triangle ABE} = \frac{1}{2}AE \cdot BE = \frac{1}{2}AB \cdot EF$, it is obtained that $EF = \frac{5}{9}\sqrt{56}$, so

$$BF = \sqrt{BE^2 - EF^2} = \sqrt{5^2 - \frac{5^2 \times 56}{9^2}} = \frac{25}{9}.$$

Because $GH \perp AB$, $EF \perp AB$, $DG = GE$, we obtain that $GH \parallel EF$, and $DH = HF$, then

$$BH = HF + BF = \frac{DF}{2} + BF = \frac{DF + BF}{2} + \frac{BF}{2} = \frac{BD}{2} + \frac{BF}{2} = \frac{5 \times 7}{9}.$$

Therefore, $BG^2 = GH^2 + BH^2 = \frac{5^2 \times 7}{9}$, that is, $BG = \frac{5\sqrt{7}}{3}$.

Join AC. Because $AC \perp BC$, $DG \perp BC$, and $AB = 9$, $BD = 5$, we get $\frac{BG}{BC} = \frac{BD}{AB}$, then $BC = \frac{9}{5} \times BG = 3\sqrt{7}$.

So the answer is A.

5. Join AD. It is known from the given condition that $\triangle CDE \backsim \triangle ABE$, then $S_{\triangle CDE} : S_{\triangle ABE} = DE^2 : AE^2$. Since $\angle ADE = 90°$, we get $\frac{DE}{AE} = \cos\alpha$. Hence, $\frac{S_{\triangle CDE}}{S_{\triangle ABE}} = \cos^2\alpha$.
6. Construct the diameter AE through the point A, and join CE, then $\triangle ACE$ is a right triangle. Because $AC = b$, $AE = d$, we get that $\sin B = \sin E = \frac{b}{d}$. In Rt$\triangle BDC$,

$$CD = BC \cdot \sin B = \frac{ab}{d}.$$

7. As shown in Figure 16.2, join OD and OC, construct DE perpendicular to AB at the point E, and construct OF perpendicular to AC at the point F. Because AD is the bisector of $\angle BAC$, we have that $\angle DOB = 2\angle BAD = \angle OAC$. And since $OA = OD$, we get $\triangle AOF \cong \triangle ODE$, then $OE = AF$, $AC = 2AF = 2OE$. Let $AC = 2x$. Then $OE = AF = x$.

In Rt$\triangle ODE$, from Pythagorean Theorem, $DE = \sqrt{OD^2 - OE^2} = \sqrt{100 - x^2}$. In Rt$\triangle ADE$, $AD^2 = DE^2 + AE^2$, it is obtained that

$$(4\sqrt{15})^2 = (100 - x^2) + (10 + x)^2.$$

Solving the equation gives $x = 2$. Hence, $AC = 2x = 4$.

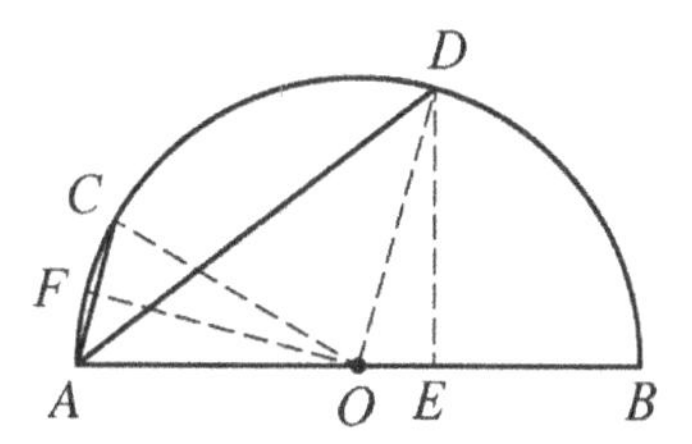

Fig. 16.2

8. From the given condition, $BC \perp AD$. Because AD is the diameter of the circle, $BC = 2\sqrt{5}$, we get $BE = EC = \frac{1}{2}BC = \sqrt{5}$.

   Since $BD \parallel FC$, and E is the midpoint of BC, we have $EF = ED$. And since F is the midpoint of OE, it is obtained that $ED = EF = FO = \frac{1}{3}OD$.

   Because $BC \perp AD$, $\angle ABD = 90°$, we know that $CD = BD$. And since $\triangle ABE \backsim \triangle BDE$, we get that $BE^2 = AE \cdot ED$.

   From $BE = \sqrt{5}$, $AE = OA + OE = OD + 2DE = 5DE$, it is obtained that $5 = 5DE^2$, $DE = 1$. Hence, $CD = BD = \sqrt{BE^2 + DE^2} = \sqrt{5+1} = \sqrt{6}$.
9. Let $C = BC = a$. Then $AB = \sqrt{2}a$. It's known from the given condition that $S_{\triangle ABC} = S_{\text{sector } ADF}$, so $\frac{\pi}{8}AD^2 = \frac{1}{2}a^2$, i.e. $AD = \frac{2a}{\sqrt{\pi}}$. Since

$$AD : DB = AD : (AB - DA),$$

   we get that $AD : DB = (\sqrt{2\pi} + 2) : (\pi - 2)$.
10. As shown in Figure 16.3, let $\odot O$ be the minimum circle that meets the requirements. Take M as the midpoint of BC. It is known from symmetry that $EM \perp BC$, and the center O is on EM. Since both OE and OB are the radius of $\odot O$, we obtain $OE = OB = R$. From the given condition, the height of the regular triangle ADE is $\frac{\sqrt{3}}{2} \times \sqrt{3} = \frac{3}{2}$, then $EM = \frac{3}{2} + 1 = \frac{5}{2}$. Therefore, $OM = EM - OE = \frac{5}{2} - R$. And $BM = \frac{\sqrt{3}}{2}$,

from Pythagorean Theorem, we get

$$\left(\frac{5}{2}-R\right)^2+\left(\frac{\sqrt{3}}{2}\right)^2=R^2.$$

Solving the equation gives $R=\frac{7}{5}$. Hence, the minimum value of R is $\frac{7}{5}$.

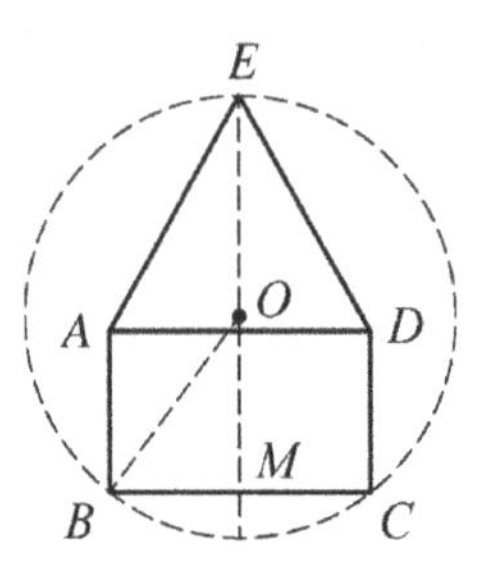

Fig. 16.3

11. Construct a line BE perpendicular to BP which intersects PA at the point E, and construct a line CF perpendicular to PC which intersects PD at the point F. Since $BP \perp CP$, then $BE \parallel CP$, $CF \parallel BP$, it is obtained that

$$\tan\angle APB\cdot\tan\angle CPD=\frac{BE}{BP}\cdot\frac{CF}{CP}=\frac{BE}{CP}\cdot\frac{CF}{BP}=\frac{AB}{AC}\cdot\frac{CD}{BD}=\frac{1}{4}.$$

12. Draw a symmetry point D' of the point D about AB, D' is on the circle O, then $CP+PD=PC+PD'$. Extend CO to intersect $\odot O$ at the point E, then $\angle D'OE=60°$. Thus we get that $\angle ECD'=30°$, $\angle CD'E=90°$. Therefore, $CD'=2\cdot\frac{\sqrt{3}}{2}=\sqrt{3}$. So the minimum value of $PC+PD$ is $\sqrt{3}$.

Solution 17

Positional Relation Between Line and Circle

1. D.
2. As shown in Figure 17.1, join OD and OA, then $OD \perp BC$. Construct AE perpendicular to OD through the point A which intersects OD at the point E, then we have that $OA = OD = R$, $OA^2 = AE^2 + OE^2$, $AE = BD = b$, $DE = AB = a$, $OE = R - a$. Therefore, $R^2 = b^2 + (R - a)^2$, that is, $R = \frac{a^2+b^2}{2a}$.

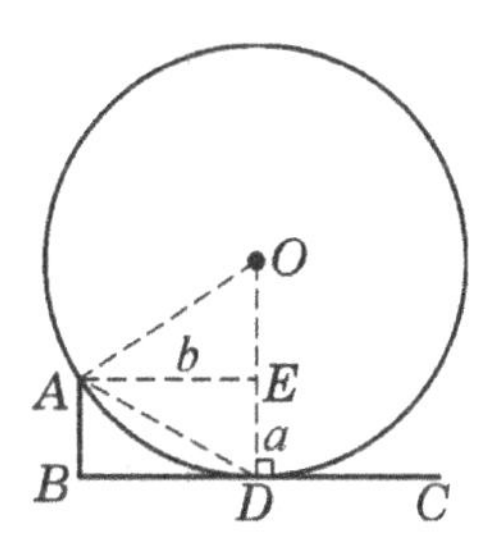

Fig. 17.1

3. Join OA and OC. From the properties of a tangent, $OC \perp AB$. From the vertical diameter theorem, $AC = BC = 4\,\text{cm}$. In Rt$\triangle OAC$, $AC^2 = OA^2 - OC^2$. So the area of the shaded part is

$$S = \pi \cdot OA^2 - \pi \cdot OC^2 = \pi(OA^2 - OC^2) = \pi \cdot AC^2 = 16\pi(\text{cm}^2).$$

4. It is known that $\angle CED = 50°$, $\overset{\frown}{CD} = 50°$, then the degree of the major arc $\overset{\frown}{AD}$ is $360° - 150° = 210°$, and $\angle ACD = 105°$. So the answer is D.

5. As shown in Figure 17.2, join AK and EK. Let the intersection of AK and $\odot O$ be H, then AH is the distance between the point A and $\odot K$. Since $AK = \sqrt{EK^2 + AE^2} = 10\,\text{cm}$, it is obtained that $AH = 4\,\text{cm}$. So the answer is A.

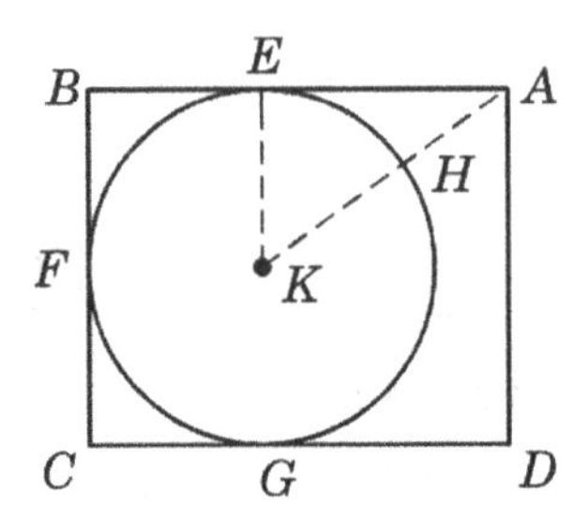

Fig. 17.2

6. As shown in Figure 17.3, join BP and CP, and construct PD perpendicular to BC through the point P which intersects BC at the point D, then it is easy to know that $\triangle BEP \backsim \triangle CDP$, $\triangle PDB \backsim \triangle PFC$. Then we have $\frac{PD}{6} = \frac{4}{PD}$, that is, $PD = 2\sqrt{6}\,\text{cm}$.

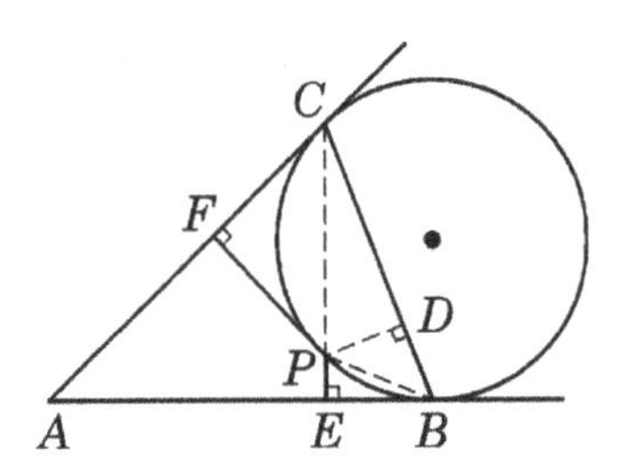

Fig. 17.3

7. B.
8. As shown in Figure 17.4, extend BC to the point G so that BC intersects $\odot O$ at the point G, draw the diameter EF of $\odot O$ through the point C which intersects $\odot O$ at the points E and F, then we get $EF \parallel AB$. Let CA intersect $\odot O$ at the point M, and connect MG which intersects EF at the point P. Then $\angle ECM = \angle ECG = 60°$. Because EF is the diameter, it is obtained from the axial symmetry that $EG = EM$, $GM \perp CE$. Hence, $\angle CGM = 30°$, $\angle MON = 60°$. So the answer is E.

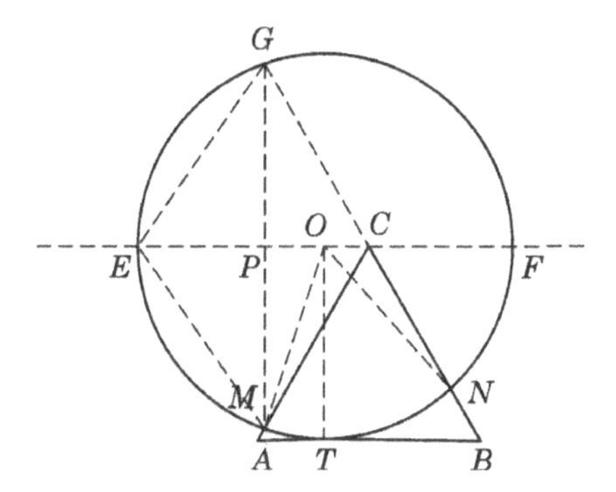

Fig. 17.4

9. Consider the position of the point E in different situations: the point E can be on the extended line of OA; the point E can be on the line segment OA (excluding point O); the point E can be on the extended line of OB; but the point E cannot be on the line segment OB (excluding point O). Hence, there are 3 points E in total.
10. Join CD and BE, and let $CD = a$, $BE = b$. Then $CD \perp AB$, $BE \perp AC$. And since $\angle A = 30°$, we get that $AC = 2a$, $AD = \sqrt{3}a$, $AB = 2b$, $AE = \sqrt{3}b$. Therefore, we know that

$$S_1 = S_{\triangle ADE} = \frac{1}{2} \cdot AD \cdot AE \cdot \sin \angle A = \frac{3}{4}ab,$$

$$S_2 = S_{BCED} = S_{\triangle ABC} - S_{\triangle ADE} = \frac{1}{2}AC \cdot BE - S_{\triangle ADE}$$
$$= \frac{1}{2} \cdot 2a \cdot b - \frac{3}{4}ab = \frac{1}{4}ab.$$

Hence, $S_1 : S_2 = \frac{3}{4}ab : \frac{1}{4}ab = 3 : 1$.
11. Join OT and AT, construct AM perpendicular to DC which intersects DC at the point M, then $OT \parallel AM \parallel BC$. And if A is the midpoint of OB, then M is the midpoint of TC, so we get that $\triangle ATM \cong \triangle ACM$, $\angle ACB = \angle CAM = \angle TAM = \angle OTA = \angle TAO$. Hence, $\angle ACB = \frac{1}{3}\angle CAO$.
12. As shown in Figure 17.5, quadrilateral $ABCD$ is inscribed in the circle O, the diagonals AC and BD intersect at the point E, $AC \perp BD$, and $CF = DF$. Join EF which intersects AB at the point G, construct line segment OH perpendicular to AB which intersects AB at the point H, and join OF.

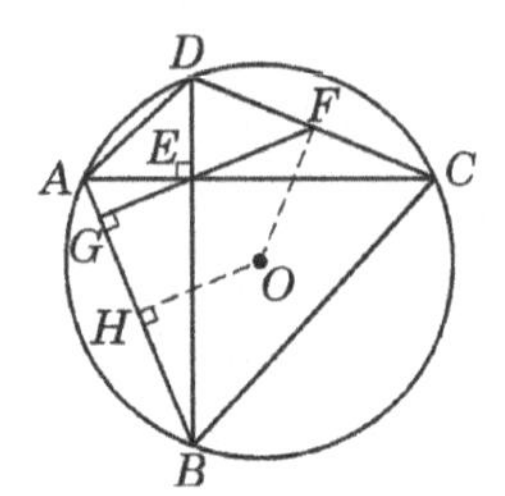

Fig. 17.5

(1) From $AC \perp BD$, $CF = DF$, it is obtained that $EF = CF$, then $\angle CEF = \angle FCE$. Since $\angle ABE = \angle FCE$, $\angle AEG = \angle FEC$, we get $\angle ABE = \angle AEG$. And since $\angle BAE + \angle ABE = \angle GAE + \angle AEG = 90°$, we obtain that $FG \perp AB$.

(2) It is known from (1) that F is the midpoint of DC, and $FG \perp AB$. Similarly, if H is the midpoint of AB, join HE which intersects DC at the point I, then $HI \perp DC$. Because $OH \perp AB$, $OF \perp DC$, $OHEF$ is a parallelogram. Hence,

$$OH = EF = CF = \frac{1}{2}CD.$$

Solution 18

Positional Relation of Two Circles

1. The common chord is the diameter of the small circle, so the answer is D.
2. Let the radius of the large semicircle be R, and the radius of the small semicircle be r. Then $2R = n \cdot 2r$, $R = nr$. Hence, $l_1 = \pi R$, $l_2 = n \cdot \pi r = \pi R$. So the answer is A.
3. Construct $O_1C \perp MA$, $O_2D \perp AN$, and the points C and D are the perpendicular feet. From $O_1C \parallel PA \parallel O_2D$, it is obtained that $CA = AD$. Hence, $AM = AN$.
4. Join QP, then $\angle APQ = \angle CDQ$, $\angle BPQ = \angle DCQ$. Hence, $\angle CQD + \angle APB = \angle CQD + \angle CDQ + \angle DCQ = 180°$.
5. Because the original equation can be factorized into $(x-1)(x-2)(x-3) = 0$, we get that $x_1 = 1$, $x_2 = 2$, $x_3 = 3$, then r_1, r_2 and $(r_1 - r_2)$ are 3, 2, 1 or 3, 1, 2 respectively. Since $d = 5$, the two circles touch externally or one circle lies completely outside the other. So the answer is E.
6. As shown in Figure 18.1, make diameter CF of $\odot O_1$, extend CF to the point G, and join AB and AF. Then $\angle AFC = \angle ABC = \angle D$, $\angle ACF = \angle GCD$. So we get $\triangle ACF \backsim \triangle GCD$, then $\angle CGD = \angle CAF = 90°$. Therefore, $CG = \frac{2S_{\triangle CDE}}{DE} = 16$. From $\frac{CF}{AC} = \frac{CD}{CG}$, we obtain $\frac{2O_1C}{8} = \frac{20}{16}$, that is, $O_1C = 5$. Hence, the radius of $\odot O_1$ is 5. So the answer is A.

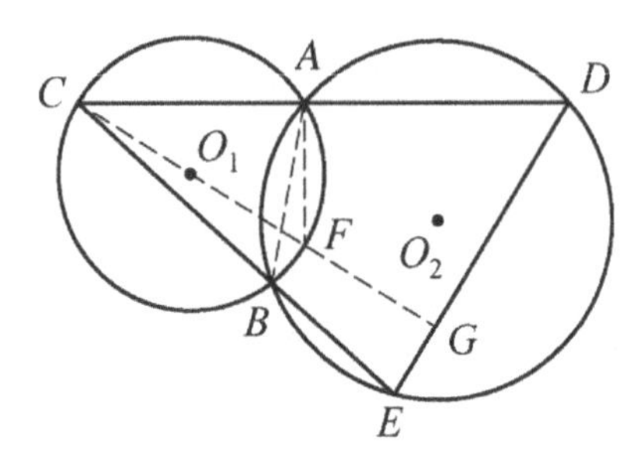

Fig. 18.1

7. Join CD, then $CD \perp AB$. Let $CD = h$, then we have

$$S = ah - \frac{\pi}{4}h^2 = -\frac{\pi}{4}\left(h - \frac{2a}{\pi}\right)^2 + \frac{a^2}{\pi}.$$

   Since $0 < h \leq a$, $0 < \frac{2a}{\pi} < a$, it is obtained that the maximum value of S is $\frac{a^2}{\pi}$ when $h = \frac{2a}{\pi}$.

8. Connecting AE and AF, then we have that $\angle AEC = 90°$, $\cos\angle ACE = \frac{CE}{AC} = \frac{1}{2}$, so $\angle ACE = 60°$. It is easy to prove that $\triangle ACD \backsim \triangle AEF$. Since AM and AN are the corresponding medians of $\triangle ACD$ and $\triangle AEF$ respectively, $\triangle ACM \backsim \triangle AEN$. Thus we get $\angle CAM = \angle EAN$, then $\angle CAE = \angle MAN$. Hence, $\triangle ACE \backsim \triangle AMN$, $\angle AMN = \angle ACE = 60°$.

9. As shown in Figure 18.2, the secant parallel to OO' which is the line of centers is CAB, and the secant not parallel to OO' is EAD. Construct the perpendicular lines of DE through the points O and O', and the perpendicular feet are the points M and N respectively. Make the perpendicular line of OM from the point O' which intersects OM at the point H, then $O'NMH$ is a rectangle.

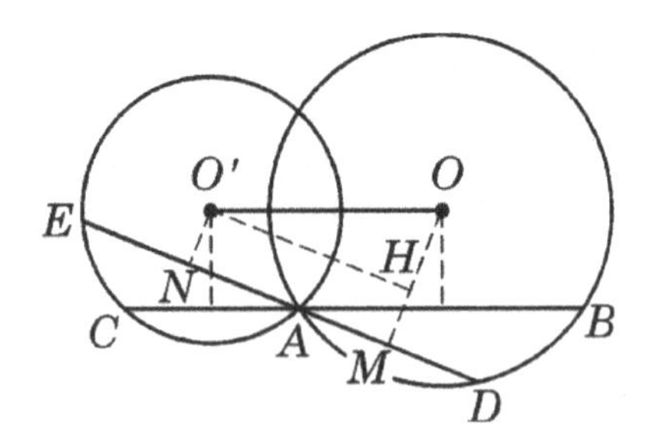

Fig. 18.2

Since M is the midpoint of AD, N is the midpoint of AE, we get that $O'H = MN = \frac{1}{2}\ DE$. In the right triangle $O'HO$, $O'H < OO'$, and $OO' = \frac{1}{2}BC$, so $DE < BC$. That is, the proposition holds.

10. As shown in Figure 18.3, extend AP to the point Q where AP and $\odot O_2$ intersect, and join AH, BD, QC, QH and BQ.

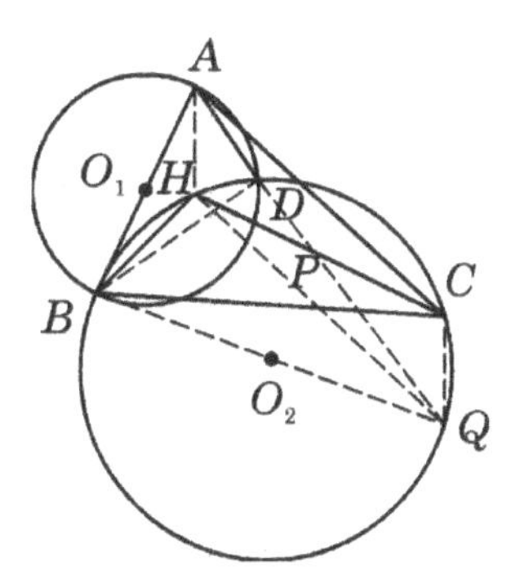

Fig. 18.3

Because AB is the diameter, $\angle ADB = \angle BDQ = 90°$, then BQ is the diameter of $\odot O_2$. Hence, $CQ \perp BC$, $BH \perp HQ$.

Since H is the orthocenter of $\triangle ABC$, $AH \perp BC$, $BH \perp AC$. It is obtained that $AH \parallel CQ$, $AC \parallel HQ$, the quadrilateral $ACHQ$ is a parallelogram.

So the point P is the midpoint of CH.

11. Let $AM = k_1 AB$, $CN = k_2 AC$, and $CL = k_3 BC$, then we need to prove that $k_1 = k_2 = k_3$. Since $\triangle ABD \backsim \triangle CAD \backsim \triangle CBA$, AM, CN and CL is a group of corresponding line segments, we get $\frac{AB}{AM} = \frac{AC}{CN} = \frac{BC}{CL}$, that is, $k_1 = k_2 = k_3$. Hence,

$$AM \cdot AB + CN \cdot AC = k_1 AB^2 + k_2 AC^2 = k_3 BC^2 = CL \cdot BC.$$

Solution 19

Power of A Point Theorem

1. Join BE. Because AB is the diameter, $BE \perp AD$. Thus we get that C and E are on the circle with the diameter of BD, then $AB \cdot AC = AD \cdot AE$. So the answer is A.
2. Extend PO to the point F where PO and $\odot O$ intersect, and join OA. Let the radius of $\odot O$ be r. From the secants and tangents theorem, we get $PA^2 = PC \cdot PF$, that is, $r = \frac{15}{2}$ cm. Since $CE \parallel OA$, it is obtained that $\frac{CE}{OA} = \frac{PC}{PO}$. Hence, $CE = 3$ cm.
3. Since CM is the bisector of $\angle C$, we get $\frac{BC}{BM} = \frac{AC}{AM}$. From Segments of Secants Theorem, $BM \cdot BA = BN \cdot BC$. Thus we have $\frac{AB}{BN} = \frac{AC}{AM}$. Because $AC = \frac{1}{2}AB$, it is obtained that $BN = 2AM$.
4. From Segments of Secants and Tangents Theorem, we get that $BM^2 = BE \cdot BF$, $CN^2 = CF \cdot CE$. Since $BE = EF = CF$, we know that $BM = CN$. Because $AM = AN$, $AB = AC$, that is, $\angle B = \angle C$.
5. From Segments of Chords Theorem, we get

$$OH = \frac{1}{2}\sqrt{(AB + CD)(AB - CD)}.$$

Let $AB = 10a + b$, $CD = 10b + a$. Then $AB^2 - CD^2 = [11(a+b)] \cdot [9(a-b)] = 11 \times 9(a^2 - b^2)$.

Since both a and b are numbers from 0 to 9, $a > b$, and OH is a rational number, we know that $a^2 - b^2$ must be a multiple of 11, then $a^2 - b^2 = 11 \cdot k^2$. Thus we get that $a = 6$, $b = 5$. Hence, $AB = 65$.
6. It is easy to know that $\frac{PB}{PC} = \frac{AB}{DC} = \frac{1}{9}$, $\frac{PB}{PA} = \frac{BC}{AD} = \frac{9}{8}$. Let $PB = 9x$. Then $PC = 81x$, $PA = 8x$. So we get that $PD = \frac{PA \cdot PC}{PB} = \frac{8x \cdot 81x}{9x} = 72x$. Hence,

$$S_{\triangle PAB} : S_{\triangle PBC} : S_{\triangle PCD} : S_{\triangle PDA} = 8 : 81 : 648 : 64.$$

7. From Segments of Secants and Tangents Theorem, we get that $AM^2 = BM^2 = MC \cdot MD$, i.e. $\frac{AM}{MD} = \frac{MC}{AM}$. And since $\angle AMC = \angle DMA$, it is obtained that $\triangle AMC \backsim \triangle DMA$. Hence, $\angle CAM = \angle D$. Since $\angle D = \angle CEF$, we obtain that $\angle CAM = \angle CEF$, then $EF \parallel AB$.
8. Joining OB and OF, then we have $BO \perp AB$. In the right triangle AOB, it is obtained from the projection theorem that $AB^2 = AC \cdot AO$. From the secants and tangents theorem, we get $AB^2 = AE \cdot AF$. Therefore, $AC \cdot AO = AE \cdot AF$, then the four points C, O, F and E are concyclic. From the intersecting chords theorem, we get that $CG \cdot GF = OG \cdot GE$, so

$$GE = \frac{1 \times 4}{\frac{7}{5}} = \frac{20}{7},$$

$$OE = OG + GE = \frac{7}{5} + \frac{20}{7} = \frac{149}{35}.$$

That is, the radius of the circle O is $\frac{149}{35}$.

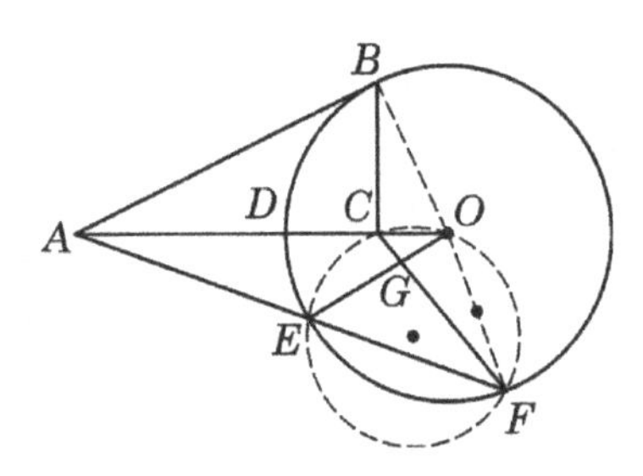

Fig. 19.1

9. From Segments of Secants Theorem, we have that $AH \cdot AJ = AG \cdot AF$, that is, $AH \cdot (AH + 7) = 2 \times (2 + 13) = 30$, so $AH = 3$. Hence, $BJ = AC - AJ = (2 + 13 + 1) - (3 + 7) = 6$, $BH = 6 + 7 = 13$. And from $BJ \cdot BH = BD \cdot BE$, $CF \cdot CG = CE \cdot CD$, we get that

$$\begin{cases} 6 \times 13 = BD \cdot (BD + DE), \\ 1 \times 14 = CE \cdot (CE + DE), \\ BD + CE + DE = 16. \end{cases}$$

Solving the system of equations gives $DE = 2\sqrt{22}$.

10. From $AD \parallel BC$, it is obtained that $\triangle AED \backsim \triangle CEF$, then $DE : EF = AE : EC$. Since $AP \parallel DC$, we get that $\triangle AEP \backsim \triangle CED$, then $AE : EC = EP : DE$. Hence, $DE : EF = EP : DE$, that is, $DE^2 = EF \cdot EP$. EG is the tangent line of the circle passing through the points B, F and P, and EFP is the secant of the circle, then $EG^2 = EF \cdot EP$. Therefore, $DE^2 = EG^2$, i.e. $DE = EG$.

Solution 20

Four Concyclic Points

1. Because the quadrilateral $ABPC$ is inscribed in a circle, from Ptolemy's Theorem and $AB = AC = BC$, it is obtained that $PA \cdot BC = PB \cdot AC + PC \cdot AB$, that is, $PA = PB + PC$.
2. Join OB and OC. Because $OF \perp AB$, $OA = OB$, it is obtained that $\angle AOF = \angle BOF$, then $\angle BOF = \angle ACB$. And since $\angle BOF + \angle POB = \angle ACB + \angle PCQ = 180^\circ$, we get that $\angle POB = \angle PCQ$, then four points O, B, Q and C are concyclic. Thus we obtain that $\angle OBC = \angle Q$, then $\angle OCP = \angle Q$, so it is deduced that $\triangle POC \backsim \triangle COQ$. Therefore, $\frac{OP}{OC} = \frac{OC}{OQ}$, that is, $OC^2 = OP \cdot OQ$. Hence, $OA^2 = OP \cdot OQ$.
3. Let the length of the side of $\triangle ABC$ be $3a$. Joining DE, then we have $DE^2 = 3a^2$ in $\triangle DEC$. Therefore, $DE^2 + CE^2 = DC^2$, and it is deduced that $\angle DEC = 90^\circ$. From $\triangle ABD \cong \triangle BCE$, we get $\angle BDA = \angle CEB$. Thus the four points D, C, E and P are concyclic, then $\angle DPC = \angle CED = 90^\circ$, i.e. $CP \perp AD$.
4. As shown in Figure 20.1, we make a circle with the diameter of AB, where $AB = 1$. If the right triangles ACB and ADB are made on both sides of AB, so that $AC = a$, $BC = b$, $BD = x$, $AD = y$, then they meet the requirements in the question. From Ptolemy's Theorem, we have

$$AC \cdot BD + BC \cdot AD = AB \cdot CD.$$

Since $CD \leq AB = 1$, we get $ax + by \leq 1$.

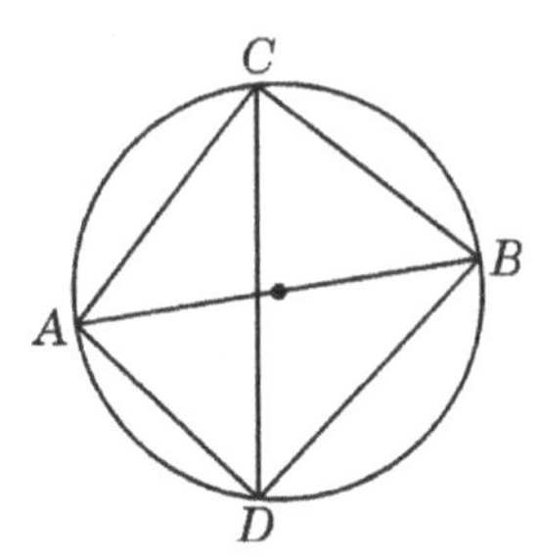

Fig. 20.1

5. It is easy to know from the given condition that $\triangle ABE \backsim \triangle AHF$, $\triangle AGE \backsim \triangle ADF$. Let the length of the side of square be a. Then $\frac{a}{AH} = \frac{AG}{a}$, i.e. $a^2 = AH \cdot AG$. Thus we know that AB is tangent to the circle passing through the points H, G and B, so the center of the circle is on BC.
6. From the fact that $ABCD$ is a parallelogram and the given conditions, it is known that $\angle ECD = \angle ACB = \angle DAF$. Since the points A, B, F and D are concyclic, we get $\angle BDC = \angle ABD = \angle AFD$. Thus it is obtained that $\triangle ECD \backsim \triangle DAF$, then $\frac{ED}{DF} = \frac{CD}{AF} = \frac{AB}{AF}$. And since $\angle EDF = \angle BDF = \angle BAF$, $\triangle EDF \backsim \triangle BAF$ is obtained. Hence, $\angle DFE = \angle AFB$.

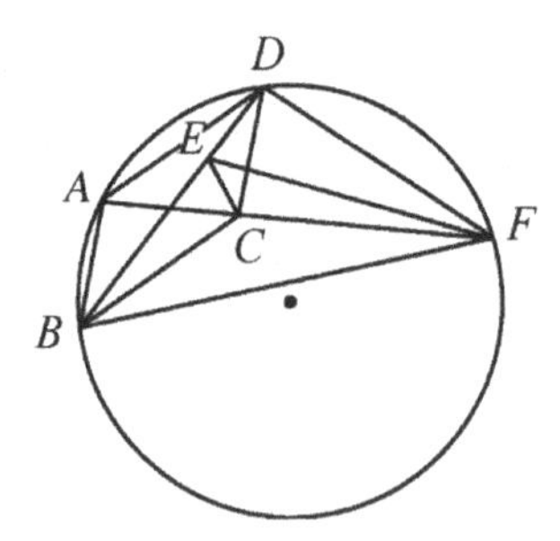

Fig. 20.2

7. From Ptolemy's Theorem, we have that $AB \cdot PC = PA \cdot BC + PB \cdot AC$, then $PC = PA + PB$. And since $\angle APC = \angle BPC = 60°$, it is obtained that $\triangle PAD \backsim \triangle PCB$, and $PA \cdot PB = PD \cdot PC$. While $PD \cdot PC = PD \cdot (PA + PB)$, it is deduced that $\frac{1}{PD} = \frac{1}{PA} + \frac{1}{PB}$.

8. Because $a^2 = b(b+c)$, and $b+c > a$, we know that $b < a$. Draw an arc with the center of A and the radius of BC, which intersects the circumscribed circle of $\triangle ABC$ at the point D, so that the minor arc $\widehat{AD}$ contains the point C. Join BD, DA and DC, then $\angle ABD = \angle BAC$, thus we have $\widehat{BD} = \widehat{AC}$. Therefore, $BD = AC = b$, $AD = BC = a$. From Ptolemy'sTheorem, $BC \cdot AD = AB \cdot CD + BD \cdot AC$, and $a^2 = b(b+c)$, we get that $CD = b = BD$, that is, $\angle DBC = \angle DAC = \angle ABC = \angle BAD$. Hence, $\angle A = 2\angle B$.
9. Join BD and AC to intersect at the point E, then

$$\angle BAE = \angle CAD, \angle ABE = \angle ACD.$$

So we get that $\triangle ABE \backsim \triangle ACD$, then $\frac{AB}{AE} = \frac{AC}{AD}$, that is,

$$AB \cdot AD = AC \cdot AE.$$

Since $\angle CBE = \angle CAB$, $\angle BCE = \angle ACB$, we obtain that $\triangle CBE \backsim \triangle CAB$. Thus $\frac{CB}{CE} = \frac{CA}{CB}$, i.e. $CB^2 = CA \cdot CE$.

Therefore, $CB^2 + AB \cdot AD = CA \cdot CE + CA \cdot AE = CA^2$.

Hence, $CA^2 - CB^2 = AB \cdot AD$.

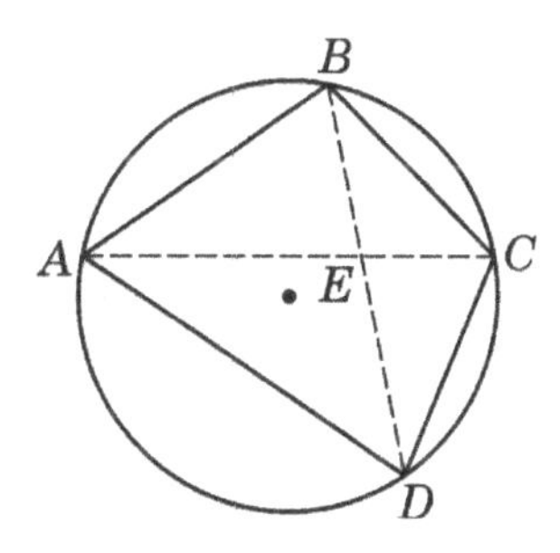

Fig. 20.3

10. As shown in Figure 20.4, join MH and EH.

Because M is the midpoint of the hypotenuse AD of Rt$\triangle AHD$, it is obtained that $MA = MH = MD$, then $\angle MHD = \angle MDH$.

Since the four points M, D, H and E are concyclic, we get that $\angle CEH = \angle MDH$, then

$$\angle MHD = \angle MDH = \angle HEC.$$

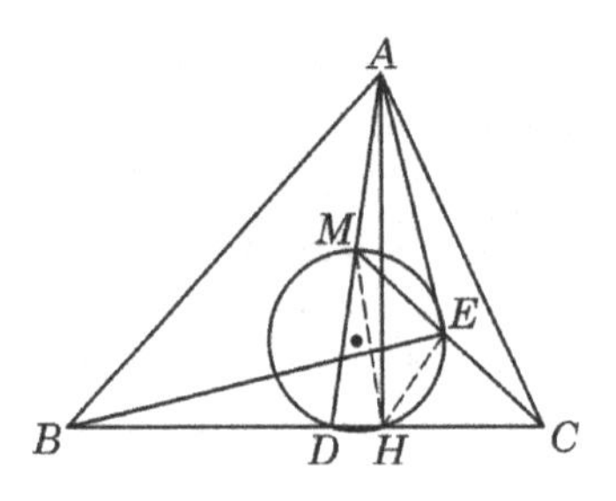

Fig. 20.4

Hence we know that

$$\angle MHC = 180^\circ - \angle MHD = 180^\circ - \angle HEC = \angle MEH.$$

Because $\angle CMH = \angle HME$, we obtain that $\triangle CMH \backsim \triangle HME$, then $\frac{MH}{MC} = \frac{ME}{MH}$, i.e. $MH^2 = ME \cdot MC$. Hence, $MA^2 = ME \cdot MC$.

And since $\angle CMA = \angle AME$, it is obtained that $\triangle CMA \backsim \triangle AME$, then $\angle MCA = \angle MAE$. Therefore,

$$\begin{aligned}
\angle BHE + \angle BAE &= \angle DHE + \angle BAD + \angle MAE \\
&= \angle DHE + \angle MAC + \angle MCA \\
&= \angle DHE + \angle DME \\
&= 180^\circ.
\end{aligned}$$

Hence, the four points A, B, H and E are concyclic, then $\angle AEB = \angle AHB = 90^\circ$.

Solution 21

Problems of Geometric Fixed Value

1. From the given condition, $MN \parallel BC$. Suppose that MN intersects AP at the point Q, then $AQ = QP$. Since A and P are the fixed points, Q is also a fixed point. Then the straight line MN passes through the fixed point Q.
2. (1) Because $DC = DA$, $EC = EB$, we get

$$PD + PE + DE = PD + DA + PE + BE = PA + PB.$$

   Hence the perimeter of $\triangle PDE$ is a fixed value.

   (2) Because $\angle DOE = \frac{1}{2}\angle AOB$, we know that the size of $\angle DOE$ is a fixed value.
3. Construct the line segment PE perpendicular to OA with the foot point of E, and construct the line segment PF perpendicular to OB with the foot point of F. Join PC, PD, then we have $\triangle PCE \cong \triangle PDF$. Therefore, it is obtained that $CE = DF$.

   Hence, $OC + OD = (OE - CE) + (OF + FD) = OE + OF$ is a fixed value.
4. Because $\triangle BPQ \backsim \triangle BDA$, $\triangle PCR \backsim \triangle DCA$, we have that $\frac{PQ}{DA} = \frac{BP}{BD}$, $\frac{PR}{DA} = \frac{PC}{DC}$. Then from $BD = DC$, $BP + PC = BC = 2BD$, it is obtained that $\frac{PQ+PR}{DA} = 2$, i.e. $PQ + PR = 2DA$.
5. Suppose that $PM \perp AB$ with the foot point of M, $PN \perp CD$ with the foot point of N. It's easy to know that the four points P, M, O and N are concyclic, and no matter what the point P is, the diameter of the

circle passing through these four point is equal to the radius of $\odot O$, which is a fixed value (when the point P coincides with one of A, B, C and D, it is the circumscribed circle of the right triangle). And since $\angle MPN = \angle AOC$ is a fixed angle, we know that MN is a fixed value.

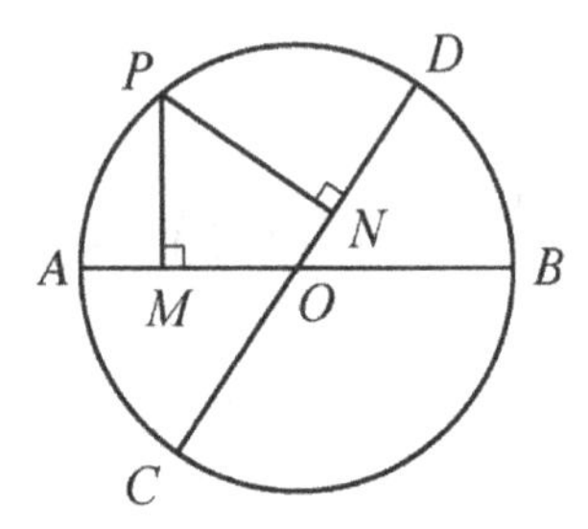

Fig. 21.1

6. Join OD, OM, OF, ON and OE, then $\angle BOM = \angle CNO$, $\angle MBO = \angle OCN$. So we get that $\triangle BOM \backsim \triangle CNO$, then $\frac{BM}{CO} = \frac{BO}{CN}$.
   Hence, $BM \cdot CN = BO^2 = \frac{1}{4}BC^2$ is a fixed value.
7. Because $CA^2 = CD \cdot CE$, that is, $BC^2 = CD \cdot CE$, we get that $\triangle BCD \backsim \triangle ECB$, then $\angle CBD = \angle CEB$. Since $\angle DEQ = \angle DPQ$, we obtain that $\angle CBP = \angle BPQ$, then $PQ \parallel AB$. Since AB is the tangent passing through the fixed point A on the fixed circle O, and its direction is certain, the conclusion holds.
8. Let E and F be the midpoints of AB and CD respectively. Then EF also bisects the area of trapezoid. Let the intersection of MN and EF be O. Then $S_{\triangle OME} = S_{\triangle ONF}$. It's easy to know that $\triangle OME \backsim \triangle ONF$, then $1 = S_{\triangle OME}:S_{\triangle ONF} = OE^2:OF^2$, that is, O is the midpoint of EF. Hence, we know that MN must pass through the fixed point O.
9. Constructing the line segment AC parallel to OQ through the point A, which intersects OP at the point C, then we have that $\frac{CP}{OC} = \frac{PA}{AQ}$, $\frac{OC}{OP} = \frac{AQ}{PQ}$, $\frac{OQ}{AC} = \frac{OP}{CP}$. Since OA is the bisector of $\angle XOY$, it is obtained that $AC = OC$. Thus we get that $\frac{OQ}{OP} = \frac{OC}{CP} = \frac{AQ}{AP}$, then $\frac{OQ}{OP+OQ} = \frac{AQ}{PQ}$. Therefore, $\frac{1}{OP} + \frac{1}{OQ} = \frac{1}{OC}$. Because A is a fixed point, OY is a fixed line, C is a certain point and OC is a fixed value, the conclusion holds.
10. Suppose that $EH \parallel FG \parallel BD$, $EF \parallel HG \parallel AC$, which intersect AB, BC, CD and DA at the points E, F, G and H, then $EFGH$ is a

parallelogram, and $\frac{EH}{BD} = \frac{AE}{AB}$, $\frac{EF}{AC} = \frac{BE}{AB}$, thus we have $EF + EH = AC$. Hence, the perimeter of $\square EFGH$ is $2(EF + EH) = 2AC$, which is a fixed value.

11. Join AM and BN. Because $\angle ANM = \angle ABM$, we get that $\triangle PAN \backsim \triangle PMB$, then $\frac{BM}{AN} = \frac{PB}{PN}$. Since $\triangle PAM \backsim \triangle PNB$, we obtain $\frac{AM}{BN} = \frac{PA}{PN}$. Hence, $\frac{AM \cdot BM}{AN \cdot BN} = \frac{PA \cdot PB}{PN^2} = \frac{PM}{PN} = \frac{a}{a+2R}$ is a fixed value, where $a = PM$ (fixed length) and R is the radius of circle O.

Solution 22

Five Centers of A Triangle

1. Construct any circle P, so that a point Q is inside $\odot P$. Take any point A on $\odot P$, and the ray AQ intersects $\odot P$ at D.

   Draw an arc with the center of D and the radius of DQ, and the arc intersects $\odot P$ at the points B and C, then $\triangle ABC$ is the required triangle. Since $\odot P$ can be large or small, and the point A is taken arbitrarily, there are countless triangles. So the answer is D.
2. Let the radius of the circumscribed circle of $\triangle ABC$ be R. Then $m = \sqrt{R^2 - \left(\frac{a}{2}\right)^2} = R\cos A$, $n = R\cos B$, $p = R\cos C$. So the answer is C.
3. It's easy to know that $\frac{EF}{BC} + \frac{PQ}{CA} + \frac{HG}{AB} = \frac{EF}{BC} + \frac{BP}{BC} + \frac{GC}{BC} = 2$, then $\frac{EF}{BC} = \frac{PQ}{CA} = \frac{HG}{AB} = \frac{2}{3}$, so the barycenter of $\triangle ABC$ is on EF, PQ and HG at the same time. Thus we know that the point O must be the barycenter of $\triangle ABC$. So the answer is B.
4. 10 points.
5. Construct the median AM on the side BC, then AM passes through the point G. Thus we have

$$S_{\triangle GBM} = S_{\triangle GCM} = S_{\triangle GDA} = S_{\triangle GEA} = S_{\triangle GDC} = S_{\triangle GEB}.$$

   Hence, $S_{\triangle GBC} = S_{ADGE} = \frac{1}{3}S_{\triangle ABC}$.
6. The orthocenter H of an acute angle triangle ABC is inside the triangle, as shown in Figure 22.1. Let the circumcenter of $\triangle ABC$ be O, and the midpoint of BC be D. The extended line of BO intersects $\odot O$ at the point E. Join CE, AE.

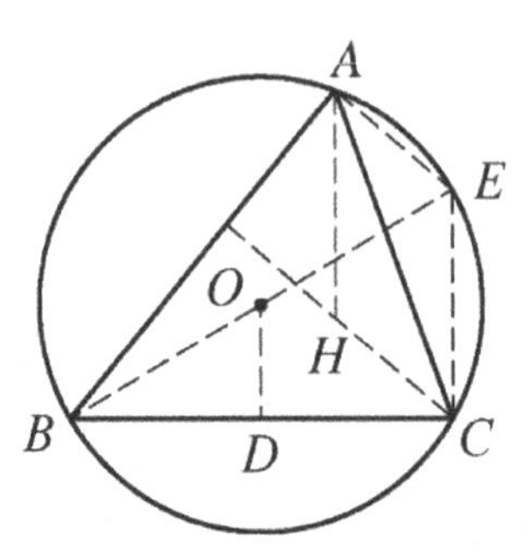

Fig. 22.1

Because BE is the diameter, $\angle BCE = 90°$, that is, $EC \perp BC$. Since H is the orthocenter of $\triangle ABC$, $AH \perp BC$. Therefore, we get $EC \parallel AH$. Similarly, $AE \parallel HC$.

Thus, it is deduced that the quadrilateral $AECH$ is a parallelogram, then $AH = EC$.

Hence, it is obtained from the given condition that $OB = AH = CE = 2OD$, then $\angle OBD = 30°$, $\angle BOD = 60°$.

So $\angle BAC = \angle BEC = \angle BOD = 60°$.

7. Let the point of tangency of AB and circle O be D, and the point of tangency of BC and circle O be E. Then the points A, O and E are collinear, both $\triangle ADO$ and $\triangle AEB$ are right triangles, and $\triangle ADO \backsim \triangle AEB$. Suppose that $AD = x$, then $OA = \sqrt{x^2+1}$, so we have $\frac{1}{2} = \frac{\sqrt{x^2+1}}{2+x}$. Solving it gives $x = \frac{4}{3}$.

   Hence, $AB = AD + BD = \frac{4}{3} + 2 = \frac{10}{3}$.

8. Join BK. Since the four points H, E, C and D are concyclic, we know $\angle BHD = \angle BCA$. And since $\angle BCA = \angle BKH$, it is obtained that $\angle BKH = \angle BHK$, then $BH = BK$. Since $BD \perp HK$, $HD = DK$.

9. As shown in Figure 22.2, construct $EM \perp BC$, $FN \perp BC$, $FP \perp EM$, and the foot points are M, N and P respectively.

   Because E and F are the circumcenters of $\triangle ABD$ and $\triangle ACD$ respectively, M and N are the midpoints of BD and CD respectively. Since $EF = BC$, then $PF = MN = \frac{1}{2}BC = \frac{1}{2}EF$, we get $\angle PEF = 30°$.

   From $EF \perp AD$, $EM \perp BC$, we know that $\angle ADC = \angle PEF = 30°$.

   Since $\angle ADC = \angle B + \angle BAD = \angle B + \frac{1}{2}(180° - 2\angle C) = 90° + \angle B - \angle C$, it is obtained that $\angle C - \angle B = 90° - \angle ADC = 60°$.

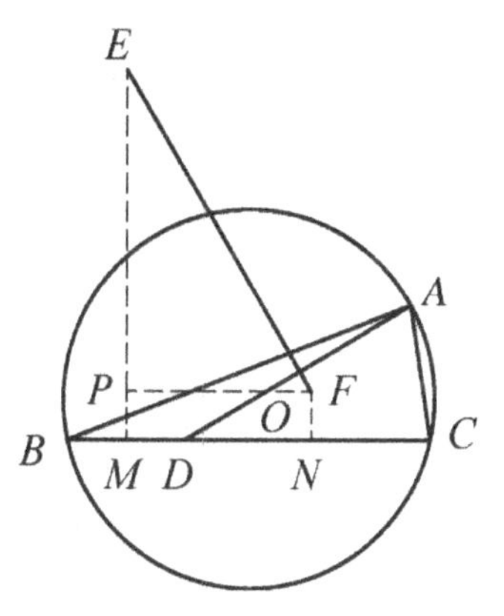

Fig. 22.2

10. In the case of Figure 22.23(2) of Chapter 22, the conclusion $BE+CF = AD$ holds.

    Join AO and extend it to the point G where AO and BC intersect. Construct $GH \perp EF$ with the foot point of H.

    Because $AD \parallel GH$, we know $\triangle ADO \backsim \triangle GHO$. From $AO = 2OG$, it is obtained that $AD = 2GH$.

    Join FG and extend it to the point M where FG and the extension line of EB intersect. Then $\triangle BMG \cong \triangle CFG$. So we get that $BM = CF$, $MG = FG$.

    Because $GH \parallel EM$, we get that $\triangle FHG \backsim \triangle FEM$, then $BE + BM = 2GH$. Hence, $BE + CF = AD$.

    Similarly, in the case of Figure 22.23(3) of Chapter 22, the conclusion $CF - BE = AD$ holds.

11. Extend KD, KE and KF respectively, and they intersect three sides AB, BC, CA of $\triangle ABC$ at the points M, N and P respectively. Since D, E and F are the barycenter of $\triangle KAB$, $\triangle KBC$ and $\triangle KCA$ respectively, it is easy to know that M, N, P are the midpoints of AB, BC and CA respectively. Then we get $S_{\triangle MNP} = \frac{1}{4}S_{\triangle ABC}$.

    It is easy to prove that $\triangle DEF \backsim \triangle MNP$, and the similarity ratio is 2:3, so

$$S_{\triangle DEF} = \left(\frac{2}{3}\right)^2 S_{\triangle MNP} = \frac{4}{9} \cdot \frac{1}{4} S_{\triangle ABC} = \frac{1}{9} S_{\triangle ABC}.$$

Therefore, $S_{\triangle DEF}:S_{\triangle ABC} = 1:9$. So the answer is A.

12. It is seen from the given condition that O and I are inside $\triangle ABC$. Since $\angle BOC = 2\angle A$, $\angle IBC = \frac{\angle B}{2}$, $\angle ICB = \frac{\angle C}{2}$, $\angle A > \angle B > \angle C$, we get that

$$\angle OBC = \frac{\angle C + (\angle B - \angle A)}{2} < \frac{\angle C}{2} = \angle ICB < \frac{\angle B}{2} = \angle IBC.$$

Then we have that $\angle OBC < \angle IBC$, $\angle OCB < \angle ICB$. It can be known that O is inside $\triangle BCI$, then IO intersects the line segment BC. And since $\angle OCB < \angle ICB$, it is obtained that $\angle OBA > \angle IBA$. Then from

$$\angle OAC = \frac{\angle A + \angle C - \angle B}{2} < \frac{\angle A}{2} = \angle IAC,$$

we get $\angle OAB > \angle IAB$, thus the point I is inside $\triangle AOB$. Hence, IO intersects AB.

13. Join AE, BD, CF, MB and BN. Because $\triangle ACE \cong \triangle BCD$, we get $BD = AE$. Similarly, $FC = BD$. It is easy to prove that $\triangle BMN \backsim \triangle BCF$, so $\frac{MN}{FC} = \frac{BM}{BC} = \frac{1}{\sqrt{3}}$. Similarly, it is obtained that $\frac{PN}{BD} = \frac{1}{\sqrt{3}}$, $\frac{MP}{AE} = \frac{1}{\sqrt{3}}$. So $MN = NP = PM$, that is, $\triangle MNP$ is a regular triangle.

Solution 23

Geometric Inequalities

1. Since one of $\angle ADB$ and $\angle ADC$ must be greater than or equal to 90°, might as well let $\angle ADB \geq 90^\circ$. Then $AB > AD$. So AD is less than the longest side of $\triangle ABC$.
2. Consider that P and Q are in $\triangle ABC$. Join AP, BP, CP. Then Q must be in $\triangle APC$, $\triangle BPC$ or $\triangle APB$, might as well let Q be in $\triangle APC$. Extend AP to the point D where AP intersects BC, then it is known from the conclusion of Exercise 1 that AP is less than the longest side of $\triangle ABC$. Similarly, BP and CP are less than the longest side of $\triangle ABC$, and PQ is less than the longest side of $\triangle APC$. Therefore, PQ is less than the longest side of $\triangle ABC$. In other cases, for the same reason, the conclusion can be obtained.
3. Extend the median by one time to form a parallelogram. By using the theorem that the sum of two sides of a triangle is greater than the third side, we obtain the conclusion.
4. It is easy to know that the conclusion holds by using Pappus Law.
5. Since $\angle BAC > \angle ABC = \angle ACB$, $\triangle FBD \cong \triangle DCE$, then $DF = DE$, $\angle FDE = \angle ABC$, so $2\angle DEF + \angle FDE = 180^\circ$. Thus we have $60^\circ < \angle DEF < 90^\circ$. So the answer is C.
6. Construct $AD \perp BC$, $BE \perp AC$, $CF \perp AB$ and the foot points are D, E and F respectively. Then $AD = h_a$, $BE = h_b$, $CF = h_c$. Thus we get that $h_a + CD > b$, $h_a + BD > c$, $h_b + CE > a$, $h_b + AE > c$, $h_c + AF > b$, $h_c + BF > a$. So $2(h_a + h_b + h_c) + (a + b + c) > 2(a + b + c)$, that is, $\frac{h_a+h_b+h_c}{a+b+c} > \frac{1}{2}$. In the right-angled $\triangle ACD$, $\triangle BAE$ and $\triangle CBF$, $h_a < b$, $h_b < c$, $h_c < a$, then we have $h_a + h_b + h_c < a + b + c$. Hence, the conclusion holds.

7. Because $b^2 + c^2 = 2m_a^2 + \frac{a^2}{2}$, we get that

$$m_a^2 = \frac{(b-c)^2}{2} + bc - \frac{a^2}{4} \geq bc - \frac{a^2}{4}.$$

And since $m_a^2 = bc + \frac{(b-c+a)(b-c-a)}{2} + \frac{a^2}{4} < bc + \frac{a^2}{4}$, the first conclusion holds. In the same way, it is obtained that the other two conclusions hold.

8. Construct a line segment DE perpendicular to the desktop at the point D on the desktop, so that $DE = BC$. When the points A, C and E are collinear, BD reaches its minimum value, that is, the light bulb B is closest to the point D on the desktop at this time.

9. Let $\triangle ABC$ be an isosceles triangle. So $AB = AC$, and $\triangle A'BC$ be any isosceles triangle with the same base and the same area as $\triangle ABC$ (might as well let A and A' be on the same side of BC). It is easy to know that $AA' \parallel BC$. Extend BA to the point C' so that $AC' = AC$, then we have $\triangle AC'A' \cong \triangle ACA'$. So $A'B + A'C = A'B + A'C' > C'B = AB + AC$.

10. Make $\angle ECF = \angle ABD = \frac{1}{2}\angle ABC$, and CF intersects BD at the point F. Because CF is inside $\angle DCE$, and F is on the line DB, we get that the four points B, C, F and E are concyclic. From $\angle ACB > \angle ABC$, it is obtained that $\angle EBC < \angle FCB$. So $CE < BF < BD$.

11. (1) Because D and G are the midpoints of AB and AC respectively, we get that $DG \parallel BC$ and $DG = \frac{1}{2}BC$.

    Construct $AM \perp BC$, $DN \perp BC$ respectively through the points A and D, and the perpendicular feet are the points M and N respectively, then $\triangle DNB \backsim \triangle AMB$, so $DN = \frac{1}{2}AM$.

    Hence, $T = DG \cdot DN = \frac{1}{2}BC \cdot \frac{1}{2}AM = \frac{1}{2}S$.

    (2) Construct $GH \parallel AB$ through the point G, and GH intersects BC at the point H, then $\angle B = \angle GHF$.

    Because $DE = GF$, $DE \parallel GF$, we get that $\angle DEB = \angle GFH$, then $\triangle DBE \cong \triangle GHF$.

    Because $DG \parallel BC$, we obtain $\triangle ADG \backsim \triangle ABC$. Suppose that $AD = kAB (0 < k < 1)$, then $S_{\triangle ADG} = k^2 S$.

    It can be obtained from $\triangle GHC \backsim \triangle ABC$ that $S_{\triangle GHC} = (1-k)^2 S$. Therefore,

$$\begin{aligned} T &= S - S_{\triangle ADG} - S_{\triangle GHC} = [1 - k^2 - (1-k)^2]S \\ &= -2\left(k - \frac{1}{2}\right)^2 S + \frac{1}{2}S \leq \frac{1}{2}S. \end{aligned}$$

(3) The discussion is divided into the following four cases:
The first case: if the parallelogram obtained by cutting out has three vertices on the edge of the triangle, and the fourth vertex is not on the edge of the triangle.

(1) When two of the three vertices are on the same edge of the triangle, as shown in Figure 23.1, extend DG to the point G' where BD and AC intersect, construct $G'F' \parallel GF$ through the point G' and $G'F'$ intersects BC at the point F'. It is easy to know that the quadrilateral $DEF'G'$ is a parallelogram, then $T \leq S_{\square DEF'G'}$. It is known from (2) that $S_{\square DEF'G'} \leq \frac{1}{2}S$. Hence, $T \leq \frac{1}{2}S$.

(2) When these three vertices are on three edges of the triangle, as shown in Figure 23.2, construct $AH \parallel DE$ through the point A and AH intersects EF and DG at the points F' and G' respectively, the question is transformed into two cases of (2)

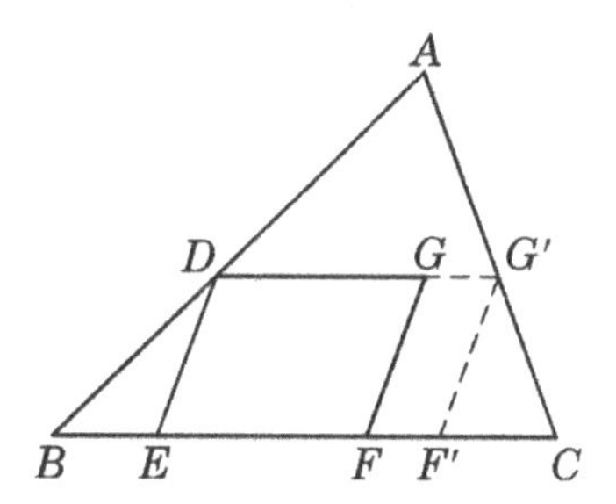

Fig. 23.1

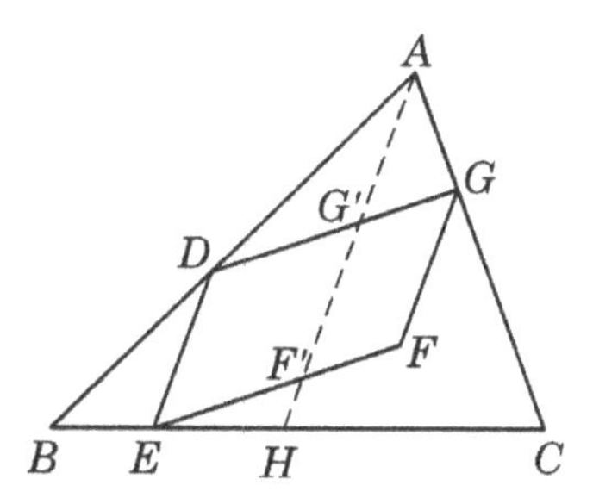

Fig. 23.2

and Figure 23.1, then

$$T = S_{\square DEF'G'} + S_{\square F'G'GF} \leq \frac{1}{2}S_{\triangle ABH} + \frac{1}{2}S_{\triangle AHC} = \frac{1}{2}S.$$

The second case: if the parallelogram obtained by cutting out has two vertices on the edge of the triangle, and the other two vertices are not on the edge of the triangle.

(1) When these two vertices are on the same edge of the triangle, as shown in Figure 23.3, extend DG which intersects two sides AB and AC of the triangle at the points L and K respectively. Construct a parallelogram $MNKL$. The question is transformed into (2). Then

$$T = S_{\square DEFG} \leq S_{\square MNKL} \leq \frac{1}{2}S.$$

(2) When these two vertices are on two edges of the triangle, as shown in Figure 23.4, extend DE and GF which intersect BC at the points K and M respectively. Construct $KN \parallel DG$ through the point K, and KN intersects GM at the point N. It is easy to know that the quadrilateral $DKNG$ is a parallelogram. Thus the question is transformed into the

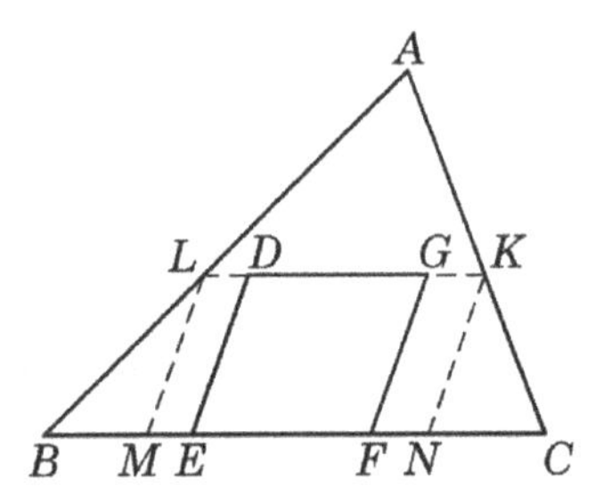

Fig. 23.3

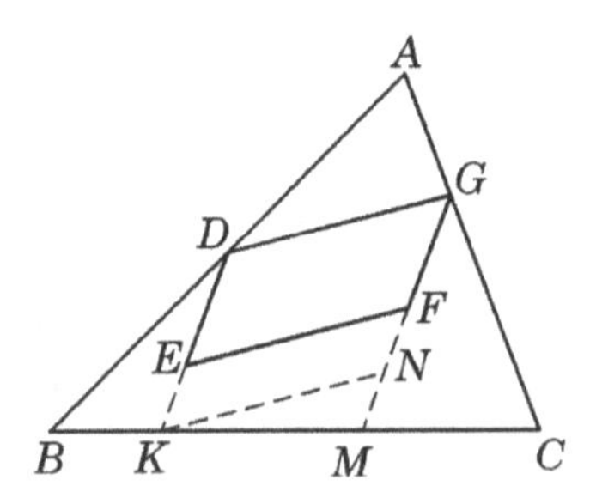

Fig. 23.4

case of Figure 23.2, then

$$T = S_{\square DEFG} \leq S_{\square DKNG} \leq \frac{1}{2}S.$$

The third case: if the parallelogram obtained by cutting out has only one vertex on the edge of the triangle, and the other three vertices are not on the edge of the triangle.

As shown in Figure 23.5, extend ED and FG which intersect AB and AC at the points K and M respectively. Construct $KN \parallel DG$ through the point K, and KN intersects FM at the point N. It is easy to know that the quadrilateral $EFNK$ is a parallelogram. Thus the question is transformed into the case of Figure 23.4, then

$$T = S_{\square DEFG} \leq S_{\square EFNK} \leq \frac{1}{2}S.$$

The fourth case: the parallelogram obtained by cutting out has no vertex on the edge of the triangle.

As shown in Figure 23.6, extend ED and FG which intersect AB and AC at the points K and M respectively. Construct $KN \parallel DG$

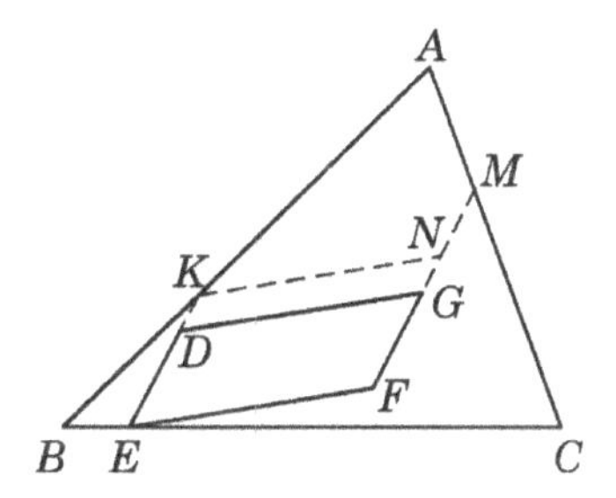

Fig. 23.5

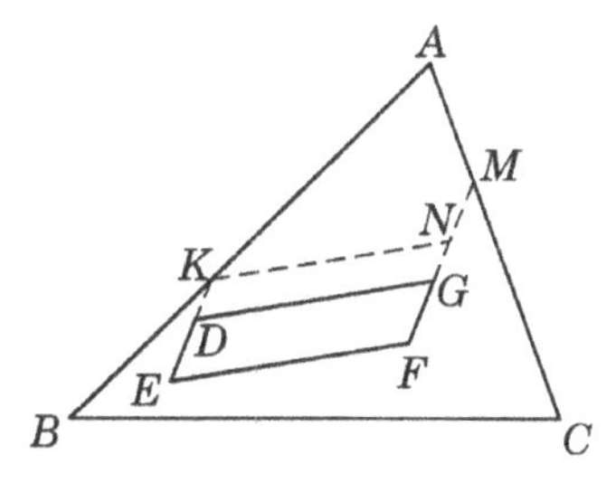

Fig. 23.6

through the point K, and KN intersects FM at the point N. It is easy to know that the quadrilateral $EFNK$ is a parallelogram. Thus the question is transformed into the case of Figure 23.5, then

$$T = S_{\square DEFG} \leq S_{\square EFNK} \leq \frac{1}{2}S.$$

To sum up, for any $\square DEFG$ obtained by cutting out, the conclusion $T \leq \frac{1}{2}S$ is true.

Solution 24
Indefinite Equations

1. From $x(x+y)=5$, we get that $\begin{cases} x=-1, \\ x+y=-5; \end{cases}$ $\begin{cases} x=-5, \\ x+y=-1; \end{cases}$ $\begin{cases} x=1, \\ x+y=5; \end{cases}$ $\begin{cases} x=5, \\ x+y=1. \end{cases}$ The integer solutions to the equation are $x_1=5$, $y_1=-4$; $x_2=1$, $y_2=4$; $x_3=-5$, $y_3=4$; $x_4=-1$, $y_4=-4$.
2. Let $\sqrt{23-a}=m$, $\sqrt{6-a}=n$. Then $m\geq 0$, $n\geq 0$, and $m^2-n^2=17$. Thus from $m+n\geq m-n$, we get $\begin{cases} m+n=17, \\ m-n=1. \end{cases}$ Solving the simultaneous inequalities gives $\begin{cases} m=9, \\ n=8. \end{cases}$

   Hence, $a=-58$.
3. From the given condition, we get the simultaneous equations $\begin{cases} y=\frac{6}{x}, \\ y=kx-4. \end{cases}$ Eliminating y gives $x(kx-4)=6$.

   Because x, y and k are integers, we have that

$$\begin{cases} x=1, \\ k-4=6, \end{cases} \begin{cases} x=-1, \\ -k-4=-6, \end{cases} \begin{cases} x=2, \\ 2k-4=3, \end{cases} \begin{cases} x=-2, \\ -2k-4=-3, \end{cases}$$

$$\begin{cases} x=3, \\ 3k-4=2, \end{cases} \begin{cases} x=-3, \\ -3k-4=-2, \end{cases} \begin{cases} x=6, \\ 6k-4=1, \end{cases} \text{or} \begin{cases} x=-6, \\ -6k-4=-1. \end{cases}$$

   Then the solutions are

$$\begin{cases} x=1, \\ k=10, \end{cases} \begin{cases} x=-1, \\ k=2, \end{cases} \begin{cases} x=3, \\ k=2. \end{cases}$$

When $k = 0$, $y = kx - 4 = -4$, but $-4x = 6$, $x = \frac{3}{2}$ is not an integer, which doesn't satisfy the requirements. Hence, the values of k are 10 and 2. So the answer is B.

4. It is known from $x^2 - 2y^2 = 5$ that x is an odd number. Let $x = 2k+1$, where k is a nonnegative integer. Then we have that $4k^2+4k-2y^2 = 4$, i.e. $2k^2 + 2k - y^2 = 2$, so y is an even number. Let $y = 2s$, where s is a positive integer. Thus we have $k^2 + k - 2s^2 = 1$, that is,

$$k(k+1) - 2s^2 = 1.$$

Since $k(k+1)$ is the product of two consecutive integers and it is even, the difference between two even numbers is still even and cannot be equal to 1. Hence, the number of positive integer solutions (x, y) is 0. So the answer is A.

5. Since the equation is $(x+y)^2+2y^2 = 34$, it is obvious that $x+y$ must be even, assume that $x+y = 2t$, then the original equation is transformed into $2t^2+y^2 = 17$. Hence, $2t^2 \leq 17$, we get that $0 \leq t^2 \leq 8$, then $t = 0$, ±1, ±2. Substituting them into $2t^2 + y^2 = 17$ we know that $t = 0$ or ±1 which do not satisfy the conditions. So its integer solutions are $\begin{cases} t = \pm2, \\ y = \pm3. \end{cases}$ Thus we obtain that the integer solutions to the original equation are $(x, y) = (-7, 3), (1, 3), (7, -3), (-1, -3)$, 4 groups in total. So the answer is B.

6. (1) The equation is transformed into $x = 2 + \frac{4}{y-2}$. Then it is known from $y > 2$ that $y - 2 = 1, 2, 4$, that is, $y = 3, 4, 6$. Accordingly, the positive integer solutions to the equation are $x_1 = 6$, $y_1 = 3$; $x_2 = 4$, $y_2 = 4$; $x_3 = 3$, $y_3 = 6$.

   (2) Might as well let $x \geq y > 0$. Then $\frac{1}{x} \leq \frac{1}{y}$. So $\frac{3}{7} \leq \frac{2}{y}$, it is obtained that $y \leq 4$. Thus $y = 1, 2, 3, 4$. Substituting them into $\frac{1}{x} + \frac{1}{y} = \frac{3}{7}$ gives that there is no positive integer x satisfying the conditions. That is, the number of positive integer solutions (x, y) is 0.

7. From the given condition, $(x - 10)(y - 10) = 101$, then we have that

$$\begin{cases} x - 10 = 1, \\ y - 10 = 101; \end{cases} \begin{cases} x - 10 = -1, \\ y - 10 = -101; \end{cases} \begin{cases} x - 10 = 101, \\ y - 10 = 1; \end{cases} \begin{cases} x - 10 = -101, \\ y - 10 = -1. \end{cases}$$

Hence, all the integer solutions are $x_1 = 111$, $y_1 = 11$; $x_2 = 11$, $y_2 = 111$; $x_3 = -91$, $y_3 = 9$; $x_4 = 9$, $y_4 = -91$.

8. It is known from $z \geq 3$ that $\frac{1}{x}+\frac{1}{y} = \frac{1}{2}+\frac{1}{z} > \frac{1}{2}$. Might as well let $x \leq y$. Then we know that $\frac{1}{x} + \frac{1}{y} \leq \frac{2}{x}$. Hence, it is obtained from $\frac{1}{2} \leq \frac{2}{x}$ that

$x \leq 4$. However, $3 \leq x$. Then we get that $x = 3$ or 4. When $x = 3$, $\frac{1}{y} - \frac{1}{z} = \frac{1}{6}$; when $x = 4$, $\frac{1}{y} - \frac{1}{z} = \frac{1}{4}$. So the integer solutions satisfying the original equation are $x_1 = 3$, $y_1 = 3$, $z_1 = 6$; $x_2 = 3$, $y_2 = 4$, $z_2 = 12$; $x_3 = 3$, $y_3 = 5$, $z_3 = 30$; $x_4 = 4$, $y_4 = 3$, $z_4 = 12$; $x_5 = 5$, $y_5 = 3$, $z_5 = 30$.

9. Since the equation $14x^2 + (4 - 24y)x + 21y^2 - 12y - 18 = 0$ about x has integer solutions, then $\Delta_x \geq 0$, that is, $(4 - 24y)^2 - 4 \times 14 \times (21y^2 - 12y - 18) \geq 0$. Solving the inequality gives $\frac{12 - 2\sqrt{132}}{15} \leq y \leq \frac{12 + 2\sqrt{132}}{15}$. Then we get that $y = 0$ or 1. When $y = 0$, $x = 1$; when $y = 1$, x has no integer solution. So the solutions to the equation are $x = 1$, $y = 0$.
10. Since x and $x + 1$ are coprime and $y^3 = x(x + 1)$, we know that there must exist k_1 and k_2, so that $x = k_1^3$, $x + 1 = k_2^3$, $y = k_1k_2$, then $k_2^3 - k_1^3 = 1$. Since k_2 and k_1 are integers, $k_2 \neq k_1$, $k_1^2 + k_1k_2 + k_2^2 > 0$ and $(k_2 - k_1)(k_1^2 + k_1k_2 + k_2^2) = 1$, we get that $k_2 - k_1 = 1$, $k_1^2 + k_1k_2 + k_2^2 = 1$. Thus $k_1k_2 = 0$, it is obtained that $k_2 = 0$, $k_1 = -1$; $k_2 = 1$, $k_1 = 0$. Hence, the integer solutions to the original equation are $x_1 = 0$, $y_1 = 0$; $x_2 = -1$, $y_2 = 0$.
11. (1) From the given condition,

$$(x - 1)(x^2 + kx - 1) = x^3 + (k - 1)x^2 - (k + 1)x + 1,$$

it implies that

$$x^3 - x^2 - x + 1 = x^3 + (k - 1)x^2 - (k + 1)x + 1.$$

Then we have that $\begin{cases} k - 1 = -1, \\ -k - 1 = -1. \end{cases}$ Solving the equations gives $k = 0$.

(2) $\dfrac{x^3 - x^2 - x + 1}{x^2 - 2x + 1} = \dfrac{(x - 1)(x^2 - 1)}{(x - 1)^2} = x + 1.$

Because x is an integer, we know that $x + 1$ is an integer. Hence, $\dfrac{x^3 - x^2 - x + 1}{x^2 - 2x + 1}$ is an integer.

12. Because $(m - n)^2 + 13n^2 = 217$, we get $n^2 \leq \frac{217}{13}$, then $n^2 \leq 16$, that is, $n = 1, 2, 3, 4$. Thus we know that $n = 3$, $m = 13$; $n = 4$, $m = 1$; $n = 4$, $m = 7$.
13. Because $x^3 - y^3 - z^3 - 3xyz = (x - y - z) \cdot \frac{1}{2}[(x + y)^2 + (x + z)^2 + (y - z)^2] = 0$, we know that $x = y + z$ or $x = -y = -z$. When $x = y + z$, $x^2 = 2x$, and $x \geq 1$, it is obtained that $x = 2$, then $y = z = 1$. When $x = -y = -z$, $x^2 = -4x$, and $x \geq 1$, then there is no positive integer solution. So the positive integer solutions to the simultaneous equations are $x = 2$, $y = z = 1$.

14. From the given codition, $\frac{1^2+2^2+\cdots+n^2}{n}=m^2$, then we have

$$\frac{(n+1)(2n+1)}{6}=m^2. \quad (*)$$

Since $n+1$ and $2n+1$ are coprime, and $2n+1$ is an odd number, we get that $n+1$ must be even, that is, n is an add number. Then it is obtained from $(*)$ that $\begin{cases} n+1=2k_1^2, \\ 2n+1=3k_2^2; \end{cases}$ or $\begin{cases} n+1=6k_1^2, \\ 2n+1=k_2^2. \end{cases}$ It is known from $\begin{cases} n+1=2k_1^2, \\ 2n+1=3k_2^2 \end{cases}$ that $4k_1^2=1+3k_2^2$, and k_2 is odd. Through the checking calculation, when $k_2=15$, the minimum value of k_1 is 13, and the minimum value of n is 337. It is known from $\begin{cases} n+1=6k_1^2, \\ 2n+1=k_2^2 \end{cases}$ that $12k_1^2=k_2^2+1$, and k_2 is odd. Because $k_2^2\equiv 1(\text{mod}4)$, we get that $k_2^2+1\equiv 2(\text{mod}4)$, but $12k_1^2\equiv 0(\text{mod}4)$, that is, $12k_1^2=k_2^2+1$ has no solution. So the minimum value of n is 337.

15. Because 208 is a multiple of 4, the remainder of the square of an even number divided by 4 is 0, and the remainder of the square of an odd number divided by 4 is 1, we get that x and y are both even numbers.

Let $x=2a$, $y=2b$. Then $a^2+b^2=104(a-b)$.

Same as above, a and b are both even numbers.

Let $a=2c$, $b=2d$. Then $c^2+d^2=52(c-d)$, so c and d are both even numbers.

Let $c=2s$, $d=2t$. Then $s^2+t^2=26(s-t)$.

Hence, $(s-13)^2+(t+13)^2=2\times 13^2$, where s and t are both even numbers. Thus we know that

$$(s-13)^2=2\times 13^2-(t+13)^2\leq 2\times 13^2-15^2<11^2,$$

then $|s-13|$ may be 1, 3, 5, 7, 9, and $(t+13)^2$ is 337, 329, 313, 289, 257. So it can only be $(t+13)^2=289$, then $|s-13|=7$. Solving the equation gives $\begin{cases} s=6, \\ t=4, \end{cases}$ or $\begin{cases} s=20, \\ t=4. \end{cases}$ So the integer solutions to the original equation are $\begin{cases} x=48, \\ y=32, \end{cases}$ or $\begin{cases} x=160, \\ y=32. \end{cases}$

Another solution Since $(x-104)^2+(y+104)^2=2\times 104^2=21632$, we have that $(y+104)^2\leq 21632$. Since y is a positive integer, $1\leq y\leq 43$.

Let $a=|x-104|$, $b=|y+104|$. Then $a^2+b^2=21632$.

Since the units digits of any perfect square number are 1, 4, 5, 6, 9, it is known from $a^2 + b^2 = 21632$ that the units digits of a^2 and b^2 can only be 1 and 1, or 6 and 6.

When the units digits of a^2 and b^2 are 1 and 1, the units digits of a and b can be 1 or 9, but the tens digits of the square number of the numbers with the units digits of 1 and 9 are even, which contradicts the condition that the tens digits of $a^2 + b^2$ is 3.

When the units digits of a^2 and b^2 are 6 and 6, the units digits of a and b can be 4 or 6.

From $105 \leq b \leq 147$, we can take the numbers $b = 106, 114, 116, 124, 126, 134, 136, 144, 146$. Substituting them into $a^2 + b^2 = 21632$ one by one gives that only when $b = 136$, $a = 56$, that is, $\begin{cases} |x - 104| = 56, \\ |y + 104| = 136 \end{cases}$ satisfy the given conditions. Solving the equations gives $\begin{cases} x = 48, \\ y = 32, \end{cases}$ $\begin{cases} x = 160, \\ y = 32. \end{cases}$

16. (1) Let $x_1, x_2, x_3, \ldots, x_{1007}$ be 1007 numbers taken out arbitrarily from $1, 2, 3, \ldots, 2008$.

First of all, numbers $1, 2, 3, \ldots, 2008$ are divided into 1004 pairs, so that the sum of each pair is 2009. Each pair of numbers is recorded as $(m, 2009 - m)$, where $m = 1, 2, 3, \ldots, 1004$.

Because there are 1001 numbers left after taking 1007 numbers out of 2008 numbers, the number pairs satisfying the condition that at least one number is one of these 1001 numbers are at most 1001. Therefore, at least three pairs of numbers are 6 of x_1, x_2, x_3, $\ldots$, x_{1007}. Might as well record them as $(m_1, 2009 - m_1)$, $(m_2, 2009 - m_2)$ and $(m_3, 2009 - m_3)$ (where m_1, m_2 and m_3 are not equal to each other).

Secondly, 2006 of these 2008 numbers (except 1004 and 2008) are divided into 1003 pairs, so that the sum of each pair is 2008. Each pair of numbers is recorded as $(k, 2008 - k)$, where $k = 1, 2, \ldots, 1003$.

At least 1005 of the 2006 numbers are taken out, so there are at most 1001 numbers among these 2006 numbers except the numbers taken out. In these 1003 pairs of numbers, at least two pairs of numbers are 4 of $x_1, x_2, x_3, \ldots, x_{1007}$, might as well record one of the pairs as $(k_1, 2008 - k_1)$.

In three pairs of numbers $(m_1,\ 2009 - m_1)$, $(m_2,\ 2009 - m_2)$ and $(m_3,\ 2009 - m_3)$ (where m_1, m_2 and m_3 are not equal to each other), there are at least two numbers in a pair that are different from two numbers in $(k_1,\ 2008 - k_1)$. Might as well let this pair of numbers be $(m_1,\ 2009 - m_1)$. Then $m_1 + 2009 - m_1 + k_1 + 2008 - k_1 = 4017$.

(2) When $n \leq 1006$, the conclusion doesn't hold.

When $n = 1006$, might as well take out the posterior 1006 numbers from $1, 2, \ldots, 2008$ as follow: $1003, 1004, \ldots, 2008$, then the sum of any four different numbers of them is not be less than $1003 + 1004 + 1005 + 1006 = 4018 > 4017$.

When $n < 1006$, take out the posterior n numbers from $1, 2, \ldots, 2008$, where the sum of any four numbers is greater than $1003 + 1004 + 1005 + 1006 = 4018 > 4017$.

Hence, the conclusions doesn't hold when $n \leq 1006$.

Solution 25

Reductio ad Absurdum

1. From the given condition, $(m+n)(p+q) = 1$, that is, $(mp+nq) + (mq+np) = 1$. Since $mp + nq > 1$, we know $mq + np < 0$. Suppose that all of m, n, p and q are more than or equal to 0, then $mq + np \geq 0$, which contradicts the conclusion that $mq + np < 0$. So the hypothesis does not hold. Hence, the original proposition holds.
2. Since $\sqrt{24} = 2\sqrt{6}$, it is only necessary to prove that $\sqrt{6}$ is an irrational number.

   Prove it by the reductio ad absurdum. Suppose that $\sqrt{6}$ is a rational number, then there exist integers p and q which are coprime, and $p > 0$, so that $\sqrt{6} = \frac{q}{p}$. Hence, $6p^2 = q^2$, then we get $2|q^2$. Since 2 is prime, $2|q$. Let $q = 2r$, where r is an integer.

   Substituting it into $6p^2 = q^2$ gives $3p^2 = 2r^2$. Since 2 and 3 are coprime, we get $2|p^2$. Similarly, it is obtained that $p = 2s$, where s is an integer. Therefore, p and q have at least one common divisor $2 > 1$, which contradicts the hypothesis that p and q are coprime. So the hypothesis does not hold, that is, $\sqrt{6}$ is an irrational number. Hence, $\sqrt{24}$ is an irrational number.
3. Suppose that there are at least four acute angles, then their exterior angles are all greater than $90°$, so the sum of their exterior angles is greater than $4 \times 90° = 360°$, which contradicts the theorem that the sum of the exterior angles of a convex polygon is equal to $360°$. Therefore, the original proposition holds.
4. Suppose that $AB > AC$, then $\angle ACB > \angle ABC$. Because the quadrilateral $ABCD$ is a convex quadrilateral, it is obtained that $\angle BCD > \angle ACB$, $\angle ABC > \angle DBC$, then $\angle BCD > \angle DBC$, $BD > CD$.

Therefore, we get that $AB + BD > AC + CD$, which contradicts the given conditions. So the hypothesis does not hold, that is, $AB \leq AC$.

5. If BF and CE bisect each other, then the quadrilateral $EBCF$ is a parallelogram, so $BE \parallel CF$, which contradicts the given condition.
6. If $\angle BAP \geq \angle CAP$, then $PB \geq PC$, $\angle PCB \geq \angle PBC$. Because $\angle ABC = \angle ACB$, we get $\angle ACP \leq \angle ABP$. Further, it is deduced that $\angle APC > \angle APB$, which contradicts the given condition.
7. Suppose that AB and CD bisect each other at the point P. Join OP, then $OP \perp AB, OP \perp CD$. AB and CD are two different chords, that is, there are two straight lines passing through the point P perpendicular to OP, which contradicts the theorem that there is only one perpendicular line passing through a point P on the line. Hence, AB and CD cannot bisect each other.
8. The discriminants of the given equations are

$$\Delta_1 = (\sqrt{p})^2 - 4 \times \frac{q}{8}, \quad \Delta_2 = (\sqrt{q})^2 - 4 \times \frac{r}{8}, \quad \Delta_3 = (\sqrt{r})^2 - 4 \times \frac{p}{8}.$$

Then we get that $\Delta_1 + \Delta_2 + \Delta_3 = p - \frac{q}{2} + q - \frac{r}{2} + r - \frac{p}{2} = \frac{1}{2}(p + q + r) > 0$. So there must be an equation whose discriminant is greater than 0, thus we know that this equation must have two distinct roots. Then from the relationship between roots and coefficients and p, q, $r > 0$, it is known that these two roots are positive.

9. Suppose that there exist four points A, B, C and D so that all of $\triangle ABC$, $\triangle BCD$, $\triangle CDA$ and $\triangle DAB$ are acute triangles.

If the four points A, B, C and D form a convex quadrilateral $ABCD$, since the sum of interior angles of a convex quadrilateral is 360°, at least one interior angle shall not be less than 90° (otherwise, if all the four interior angles are less than 90°, then the sum of interior anglesis less than 360°, which is impossible). Therefore, the triangle composed of the interior angle of not less than 90° is not an acute triangle, which contradicts the hypothesis that four triangles are all acute triangles.

If four points A, B, C and D form a concave quadrilateral $ABCD$, then there must be one point (set as D) inside the triangle composed of other three points, as shown in Figure 25.1. Since the round angle is 360°, at least one angle of $\angle ADB$, $\angle BDC$ and $\angle ADC$ is not less than 120° (otherwise, the sum of the three angles is less than 360°, which is impossible). Therefore, the triangle composed of the angle of not less than 120° is not an acute triangle, which contradicts the hypothesis that four triangles are all acute triangles.

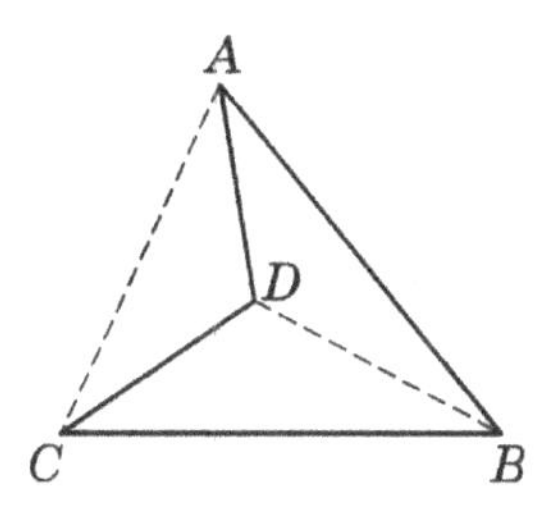

Fig. 25.1

It can be seen from the above that the four points A, B, C and D that make $\triangle ABC$, $\triangle BCD$, $\triangle CDA$ and $\triangle DAB$ be acute triangles do not exist.

10. Let the area of $\triangle AEG$, $\triangle BEF$, $\triangle CFG$ and $\triangle ABC$ be S_1, S_2, S_3 and S respectively. And $AE = a$, $BE = b$, $BF = c$, $CF = d$, $CG = e$, $AG = f$. Suppose that S_1, S_2 and S_3 are greater than $\frac{1}{4}S$. Since $S_1 = \frac{1}{2}af\sin A$, $S = \frac{1}{2}(a+b)\cdot(f+e)\cdot\sin A$, we know that $af > \frac{1}{4}(a+b)\cdot(f+e)$. Similarly, we get that

$$bc > \frac{1}{4}(a+b)\cdot(c+d), de > \frac{1}{4}(c+d)(e+f).$$

Thus we obtain that

$$abcdef > \left(\frac{1}{4}\right)^3 (a+b)^2(c+d)^2(e+f)^2. \tag{1}$$

However, $ab \leq \frac{1}{4}(a+b)^2$, $cd \leq \frac{1}{4}(c+d)^2$, $ef \leq \frac{1}{4}(e+f)^2$, so

$$abcdef \leq \left(\frac{1}{4}\right)^3 (a+b)^2(c+d)^2(e+f)^2. \tag{2}$$

Because (1) and (2) are contradictory, the hypothesis is wrong and the original conclusion is correct.

11. Proof 1 As shown in Figure 25.2, it is known that $\angle 1 = \angle 2$, $\angle 3 = \angle 4$, $BE = CF$.

Suppose that $AB \neq AC$, there are two possibilities, that is, (i) $AB > AC$; (ii) $AB < AC$. Now prove that both of them are impossible. Suppose that (i) holds, i.e. $AB > AC$, then $\angle 3 + \angle 4 > \angle 1 + \angle 2$. Because $\angle 1 = \angle 2$, $\angle 3 = \angle 4$, it is obtained that $\angle 4 > \angle 2$, $\angle 3 > \angle 1$. In $\triangle BCF$ and $\triangle BCE$, $BE = CF$, $BC = BC$, $\angle 4 > \angle 2$, so $BF > EC$. Construct $EG \parallel BF$ through the point E, and construct $FG \parallel BE$

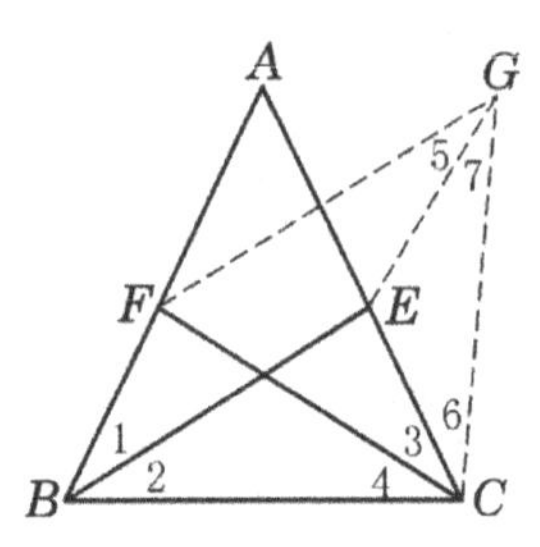

Fig. 25.2(1)

through the point F, then we get a parallelogram $BEGF$. Join GC, then $\angle 1 = \angle 5$, $BE = FG = FC$, so $\angle 3 + \angle 6 = \angle 5 + \angle 7$. Because $\angle 3 > \angle 1$, we know that $\angle 3 > \angle 5$. So $\angle 6 < \angle 7$, $EG < EC$. Since $EG = BF$, $BF < EC$. This contradicts the previous conclusion $BF > EC$, so (i) $AB > AC$ does not hold. For the same reason, it is obtained that (ii) $AB < AC$ does not hold. Therefore, $AB = AC$, the question is proved.

Remark In the proof of this question, we deduce the contradictory result from the hypothesis, so as to overturn the hypothesis and get the original proposition.

There are many ways to prove this problem. Here is another way to prove it.

Proof 2 Suppose that the two base angles are not equal, might as well let $\angle ABC < \angle ACB$, as shown in Figure 25.2(2).

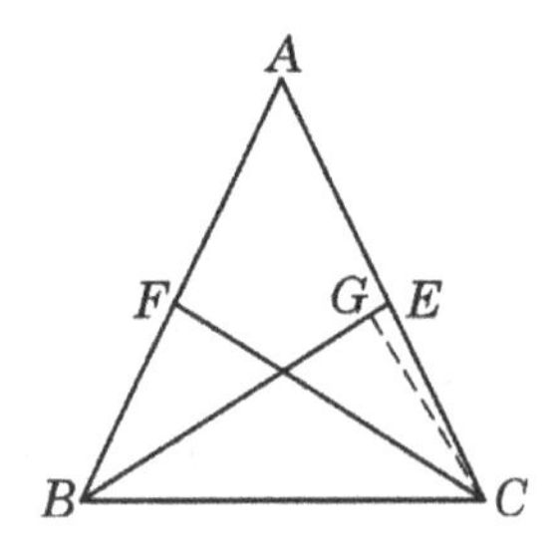

Fig. 25.2(2)

Take a point G on BE so that $\angle FCG = \angle FBG$. It is obvious that four points F, B, C and G are concyclic. Because $\angle ABC < \frac{1}{2}(\angle ABC + \angle ACB) < \frac{1}{2}(\angle A + \angle ABC + \angle ACB)$, we get $\angle CBF < \angle GCB < 90°$. But in the same circle, the chords opposite two unequal acute angles are unequal, and the chord opposite the smaller angles is shorter, that is, $CF < GB < BE$.

This contradicts the given condition (similarly, it can also negate $\angle ABC > \angle ACB$), so $\angle ABC = \angle ACB$. Thus we get $AB = AC$.

12. From the condition that p and q are positive prime numbers, it is known that the original equation cannot have a positive root. And from the property of the root of the integral coefficient equation, if the original equation has a rational root, it can only be an integer root and must be a divisor of q^3. Therefore, if the original equation has a rational root, it can only take the following numbers: $-q$, $-q^2$, -1, $-q^3$.

    (1) Suppose that the two roots of the equation are $-q$ and $-q^2$, then $p^2 = q + q^2 = q(1+q) \geq 6$. Because $q(1+q)$ is a multiple of 2, p^2 is also a multiple of 2, and $p^2 \neq 4$. Then p is a composite number, which contradicts the given conditions. So $-q$ and $-q^2$ cannot be the roots of the original equation.

    (2) Suppose that the two roots of the equation are -1 and $-q^3$, then $p^2 = 1 + q^3$ at this time. Obviously, q cannot be an odd prime number. Because if q is an odd prime number, p^2 must be an even number greater than 4, and p must be an even number greater than 2, that is, p is a composite number, which contradicts the given conditions. Therefore, q must be an even prime number. Thus we get that $q = 2$, $p = 3$.

    Hence, if and only if $p = 3$, $q = 2$, the original equation has rational roots which are -1 and -8.

13. No! Prove it by reductio ad absurdum.

    Suppose that there is a return circuit, we examine the horse's circuit in two ways:

    (1) Firstly, the small squares of $4 \times n$ chessboard are colored with black (with the oblique line) and white according to Figure 25.3(1). Obviously, the square at the beginning and ending points of each step of the horse must be in different colors, that is, the horse's return circuit presents as: white $\to$ black $\to$ white $\to$ black $\to \cdots$ (or opposite). Might as well set {the lattice to which odd steps jump} = {white lattice}; {the lattice to which even steps jump} = {black lattice}.

    (2) Then the small squares of $4 \times n$ chessboard are classified into two groups: inside and outside. Since the horse has a return circuit, it is obtained that {the number of inner lattice} = {the number of outer lattice}. When the horse takes off from the outer lattice,

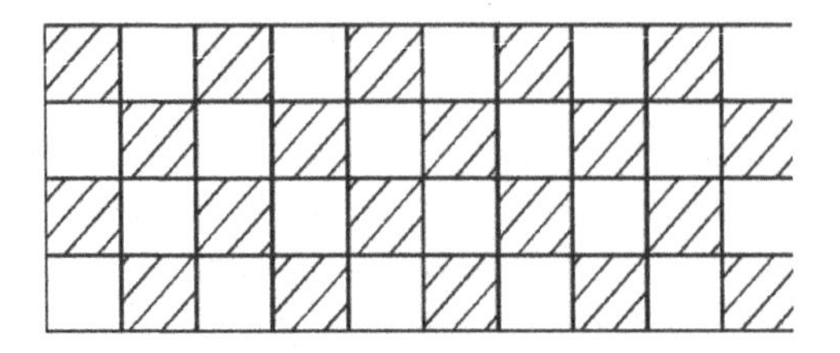

Fig. 25.3(1)

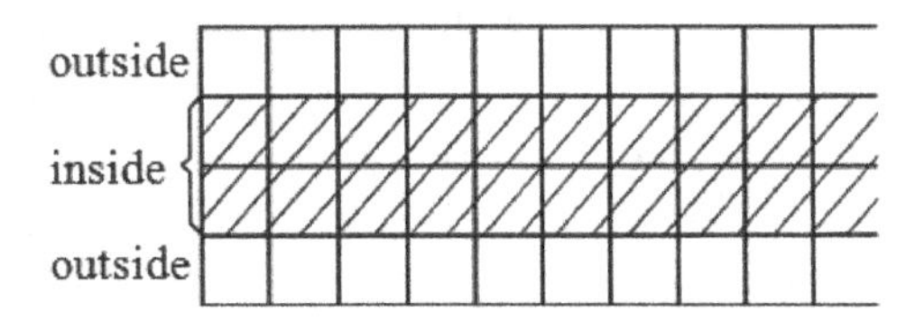

Fig. 25.3(2)

it must jump into the inner lattice. Therefore, the horse must jump into the external lattice to take off from the inner lattice. In this way, the horse's return circuit presents as: inside $\to$ outside $\to$ inside $\to$ outside $\to \cdots$ (or opposite).

Combining (1) and (2) gives that {the lattice to which odd steps jump} = {outer lattice} or {inner lattice}, that is, {white lattice} = {outer lattice} or {inner lattice}. This is obviously contradictory, because both inner and outer lattices are not of the same color. Hence, the hypothesis that the horse has a return circuit is incorrect.

14. Suppose that there is a lattice point regular triangle, as shown in Figure 25.4, which is inscribed in the rectangle $ADEF$. $\triangle ADB$ is a lattice point right triangle, then the right angle sides AD and BD are integers, $S_{\triangle ADB} = \frac{1}{2}AD \cdot BD$ (half of an integer).

Similarly, $S_{\triangle ACF}$ and $S_{\triangle CBE}$ are also half of an integer, so $S_{\triangle ABC} = S_{ADEF} - S_{\triangle ABD} - S_{\triangle BCE} - S_{\triangle ACF}$ is a rational number (half of an integer). On the other hand, $S_{\triangle ABC} = \frac{1}{2} \cdot \frac{\sqrt{3}}{2}AB \cdot AC = \frac{\sqrt{3}}{4}AB^2$ (irrational number). In this way, we get the contradiction that the rational number is equal to the irrational number! Therefore, the lattice point regular triangle does not exist.

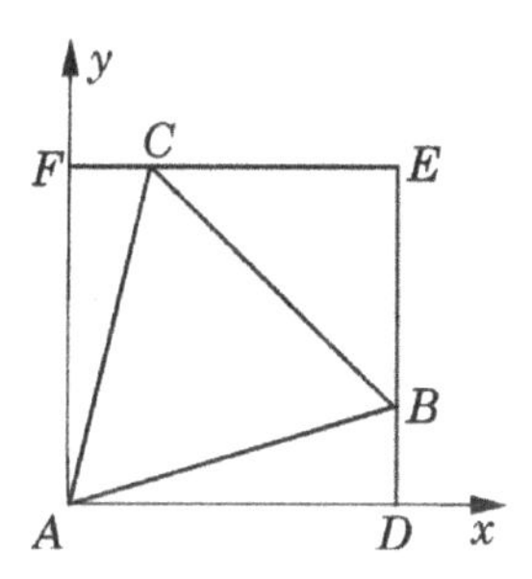

Fig. 25.4

15. Consider any question A, let it be solved by r people. If the calculation is repeated, these people will solve the other $6r$ questions.

On the other hand, the question A is paired with one of the other 27 questions, 27 pairs can be matched and each pair is just solved by two people. So each of the other 27 questions is solved by only 2 people who solve the question A at the same time, thus we have $6r = 2 \times 27$. Solving it gives $r = 9$. That is, each question is solved by 9 people. Hence, the number of participants should be $9 \times 28 \div 7 = 36$.

By using the method of reductio ad absurdum, we first make counter-assumption. Suppose that the numbers of people who solve one, two and three questions in the first test are x, y and z respectively, then we have

$$x + y + z = 36. \tag{1}$$

Suppose that there are n questions in the first test. If the number of questions solved is calculated repeatedly, then

$$x + 2y + 3z = 9n, \tag{2}$$

Pairing these n questions can form $\frac{n(n-1)}{2}$ pairs, each pair is solved by two people, and these people should be among y people and z people, so according to the number of people in each pair calculated repeatedly, it should be known that

$$y + 3z = n(n-1). \tag{3}$$

If x, y and z expressed by n are solved by (1), (2), and (3), it can be found that they are not all non-negative integers. For example, it is obtained that

$$y = -2n^2 + 29n - 108 = -2\left(n - \frac{29}{4}\right)^2 - \frac{23}{8} < 0,$$

which is impossible and leads to contradictions. Hence, the proposition is proved.

Solution 26

Extreme Principle

1. In the extreme case, no four monkeys get as many peanuts as they have. Then it needs at least $3\times$ $(0+1+2+3+\cdots+32)+33=1617$ peanuts. So there must be 4 monkeys who get as many peanuts as they have.
2. Notice that 4 is the minimum composite number. Take all the numbers of $4k+1$. It is obvious that the difference between any two numbers is a multiple of 4, which is a composite number, and this is the best possibility. Because in the consecutive 8 numbers $m, m+1, \ldots, m+7$, the difference between two numbers is $1, 2, \ldots, 7$, the difference between the numbers selected can only be the composite number 4 or 6, so at most two of the eight consecutive numbers can be taken.

   From $1 \leq 4k+1 \leq 2018$, it is obtained that $0 \leq k \leq 504$, then $n_{\max}=505$.
3. In the extreme case when PCD becomes PO, we have that

$$\angle AEP = \angle AOP = 90^\circ - \frac{1}{2}\angle APB = 70^\circ.$$

   So the answer is D.
4. The extreme case is the maximum angle. For any point P in a quadrilateral $ABCD$, the sum of $\angle APB$, $\angle BPC$, $\angle CPD$ and $\angle DPA$ is 360°. Let the maximum angle be $\angle APB$. Then $\angle APB \geq 90^\circ$, so P is in a semicircle with AB as the diameter. From the arbitrariness of the point P, we can see that the conclusion holds.
5. Take any two points to construct a straight line. Consider the distance from other points to the straight line, and take a group of straight line and the point with the minimum distance. It can be proved by the method of inverse proof that such three points (two of them are on the straight line) meet the requirements.

6. First of all, it is noticed that if the weight of each box is not more than 1 ton, the weight of the box that can be transported by each vehicle at one time will not be less than 2 tons. Otherwise, another box can be placed.

   Suppose that n cars are needed, the weights of the boxes are $a_1, a_2, \ldots, a_n$ in turn, then $2 \leq a_i \leq 3$ $(i = 1, 2, \ldots, n)$.
   Let the total weight of all the transported goods be S. Then

$$2n \leq S = a_1 + a_2 + \cdots + a_n \leq 3n.$$

   It is obtained that $2n \leq 10 \leq 3n$, i.e. $\frac{10}{3} \leq n \leq 5$. So $n = 4$ or 5.

   Then we explain that four cars are not enough.

   Suppose that there are 13 boxes, each box weighs $\frac{10}{13}$ tons. Because $\frac{10}{13} \times 4 > 3$, each car can only carry 3 boxes at most, so 4 cars can't carry all the boxes.

   As a result, it will take at least 5 vehicles to transport the boxes at one time.
7. Take the "extreme" line so that these points are on the same side of the line. Translate (or rotate) the line until there are at least two of the given points on such line l while the rest points are on the same side of l. Let A and B be the adjacent two points on l, and the given points outside l be $P_1, \ldots, P_k$. Let a point P be the point P_i which makes the largest angle among $\angle AP_1B, \ldots, \angle AP_kB$. Then there is no other point of the given points in the circle passing through the points A, B and P.
8. It can't be made. Suppose that the maximum distance between the two points among the endpoints of these n line segments is the distance between the point A and the point B. From the given condition, B is in the interior of another line segment d. Let the two endpoints of d be C and D respectively. Since $\angle ABD + \angle ABC = 180°$, there must be at least one angle of $\angle ABD$ and $\angle ABC$ not less than $90°$, might as well let $\angle ABD \geq 90°$. Thus we get $AD > AB$, which contradicts the hypothesis that AB is the maximum distance.
9. In the extreme case, AA_1 passed through O_1 and O_2, the points B_1, B and P coincide, then we have $AA_1^2 + BB_1^2 = (d_1 + d_2)^2$. Let $AP = x_1$, $A_1P = x_2$, $BP = y_1$, $B_1P = y_2$. Then $x_1^2 + y_1^2 = d_1^2$, $x_2^2 + y_2^2 = d_2^2$, and $\frac{x_1}{x_2} = \frac{y_1}{y_2} = \frac{d_1}{d_2}$. Suppose that $x_1 = d_1 t$, $x_2 = d_2 t$, $y_1 = d_1 s$, $y_2 = d_2 s$, then $s^2 + t^2 = 1$, thus we get $2(x_1x_2 + y_1y_2) = 2d_1d_2$. Hence,

$$(x_1 + x_2)^2 + (y_1 + y_2)^2 = (d_1 + d_2)^2.$$

Solution 27

Coloring Problems

1. A coloring method is shown in Figure 27.1.

1	3	3	1	1	3
2	2	2	2	2	2
3	1	1	3	3	1
3	1	1	3	3	1
2	2	2	2	2	2
1	3	3	1	1	3

Fig. 27.1

2. (1) First of all, as shown in Figure 27.2, suppose that there are two points A and B satisfying $AB = 2$ while A and B have different colors. Let the midpoint of AB be C. Might as well let it be the same color as A. Make a regular triangle ACD and a regular triangle ACE. From the given condition, D and E cannot be the same color as A and C but will be the same color as B, then $\triangle BDE$ is a regular triangle with the side length of $\sqrt{3}$ and its three vertices are homochromatic.

Secondly, take any point A in the plane, and make a circle with A as the center and 4 as the radius. If all the points in the circle have the same color as A, there is obviously a triangle with the side length of $\sqrt{3}$ and three vertices of the same color in the circle. Otherwise, there is at least one point C in the circle with different colors of A. At this time, $AC < 4$. Construct an isosceles triangle ABC, so

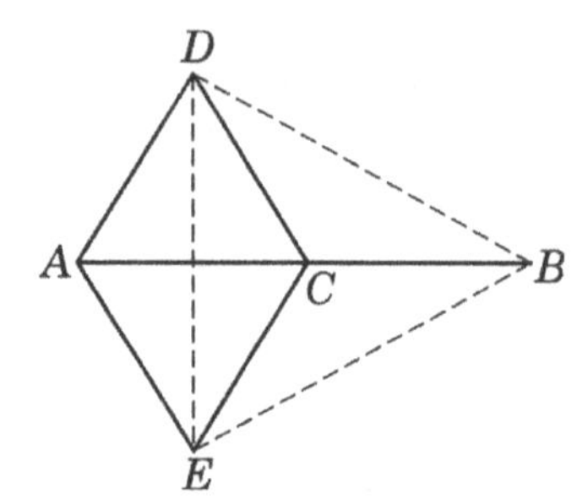

Fig. 27.2

that $AB = BC = 2$, then B has the different color from one of A and C. That is, there is a line segment with the length of 2 and with different colors at two endpoints. It is proved by the first paragraph that there is a regular triangle with the side length of $\sqrt{3}$ and three vertices of the same color.

(2) Divide the plane into horizontal belt areas with the width of $\frac{\sqrt{3}}{2}$, and each area contains a lower bound line and no upper bound line. Dye the adjacent belt areas with different colors.

3. Without loss of generality, suppose that the first ball is red and the second ball is blue. If they have different weights, the conclusion is true. If they have the same weight, such as one pound, then the third ball that weighs two pounds and one of the first or second balls must be the balls of different weights and different colors.
4. According to the dyeing rule, there are at most nine black dots. We number these 27 points in turn. It's easy to know that they can form 9 equilateral triangles (1, 10, 19), (2, 11, 20), ..., (9, 18, 27). If the number of black dots is not more than 8, there must be an equilateral triangle in which all the vertices are white.

   If the number of black dots is exactly 9, then it is known from the dyeing rule that they can only be arranged in one black and two whites, and there must also be an equilateral triangle in which all vertices are white.
5. Dye the small squares in the 1st, 4th, 7th, 10th, 13th, 16th, 19th and 22nd columns of the 23×23 square ground into black, and the rest into white. Then the number of small white squares is 15×23, which is an odd number. Because each tile with the sizes of 2×2 always covers two black squares and two white squares or four white squares, and each tile with the sizes of 3×3 always covers three black squares and six white squares, no matter how many tiles with the sizes of 2×2 and 3×3 cover

an even number of white squares, it is impossible to cover 15×23 white squares. Hence, only two kinds of ceramic tiles with the sizes of 2×2 and 3×3 can't cover the square ground with the size of 23×23.

6. Let the 9 points be A_i $(i = 1, 2, \ldots, 9)$. Then there exists a point, and there are not five lines dyed in red among the eight lines starting from this point(because $\frac{1}{2} \times 5 \times 9$ is not an integer). Might as well let this point be A_0. If 6 or more of the 8 lines starting from A_0 are red, might as well let A_0A_i $(i = 1, 2, \ldots, 6)$ be red.

   Then use two colors to paint the connection line of 6 points A_1, $A_2, \ldots, A_6$, there must be $\triangle A_iA_jA_k(1 \leq i < j < k \leq 6)$ with the same color. According to the given condition, it can only be red, and at this time, the connection lines between each of the four points A_0, A_i, A_j, A_k are red. If only 4 or less lines starting from A_0 are red, then at least four are blue. Might as well let $A_0A_i(i = 1, 2, 3, 4)$ be blue. From the given condition, the connection lines between each of the four points A_1, A_2, A_3 and A_4 are red.

7. Nine points on the circle are used to represent nine mathematicians. If two people know each other, red line will be connected between the corresponding points. If two people don't know each other, blue line will be connected between the corresponding points. The following proof is the same as Exercise 6.

8. In the extreme case, the total length of line segments is minimum.

   Considering all the n line segments connected by a blue point and a white point, then we have that the connection method of taking the minimum total length of n line segments meets the requirements. As shown in Figure 27.3, if A_1B_1 intersects A_2B_2(A represents the blue dot, B represents the white dot), connect A_1B_2 and A_2B_1 (the rest $n-2$ lines do not move), then the total length of n line segments must be reduced. So it goes on a finite number of times, and the conclusion is obtained.

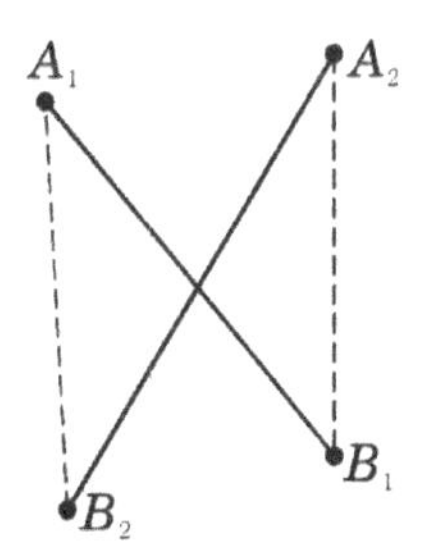

Fig. 27.3

Solution 28

Probability

1. There are six possible outcomes to select two cards from four cards randomly, and there are four possible outcomes to pick two cards with different suits. Thus the probability of getting different suits is $\frac{4}{6} = \frac{2}{3}$. So the answer is B.
2. Among the 36 pairs of possible results, there are four pairs: (1, 4),(2, 3), (2, 3), (4, 1) in which the sum of the two numbers is 5. So the probability that the sum of the two numbers of the upward face is 5 is $\frac{4}{36} = \frac{1}{9}$.
3. From the given condition, $\Delta = a^2 - 4b^2 = (a - 2b)(a + 2b) = 0$. Then we get that $a = 2b$, so the number of b can only be 1, 2, 3. While there are $6 \times 6 = 36$ possible outcomes in two throws, we obtain that the probability required is $\frac{3}{6\times6} = \frac{1}{12}$. So the answer is C.
4. By making the list or drawing the tree chart, we can see that there are 24 kinds of running sequence for the four students, among which there are 6 kinds of running sequence handed over by A to B as follow: ABCD, ABDC, CABD, DABC, CDAB and DCAB. So the probability required is $\frac{6}{24} = \frac{1}{4}$. So the answer is A.
5. Solved by enumerating method.

Number of red balls	Number of white balls	Number of black balls	Number of methods
5	2, 3, 4, 5	3, 2, 1, 0	4
4	3, 4, 5, 6	3, 2, 1, 0	4
3	4, 5, 6, 7	3, 2, 1, 0	4
2	5, 6, 7, 8	3, 2, 1, 0	4

Therefore, there are 16 species in total. So the answer is B.

6. $P_1 = 0$; $P_2 = \frac{1}{6\times6} = \frac{1}{36}$; $P_3 = \frac{1\times2}{6\times6} = \frac{2}{36}$; $P_4 = \frac{1\times2+1}{6\times6} = \frac{3}{36}$; $P_5 = \frac{1\times2+1\times2}{6\times6} = \frac{4}{36}$. Hence, $P_1 + P_2 + P_3 + P_4 + P_5 = \frac{1+2+3+4}{36} = \frac{5}{18}$.
7. The total number of basic events is $6 \times 6 = 36$, that is, 36 quadratic functions can be obtained. It is known from the given condition that $\Delta = m^2 - 4n > 0$, i.e. $m^2 > 4n$.

   It is known through enumeration that 17 pairs of m and n meet the requirements. So $P = \frac{17}{36}$, and the answer is C.
8. Use the tree diagram to list the outcomes of three people's gestures in a round.

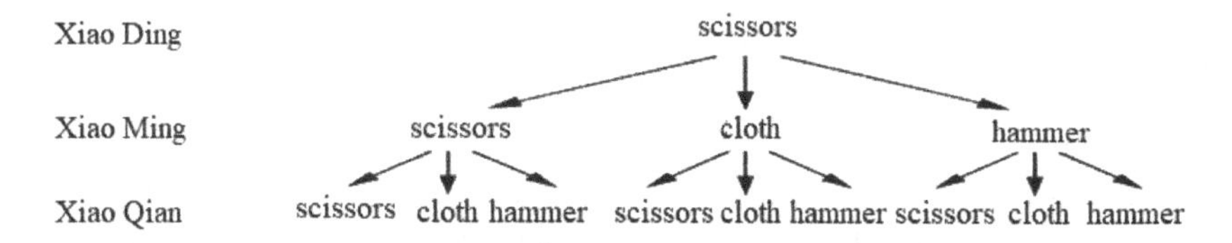

   Only a part of the tree diagram is drawn above (9 outcomes are listed). Change the "scissors" of Xiao Ding in the figure to "cloth" and repeat the above drawing, 9 more outcomes are listed. Finally, change the "scissors" to "hammer", 9 more outcomes are listed too. Therefore, there are 27 outcomes in total. So it is obtained that P (cloth, cloth, cloth) $= \frac{1}{27}$.
9. Calculate the total number N of placements: there are 16 placements for the first coin in 16 square grids, 15 placements for the second coin in the remaining 15 square grids, and 14 placements for the third coin in the remaining 14 square grids. So the total number of placements is

$$N = 16 \times 15 \times 14 = 3360.$$

   Calculate the number m of placements that meet the requirements of the topic: there are 16 placements for the first coin in 16 square grids, and there are 9 square grids in different rows and columns from it. Therefore, there are 9 placements for the second coin in different rows or different columns from the first coin. There are 4 square grids in different rows and columns from the first two coins, so there are 4 placements for the third coin in the remaining 4 square grids.

   Therefore,the number of placements that meet the requirements is

$$m = 16 \times 9 \times 4 = 576.$$

   Hence, the probability required is $P = \frac{m}{N} = \frac{16\times9\times4}{16\times15\times14} = \frac{6}{35}$.

Printed in the United States
by Baker & Taylor Publisher Services